Study Guide
Essential Biology with Physiology

Edward J. Zalisko, Ph.D.
Blackburn College

San Francisco Boston New York
Cape Town Hong Kong London Madrid Mexico City
Montreal Munich Paris Singapore Sydney Tokyo Toronto

Executive Editor: Beth Wilbur
Acquisitions Editor: Chalon Bridges
Editorial Project Manager: Ginnie Simione Jutson
Production Editor: Corinne Benson
Developmental Editor: Evelyn Dahlgren
Senior Producer, Art and Media: Russell Chun
Publishing Assistant: Nora Lally-Graves
Managing Editor, Production: Erin Gregg
Senior Marketing Manager: Josh Frost

On the cover: Photograph of a garden praying mantid (*Orthodera ministralis*). © Michele Westmorland/ The Image Bank

ISBN 0-8053-7494-9

Benjamin Cummings
1301 Sansome Street
San Francisco, CA 94111
www.aw.com/bc

5 6 7 8 9 10—PBT—06 05

Contents

Preface

As you begin . . .

Have you ever felt that you studied for an exam, but your grade was still low? Have you ever wished there was a better way to study?

Ask yourself these questions:

- How do you know what to study?
- How do you know when to study?
- How do you know what you already know?
- How do you know when you are done?

If these questions leave you wondering, you might benefit from the study tips below.

Ten Tips for Studying Biology

There are many subjects and many ways to study. These are tips that have proven successful for my biology students. Try them, refine them, and tailor them to your specific needs.

1. **Read your textbook assignments before lectures over the same material.**
   Your comprehension of the material and the quality of your notes are greatly improved if you read ahead. Think about it—if you have read ahead, the lecture material will seem familiar, your notes will make more sense, and you will need to take fewer notes because you know what is included in the textbook. *Research reveals that reading ahead is one of the most important ways to improve a student's comprehension and course grades.*

2. **Study on a regular schedule.**
   Waiting until the last minute to study usually produces disappointing results. One bad grade can really pull down an average of many good grades. Students who play sports, play an instrument, or have participated in a play all know about the need for *regular practice.* We all know that daily practice is required to perform well. The same is true in the classroom.
   - Start studying when the course begins. Don't wait until a week before an exam.
   - Establish good study habits early. Don't wait until you've received poor grades.
   - Know that a little bit of studying over a long period of time is much better than the same amount of studying over a short period of time. An hour a day for 15 days is many times better than 15 hours at one time (cramming). You can spend less time than cramming and get better grades!

3. **Take frequent and short study breaks.**
   - Many times when we study, our minds drift as we listen to someone in the hallway, a sound outside, or get distracted by people around us. There are two main methods to address this problem.
   - First, study in short bursts of 20–30 minutes each. Then take a 5–10 minute break to rest your mind. After the break, settle back in and focus intensely on your work. Repeat as necessary!
   - Second, try to study in ways that are more interactive. The additional tips below describe a highly successful interactive way to study.

4. **Identify all of the material that is addressed by the exam. Know what you need to know.**
   - After the first day of class, start identifying material for the first exam.
   - Use your course syllabus to help you determine this material.
   - If you are still unclear, talk to your instructor.
5. **Try to write many questions that address all of the information for the exam.**
   - Write questions that can be answered with short answers, 5–20 words or so.
   - Many students like to use 3 × 5 cards, with a question on the front and the answer on the back.
   - Consider using some of the questions from the study guide.
   - Begin with definitions and short lists from the material you need to study.
   - Write these questions after every lecture, keeping up with the material.
   - Writing the questions checks your understanding of the material. Use this opportunity to clear up confusion.
6. **Quiz yourself every day or so on all of the questions you have created.**
   - As you try to answer the questions, create a pile of cards with questions you got correct and a pile for those questions that you missed.
   - After quizzing yourself over all the questions, return to the pile of questions that you missed.
   - As you quiz yourself over the questions you missed, again create a pile of those you got correct and those questions you have now missed twice.
   - Look at the questions that you have missed twice. Use the Web/CD activities and your textbook to review this material again. Try to figure out ways to remember this difficult information.
   - With every new lecture, add new questions to the list of questions you have already created.
7. **Stop studying when you have correctly answered every question.**
   - Finally, a way to study that builds confidence, takes less time, and has a clear end!
   - As you continue to quiz yourself, the questions will seem easier and you should feel more confident.
   - Remember to keep up with your writing and quizzing. This process doesn't work well unless done regularly.
8. **Use short 10–30 minute periods of your day to quiz yourself.**
   - Students that manage their time well will use the short periods before, between, and after classes to review.
   - The result is more free time.
9. **Study during the times of the day when you are most alert.**
   - Many students study in the evenings, when their mind and/or body is tired.
   - Find times in your day when you are most alert and make these your study times.
   - Use times when you tend to get sleepy for other activities (for many of us, this is early afternoon).

10. **Use the night before a test to quickly review. Then get plenty of sleep.**
    - These techniques build confidence in your knowledge of the course material.
    - Many students need only a quick review of their stack of note cards the night before the test. This might take an hour or so.
    - Then get plenty of sleep. A well-rested mind always functions better.
    - Review the questions that you missed on the exam and see if the information was in your note cards. Try to improve your study methods based upon the reasons for missing these questions.

Using Your Study Guide

The purpose of this study guide is to help you organize and review the information presented in the *Essential Biology* text. Each textbook chapter corresponds to a study guide chapter of the same number. The common components of each study guide chapter are described below in the sequence that they appear. Take a few minutes to review how they can help you master the textbook information.

Studying Advice

These study tips provide specific advice for best approaching each chapter.

Organizing Tables

These tables provide the framework for organizing complex information presented in the textbook. They help you organize the information for easier review.

Content Quiz

This is the largest section of each chapter. Some students use these questions to develop their note card questions. Others prefer to quiz themselves after they have mastered their note cards. The answers to the content quiz are located in the back of the study guide. The Content Quiz includes the following types of questions:

Matching—Most chapters have at least one set of matching questions which include long lists of terms or names.

Multiple Choice—These are generally short questions with one correct answer. Consider using the correct statements in these questions for note card questions.

Correctable True/False—These true/false statements have underlined words that are to be corrected if the statement is false. These corrected statements also make good material for note card questions.

Fill-in-the-Blank—These questions check your knowledge of definitions. They also make excellent note card questions.

Figure Quiz—These questions address the information presented in the textbook figures. You will need to refer to the figures in your textbook to answer them.

Analogy Questions—These questions ask you to identify relationships that are analogous to something in biology. For example, the shape of a DNA molecule is most like the shape of a spiral staircase.

Word Roots

This list presents the meanings of the word roots that form the key terms. Learning the meaning of word roots helps you guess the meaning of new terms. For example, you might know that "milli" means one thousandth and "centi" means one hundredth because you know that a millimeter is one thousandth of a meter and a centimeter is one hundredth of a meter. Therefore, you might guess that a millipede has more legs than a centipede!

Key Terms

This list consists of all the boldfaced terms and phrases in each chapter.

Crossword Puzzle

The crossword puzzle consists of the key terms and key phrases from the chapter. This is a fun way to review your knowledge of these definitions. The crossword answers are also listed at the back of the study guide.

CHAPTER 1

Introduction: Biology Today

Studying Advice

a. If you have not already done so, read through the preceding pages to familiarize yourself with the typical components of each study guide chapter and review the advice for studying biology.

b. In addition to this study guide, spend time reviewing the media activities that are noted in your textbook and that appear on the CD-ROM included with your text. These activities illustrate and review principles addressed in your textbook.

Student Media

Biology and Society on the Web

Find out how "wonky holes" affect coral reefs.

Activities

1A The Levels of Life Card Game
1B Energy Flow and Chemical Cycling
1C DNA Molecules: Blueprints of Life
1D Classification Schemes
1E Darwin and the Galápagos Islands
1F Investigate Case Studies of Antibiotic Resistance
1G Evolution: Sea Horse Camouflage Video
1H Science and Technology: DDT

Case Studies in the Process of Science

How Do Environmental Changes Affect a Population of Leafhoppers?
How Does Acid Precipitation Affect Trees?

Evolution Connection on the Web

Examine issues surrounding the teaching of evolution.

Organizing Tables

Distinguish between prokaryotic and eukaryotic cells in the table below.

TABLE 1.1

Characteristics	Prokaryotic Cells	Eukaryotic Cells
Compare the relative size of organisms in each group.		
Indicate which groups have cells that are subdivided by internal membranes, forming organelles.		
List examples of each group.		

Identify the major groups of life in the table below.

TABLE 1.2

List the three domains of life.			
Indicate whether the members of the group have prokaryotic cells or eukaryotic cells.			

Compare the groups withing the domain Eukarya.

TABLE 1.3

	Protista	Plantae	Animalia	Fungi
Are the organisms multicellular or unicellular?				
Describe their mode of nutrition.				
Indicate examples of each group.				

Content Quiz

Directions: Identify the *one* best answer for the multiple-choice questions. For true/false questions, determine if the statement is true or false. If false, change the underlined word(s) to make the statement true. Finally, add the correct word(s) to the fill-in-the-blank questions to make the statements true.

The Scope of Biology

Life at Its Many Levels

1. The dynamics of any ecosystem depend on two main processes. These are
   A. the cycling of nutrients and the flow of energy from sunlight to producers and then to consumers.
   B. the cycling of nutrients and the movement of water through the water cycle.
   C. the movement of water through the water cycle and photosynthesis.
   D. photosynthesis and cellular respiration.

2. Which one of the following statements comparing prokaryotic and eukaryotic cells is false?
   A. Eukaryotic cells are usually larger than prokaryotic cells.
   B. Eukaryotic cells are usually more complex than prokaryotic cells.
   C. Only eukaryotic cells have DNA. Prokaryotes use RNA.
   D. Eukaryotic cells occur in animals and plants. Bacteria have prokaryotic cells.

3. **True or False?** A cell is the lowest level of structure that can perform all activities required for life, including reproduction. ____________________

4. The ____________________, or global ecosystem, is the sum of dynamic processes in all ecosystems.

5. The branch of biology that investigates the relationships between organisms and their environments is called ____________________.

6. The ____________________ is the basic unit of structure and function.

7. The entire "book" of genetic instructions an organism inherits is called its ____________________.

Life in Its Diverse Forms

Matching: Match the group on the left to its best description on the right.

_____	8. Protista	A. multicellular eukaryotes that obtain food by ingestion
_____	9. Plantae	B. eukaryotic organisms that are generally single-celled
_____	10. Fungi	C. multicellular eukaryotes that absorb nutrients by breaking down dead organisms and organic wastes
_____	11. Animalia	D. multicellular eukaryotes that produce their own sugars and other foods by photosynthesis

12. Which one of the following choices correctly lists the groups in their order of abundance, from the group with the greatest number of species to the group with the fewest species?
    A. plants, insects, vertebrates
    B. vertebrates, plants, insects
    C. plants, vertebrates, insects
    D. insects, plants, vertebrates
    E. insects, vertebrates, plants

13. Which two domains include organisms with prokaryotic cells?
    A. Bacteria and Eukarya
    B. Bacteria and Archaea
    C. Archaea and Eukarya
    D. Bacteria and Fungi

14. Look at Figure 1.8 of your text. Which one of the following groups contains the smallest organisms?
    A. domain Archaea
    B. kingdom Protista
    C. kingdom Plantae
    D. kingdom Animalia
    E. kingdom Fungi

15. **True or False?** All life is presently classified into <u>four</u> domains.

    ____

16. The branch of biology that names and classifies species is called

    ____________________.

17. The kingdoms of life can be assigned to three even-higher levels of classification called ____________________.

Evolution: Biology's Unifying Theme

The Darwinian View of Life

18. Which one of the following traits most likely occurred in the most recent common ancestor of a lizard, pigeon, turtle, and bear?
    A. fur
    B. feathers
    C. shell
    D. wings
    E. backbone

19. In his book, *The Origin of Species,* Charles Darwin developed two main points. These were that
    A. the world is very old and that evolution occurs by natural selection.
    B. evolution occurs slowly and that new species form by spontaneous generation.
    C. new species evolve by mutations and that new species don't reproduce with old species.
    D. modern species descended from ancestral species and that organisms evolve by natural selection.

20. Examine the relationships between finches in the diagram in Figure 1.13 of your text. Which one of the following pairs of finches is most closely related?
    A. warbler finch; woodpecker finch
    B. mangrove finch; woodpecker finch
    C. small tree finch; mangrove finch
    D. large cactus ground finch; small tree finch
    E. large ground finch; small tree finch

21. **True or False?** The common ancestor of bears and chipmunks <u>had</u> hair and mammary glands. ____________________

22. ____________________ is the theme that unifies all of biology.

23. In *The Origin of Species,* Darwin proposed a mechanism for descent with modification, which he called ____________________.

Natural Selection

24. Which one of the following pairs of facts was the basis for Darwin's conclusion of unequal reproductive success?
   A. individual variation among offspring; offspring are the products of a blending of the parental genetics
   B. individual variation among offspring; overproduction of offspring and struggle for existence
   C. overproduction of offspring and struggle for existence; males usually fight for a mate
   D. parents produce offspring similar to themselves; overproduction of offspring and struggle for existence
   E. males usually fight for a mate; parents produce offspring similar to themselves

25. Darwin could see that, in artificial selection, humans are substituting for
   A. individual variation.
   B. overproduction of offspring.
   C. the generation of genetic diversity.
   D. the environment.

26. Antibiotic resistance in bacteria is a good reminder that it is the ____________________ that selects and directs the evolution of life.
   A. individual
   B. species
   C. genotype
   D. environment
   E. population

27. What do artificial selection and natural selection have in common?
   A. Both result from a need or desire.
   B. Both are purposeful processes.
   C. Both rely upon individual variation.
   D. Both are directional, with selection towards a goal.

28. Evolution works most like
    A. remodeling an old home.
    B. an architect designing a new home.
    C. people in a community planning where to build a park.
    D. people voting in an election.

29. **True or False?** The product of natural selection is <u>adaptation</u>.

    ____

30. Darwin used the term ___________________ to refer to unequal reproductive success.

The Process of Science

Discovery Science

31. Which one of the following is an example of discovery science?
    A. comparing the effects of two different drugs on the healing of wounds
    B. testing to see if large doses of vitamin C will prevent the common cold
    C. describing the structure of a new dinosaur skull
    D. testing tropical plants for chemicals that might help fight cancer

32. An ecologist describing all the animals and plants in a region of a tropical rain forest is using ___________________ science.

33. The word ___________________ is derived from a Latin verb meaning "to know."

Hypothesis-Driven Science

34. Consider the following statement. "If all vertebrates have backbones, and turtles are vertebrates, then turtles have backbones." This statement is an example of
    A. a hypothesis.
    B. discovery science logic.
    C. rationalization.
    D. hypothetico-deductive reasoning.

35. **True or False?** Using <u>deduction</u>, the reasoning flows from the general to the specific. ___________________

36. A tentative answer to a question defines a(n) ___________________.

A Case Study in the Process of Science

37. In the snake experiment, the statement, "The warning coloration of the kingsnake increases its survival by deterring predators" is an example of
    A. the experimental question.
    B. an observation.
    C. an expected result.
    D. the experiment.
    E. a hypothesis.

38. In the snake experiment, the kingsnakes were using ________________, an evolutionary adaptation that makes them look poisonous.

39. In most cases, a control group and an experimental group differ by just one ________________.

The Culture of Science; Theories in Science; Science, Technology, and Society

40. Which one of the following *is not* a characteristic of modern science?
    A. repeatability of experiments
    B. dependence upon observations
    C. requirement that ideas be testable
    D. independence and isolation of researchers

41. Which one of the following *does not* apply to scientific theories? Scientific theories
    A. provide a comprehensive explanation.
    B. are the same as hypotheses.
    C. are supported by abundant evidence.
    D. are widely accepted.

42. **True or False?** Scientists usually pay little attention to other researchers currently working on the same problem. ________________

43. The one theme that continues to hold all of biology together is ________________.

44. Science and ________________ are interdependent.

Word Roots

eu = true (*eukaryote:* cell type with membrane-enclosed nucleus and other organelles)

karyon = kernel (*prokaryotic:* cell lacking a membrane-enclosed nucleus and other organelles)

pro = before (*prokaryotic:* the first cells, lacking a membrane-enclosed nucleus and other organelles)

Key Terms

artificial selection
biosphere
cell
consumers
DNA
ecosystem
eukaryotic
gene
genome
hypothesis
kingdoms
natural selection
organism
photosynthesis
producers
prokaryotic
species
theory

Crossword Puzzle

Use the Key Terms list from this chapter to fill in the crossword puzzle.

ACROSS

3. organisms that make organic food molecules from CO_2, H_20, and other inorganic raw materials
4. the process by which plants use light energy to make sugars from carbon dioxide and water
9. selective breeding of domesticated organisms by humans
13. a widely accepted explanatory idea supported by a large body of evidence
14. the taxonomic category just below genus
15. an organism's genetic material
16. a type of cell lacking a membrane-enclosed nucleus and other membrane-enclosed organelles
17. an individual living thing
18. the broad taxonomic category above phylum

DOWN

1. a way that a population of organisms can change over generations as a result of heritable differences in reproduction
2. a tentative explanation
5. a type of cell that has a membrane-enclosed nucleus and other membrane-enclosed organelles
6. a discrete unit of hereditary information
7. the global ecosystem
8. the abbreviation for deoxyribonucleic acid
10. the fundamental structural unit of life
11. a biological community and its physical environment
12. organisms that obtain their food by eating plants or by eating animals that have eaten plants

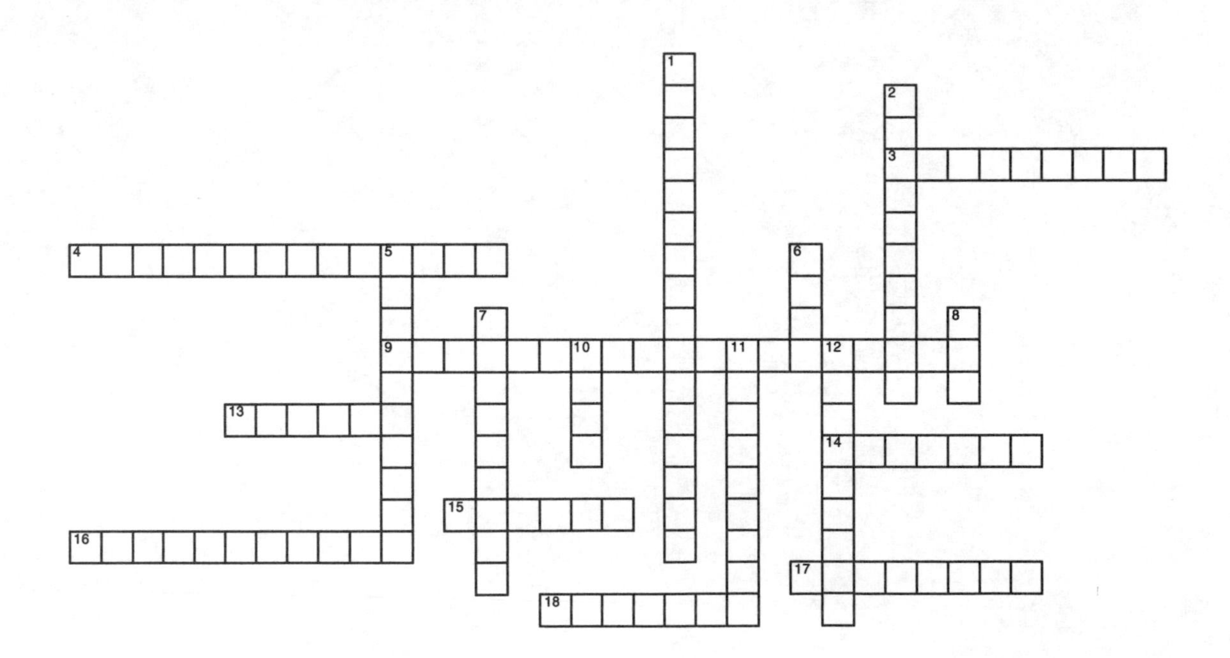

CHAPTER 2

Essential Chemistry for Biology

Studying Advice

a. If this is your first experience with chemistry, work slowly to master the terminology and basic relationships of atoms and their interactions. Before reading the chapter, look through the figures and read the chapter summary. This will give you an idea of the subjects that will be addressed in the text.

b. Understanding the terminology and ideas in this chapter is essential to understanding the content in the next few chapters. These next chapters continue to explore the molecules of life and how they interact to form cells. Your investment of energy in this chapter will pay dividends in the chapters to come!

c. Finally, think carefully about how you study. Refer to the study advice at the start of this study guide and establish good habits now.

Student Media

Biology and Society on the Web

Learn what the American Dental Association has to say about the importance of fluoride.

Activities

2A The Levels of Life Card Game
2B The Structure of Atoms
2C Electron Arrangement
2D Build an Atom
2E Ionic Bonds
2F Covalent Bonds
2G The Structure of Water
2H The Cohesion of Water in Trees
2I Acids, Bases, and pH

Case Studies in the Process of Science

How Are Space Rocks Analyzed for Signs of Life?

How Does Acid Precipitation Affect Trees?

Evolution Connection on the Web

Learn more about what Earth was like before life appeared.

Organizing Tables

Compare the three main subatomic particles.

TABLE 2.1

	Relative Size Compared to Other Parts of an Atom	Electrical Charge (if any)	Location in an Atom
Proton	Determines element	single unit of (+) charge	nucleous
Neutron	Determines Isotopes	neutral (no electrical charge)	nucleous
Electron	Determines chemical behavior	single unit (−) charge	outer-shell election

Compare the three main types of chemical bonds.

TABLE 2.2

	Nature of the Bond	Example
Ionic bonds		
Covalent bonds		
Hydrogen bonds		

Content Quiz

Directions: Identify the *one* best answer for the multiple-choice questions. For true/false questions, determine if the statement is true or false. If false, change the underlined word(s) to make the statement true. Finally, add the correct word(s) to the fill-in-the-blank questions to make the statements true.

Biology and Society: Fluoride in the Water

1. Fluoride prevents cavities by
   A. affecting the metabolism of the bacteria.
   B. promoting the replacement of lost minerals on the tooth surface.
   C. dissolving sugars.
   D. coating the teeth with a protective barrier.
   E. More than one of the above are correct.
   F. None of the above are correct.

2. **True or False?** Fluoride is a form of the element <u>fluorine</u>. ______________

3. Living organisms are, at their most basic level, ___chemical___ systems.

Tracing Life Down to the Chemical Level

4. Which one of the following is a correct sequence of levels of biological organization?
   A. population, organism, organ system, organs, tissues
   B. organs, cells, tissues, molecules, atom
   C. ecosystem, population, community, organism, organ system
   D. community, population, organ system, organism, organs
   E. organism, organ system, organs, cells, tissues

5. **True or False?** A <u>community</u> is an interacting group of just one species.
   population

6. Tissues are made of cells and molecules are made of atoms.

Some Basic Chemistry

Matter: Elements and Compounds

7. Four elements make up 96% of the human body. Which one of the following *is not* one of those four major elements?
   A. carbon
   B. nitrogen
   C. iron
   D. oxygen
   E. hydrogen

8. **True or False?** Carbon dioxide is an example of a <u>compound</u>, because it contains two or more elements in a fixed ratio. ________________

9. Elements that your body needs in very small amounts are called Trace elements.

Atoms

10. Which one of the following statements about atoms is *false*?
    A. All atoms of an element have the same number of protons.
    B. Atoms whose outer shells are not full tend to interact with other atoms.
    C. An atom is the smallest unit of matter that still retains the properties of an element.
    D. A proton and an electron are almost identical in mass.
    E. The farther an electron is from the nucleus, the greater its energy.

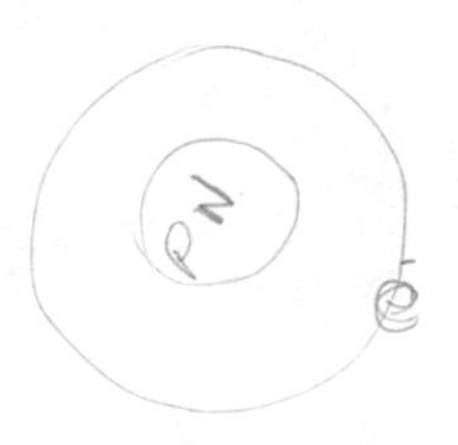

11. The way that a satellite orbits the Earth is most like the relationship between
    A. a proton and a neutron of an atom.
    B. a cell and its molecules.
    C. an electron and the nucleus of an atom.
    D. organisms in an ecosystem.
    E. a cell in a tissue.

12. Examine Figure 2.5 of your text of a helium atom. Which one of the following is a problem with the way that helium is represented?
    A. The protons and neutrons don't form the nucleus.
    B. Electrons are actually larger than protons and neutrons.
    C. Electrons don't orbit around the nucleus.
    D. The distance between the electrons and the nucleus is many times greater.
    E. Sometimes the electrons contribute to the structure of the nucleus.

13. **True or False?** Isotopes of an element have the same number of protons and electrons but different numbers of neutrons. True

14. The mass number is the sum of the numbers of protons and neutrons.

Chemical Bonding and Molecules

15. When a covalent bond occurs,
    A. protons are transferred from one atom to another.
    B. the atoms in the reaction become positively charged.
    C. an atom gives up one or more electrons to another atom.
    D. ions are formed.
    E. two atoms share electrons.

16. A person borrowing money from a bank is most like the relationship between
    A. atoms that form a covalent bond.
    B. atoms that form an ionic bond.
    C. a proton and a neutron in an atom.
    D. two isotopes of an element.
    C. an organism and its cells.

17. Examine the electrons in the atoms in Figure 2.7 of your text. How many atoms of oxygen will covalently bond to one atom of carbon?
    A. one atom of oxygen
    B. two atoms of oxygen
    C. three atoms of oxygen
    D. four atoms of oxygen
    E. Oxygen will not covalently bond to carbon.

18. **True or False?** A molecule of carbon dioxide is a compound, because it consists of two or more elements. ____________________

19. Atoms that are electrically charged as a result of gaining or losing electrons are called ions.

Chemical Reactions

20. In the chemical reaction of $2 H_2 + O_2 \rightarrow 2 H_2O$,
   A. water is a reactant.
   B. hydrogen and oxygen are the products.
   C. the number of atoms of reactants is more than the number of atoms of the products.
   D. a chemical compound is formed.

21. **True or False?** Chemical reactions cannot destroy or create matter.
   ____

22. New chemical bonds are formed in the products of a reaction.

Water and Life

The Structure of Water

23. Water molecules are attracted to each other because of
   A. covalent bonds.
   B. hydrogen bonds.
   C. atomic bonds.
   D. ionic bonds.

24. When we get out of a shower or bath, the surface of our skin is mostly wet because
   A. water is being released by our skin cells.
   B. water is a polar molecule that sticks to polar surfaces.
   C. water from the air condenses on our skin.
   D. our body continues to produce water.

25. **True or False?** In a water molecule, the electrons spend most of their time near the hydrogen atoms. Oxygen atom

26. Water's two hydrogen atoms are joined to the oxygen atom by a(n) covalent bond.

Water's Life–Supporting Properties

27. Which one of the following statements about water is *false*?
    A. Covalent bonds give water unusually high surface tension.
    B. Water absorbs and stores a large amount of heat while warming up only a few degrees.
    C. Evaporative cooling occurs because water molecules with the greatest energy vaporize first.
    D. Ice floats because water molecules move farther apart when they form a solid.
    E. Water is a common solvent inside your body.

28. Which one of the following is a result of surface tension?
    A. floating ice
    B. a water strider "standing" on the surface of water
    C. sweat cooling the surface of your skin
    D. sugar dissolving quickly in a cup of hot tea

29. It's a hot summer day, and Jill just finished jogging. As she grabs a drink, ice cubes floating at the top of the glass hit her nose and lemonade spills out. She uses her towel to wipe up drops of lemonade clinging to her cheeks. Which one of the following properties of water *was not* represented in this situation?
    A. evaporative cooling
    B. the cohesive properties of water
    C. the ability of ice to float
    D. the ability of water to resist temperature change

30. Which one of the following is represented in Figure 2.13 of your text?
    A. surface tension
    B. ability of water to moderate temperature
    C. ability of ice to float
    D. evaporative cooling
    E. versatility of water as a solvent

31. **True or False?** Because of hydrogen bonding, water has a better ability to resist temperature change than most other substances on Earth. ____________

32. A(n) solute ____________ is dissolved in a solvent to produce a(n) solution ____________.

solvent
Solution

Acids, Bases, and pH

33. The pH of a solution
    A. inside most living cells is close to 5.
    B. can be quickly changed by the addition of a buffer.
    C. can range from 0 to 14.
    D. is acidic above the level of 7.
    E. is determined by the relative amount of OH^- ions.

34. A solution with a pH of 7 contains
    A. more H^+ than OH^- ions.
    B. more OH^- than H^+ ions.
    C. only H^+ ions.
    D. only OH^- ions.
    E. equal amounts of H^+ and OH^- ions.

35. Which one of the following people functions most like a buffer?
    A. a comedian who keeps the audience laughing
    B. a funeral director who helps people in times of grief
    C. a counselor who tries to understand the hidden meaning in your dreams
    D. a coach pumping up a losing team and working hard a winning team
    E. an actor playing a serious role in a drama

36. **True or False?** A solution with a pH of 6 has <u>100 times</u> more H^+ than an equal amount of a solution with a pH of 5. ________________

37. Most biological fluids contain ________________, substances that resist changes in pH.

Evolution Connection: Earth Before Life

38. Life first evolved
    A. about 3.5–4.0 billion years ago, in an atmosphere similar to what we see today.
    B. about 3.5–4.0 million years ago, in an atmosphere similar to what we see today.
    C. about 3.5–4.0 billion years ago, in an atmosphere different from what we see today.
    D. about 3.5–4.0 million years ago, in an atmosphere different from what we see today.

39. **True or False?** The most dense layers of the Earth are at the <u>surface</u>.
    ____

40. The second atmosphere of Earth was formed by gases released from
    ________________.

Word Roots

aqua = water (*aqueous:* a type of solution in which water is the solvent)

co = together; **valent** = strength (*covalent bond:* an attraction between atoms that share one or more pairs of outer-shell electrons)

eco = house (*ecosystem:* all the organisms in a particular area and their physical environment)

iso = equal (*isotope:* an element having the same number of protons and electrons but a different number of neutrons)

neutr = neither (*neutron:* a subatomic particle with a neutral electrical charge)

Key Terms

acid
aqueous solution
atom
atomic number
base
buffer
cell
chemical bond
chemical element
chemical reaction
cohesion
community
compound
covalent bond
ecosystem
electron
evaporative cooling
heat
hydrogen bond
ion
ionic bond
isotope
mass
mass number
matter
molecule
neutron
nucleus
organ
organism
organ system
pH scale
polar molecule
population
products
proton
radioactive isotope
reactants
solute
solution
solvent
temperature
tissue
trace element

Crossword Puzzle

Use the Key Terms list from this chapter to fill in the crossword puzzle.

ACROSS

1. a substance that cannot be broken down to other substances by ordinary chemical means
5. a process leading to chemical changes in matter; involves the making and/or breaking of chemical bonds
8. a variant form of an atom that has the same number of protons as the element, but different numbers of neutrons
9. a substance that increases the hydrogen ion (H^+) concentration in a solution
12. a substance that is dissolved in a solution
13. a subatomic particle with a single positive electrical charge; found in the nucleus of an atom
16. a substance that decreases the hydrogen (H^+) concentration in a solution
17. a group of two or more atoms held together by covalent bonds
18. a measure of the relative acidity of a solution, ranging from a value from 0 (most acidic) to 14 (most basic)
20. a chemical substance that resists changes in pH by accepting hydrogen ions from a solution (when there is an excess of H^+) or donating hydrogen ions to a solution (when H^+ is depleted)
21. a measure of the intensity of heat, reflecting the average kinetic energy or speed of molecules
22. the measure of the quantity of material in an object
24. an attraction between two ions with opposite electrical charges; the electrical attraction of the opposite charges holds the ions together
25. (1) an atom's central core, containing protons and neutrons; (2) the genetic control center of a eukaryotic cell
27. the fundamental structural unit of life; a basic unit of living matter separated from its environment by a plasma membrane
28. a subatomic particle that is electrically neutral (has no electrical charge); found in the nucleus on an atom
30. a cooperative unit of many similar cells that perform a specific function within a multicellular organism
31. anything that occupies space and has mass
32. the smallest unit of matter that retains the properties of an element
34. an attraction between two atoms resulting from a sharing of outer-shell electrons or the presence of opposite charges on the atoms; the bonded atoms gain complete outer electron shells
37. an element that is essential for the survival of an organism but only in minute quantities
38. an individual living thing, such as bacterium, fungus, protist, plant, or animal
39. an atom whose nucleus decays spontaneously, giving off particles and energy
40. a structure consisting of several tissues adapted as a group to perform specific functions
41. surface cooling that results when a substance evaporates; this occurs because the "hottest" molecules evaporate first
42. a group of organs that work together in performing vital body functions

DOWN

2. the number of protons in each atom of a particular element
3. all the organisms in a given area, along with the nonliving (abiotic) factors with which they interact; a biological community and its physical environment

4. the sum of the number of protons and neutrons in an atom's nucleus
6. a type of weak chemical bond formed when the partially positive hydrogen atom in one molecule is attracted the partially negative atom of a neighboring molecule
7. an atom or molecule that has gained or lost one of more electrons, thus acquiring an electrical charge
9. a mixture of two or more substances, one of which is water
10. the molecule that has opposite charges on opposite ends
11. the amount of energy associated with the movement of the atoms and molecules in a body of matter
14. the attraction between molecules of the same kind
15. the dissolving agent in a solution; water is the most versatile one known
19. an ending material in a chemical reaction
23. a group of interacting individuals belonging to one species and living in the same geographical area
26. an attraction between atoms that share one or more pairs of outer-shell electrons; symbolized by a single line between the atoms
29. a starting material in a chemical reaction
33. a subatomic particle with a single negative electric charge; one or more of them move around the nucleus of an atom
34. all the organisms living together and potentially interacting in a particular area
35. a substance containing two or more elements in a fixed ratio, for example, table salt (NaCl) consists of one atom of the element sodium (Na) for every atom of chlorine
36. a fluid mixture of two or more substances, consisting of a dissolving agent, the solvent, and a substance that is dissolved, the solute

CHAPTER 3

The Molecules of Life

Studying Advice

a. Chapter 3 introduces the four main types of biological molecules. Table 3.1 will help you compare these categories and organize the details of each group.

b. Find a nutrition label on some packaged food around you. Notice that three of the four categories of biological molecules discussed in this chapter are sources of nutrition. These three types, carbohydrates, proteins, and lipids, are all sources of calories. The chemistry in this chapter is already something you know a little about.

Student Media

Biology and Society on the Web

Learn more about lactose intolerance.

Activities

3A Diversity of Carbon-Based Molecules
3B Functional Groups
3C Making and Breaking Polymers
3D Models of Glucose
3E Carbohydrates
3F Lipids
3G Protein Functions
3H Protein Structure
3I Nucleic Acid Functions
3J Nucleic Acid Structure

Case Study in the Process of Science

What Factors Determine the Effectiveness of Drugs?

Evolution Connection on the Web

Learn why mice can be used as models for medical research.

Organizing Table

Compare the four major classes of biological molecules by completing the table below. Some of the items are already filled in to help you complete the table.

TABLE 3.1

Group	**Monomer**	**Subgroups**	**Examples**
Carbohydrates		Monosaccharides	
		Disaccharides	
		Polysaccharides	
Lipids	*No monomer*	Unsaturated fats	
		Saturated fats	
		Steroids	
Proteins		*No subgroup*	
Nucleic acids		DNA	
		RNA	

Content Quiz

Directions: Identify the *one* best answer for the multiple-choice questions. For true/false questions, determine if the statement is true or false. If false, change the underlined word(s) to make the statement true. Finally, add the correct word(s) to the fill-in-the-blank questions to make the statements true.

Biology and Society: Got Lactose?

1. People who are lactose intolerant
   - A. produce antibodies against lactose.
   - B. are allergic to soy products.
   - C. produce insufficient amounts of lactose.
   - D. produce insufficient amounts of lactase.
   - E. None of the above are correct.

2. **True or False?** People who are unable to properly digest the main sugar found in milk can avoid the problem by consuming lactase enzyme pills when eating lactose-containing food. ________________

3. Some people are unable to properly digest ___lactose___, the main sugar found in milk.

Organic Molecules

Carbon Chemistry

4. Which one of the following statements about organic molecules is *false*?
   - A. It is possible to construct an endless diversity of carbon skeletons.
   - B. The simplest hydrocarbon is methane.
   - C. Carbon can use only one bond to attach to another carbon atom.
   - D. Carbon completes its outer shell by sharing electrons with up to four other atoms.
   - E. Carbon frequently bonds to the elements hydrogen, oxygen, and nitrogen.

5. Examine the bonds between the carbons in Figure 3.2 of your text. Find a pair of carbons with a double line between them. The double line between indicates that these two carbons:
   - A. share two electrons.
   - B. share two hydrogens.
   - C. are joined by an ionic bond.
   - D. are joined at the nucleus.
   - E. share electrons with hydrogens that we can not see.

6. **True or False?** The groups of atoms that usually participate in chemical reactions are called functional groups. ________________

7. In methane, carbon is joined to four hydrogen atoms by ________________ bonds.

Giant Molecules from Smaller Building Blocks

8. Which one of the following processes is the reverse of dehydration synthesis?
   A. hydrolysis
   B. hydration
   C. denaturation
   D. diffusion
   E. osmosis

9. **True or False?** In the process of dehydration synthesis, a molecule of <u>carbon dioxide</u> is formed. ________________

10. In the same way that a train is made by linking together railroad cars, cells make ________________ by linking together ________________.

Biological Molecules

Carbohydrates

11. Automobiles use gasoline the way that cells use
    A. cellulose.
    B. amino acids.
    C. glucose.
    D. proteins.
    E. lipids.

12. Sucrose is
    A. the main carbohydrate in plant sap.
    B. the main sweetener in soft drinks.
    C. a monosaccharide.
    D. sweeter tasting than fructose.
    E. abundant in corn syrup.

13. Plants use starch the way that animals use
    A. glucose.
    B. glycogen.
    C. fructose.
    D. cellulose.
    E. sucrose.

14. Compare the structures of glucose and fructose in Figure 3.9 of your text. How are these molecules different?
    A. Glucose has more double bonds.
    B. Fructose has more carbon atoms.
    C. Fructose has more hydrogen atoms.
    D. Glucose has more oxygen atoms.
    E. Only the shapes are different.

15. Examine the structure of glucose in Figure 3.10 of your text. How are the linear and ring structures different?
    A. The linear structure has more carbon atoms.
    B. The linear structure has more oxygen atoms.
    C. The ring structure has more hydrogen atoms.
    D. The ring structure has more double bonds.
    E. Only the shapes are different.

16. **True or False?** Starch is the most abundant organic compound on Earth.

    ____

17. **True or False?** Cellulose cannot be hydrolyzed by most animals.

    ____

18. Glucose and fructose are examples of ____________________, molecules that match in their molecular formulas.

19. Cells construct a disaccharide by joining two ____________________ in the process of ____________________.

20. Starch is a polysaccharide made entirely from ____________________ monomers.

Lipids

21. Unsaturated fats
    A. are usually solid at room temperature.
    B. contribute to cardiovascular disease.
    C. have the maximum number of hydrogen atoms attached.
    D. lack double bonds in their hydrocarbon portions.
    E. None of the above are true.

22. If we add hydrogen to an unsaturated fat, we would expect that the unsaturated fat would
    A. have more double bonds.
    B. be a much longer molecule.
    C. be stiffer at room temperature.
    D. have a lower melting point.

23. Which one of the following statements about cholesterol is *false*? Cholesterol is
    A. made of a carbon skeleton consisting of four fused rings.
    B. one of the most well known steroids.
    C. used by our bodies to make sex hormones.
    D. similar to fats in structure and function.
    E. present in cell membranes.

24. Examine the fatty acids in Figure 3.15b of your text. What causes one of the fatty acid tails to bend? The bend occurs
    A. wherever there is a double bond.
    B. at the middle of long fatty acids.
    C. in the shortest of the fatty acid tails.
    D. wherever there is no oxygen atom present.
    E. where the tail forms a ring.

25. **True or False?** A pound of fat contains more than twice as much energy as a pound of starch. ________________

26. Use of synthetic variants of testosterone, called ________________, can lead to liver damage, a reduced sex drive, and infertility.

27. Lipids are ________________, which means that they can not mix with water.

Proteins

28. Which one of the following statements about proteins is *false*?
    A. A protein's three-dimensional shape enables it to carry out its normal functions.
    B. All proteins are made from a common set of just 20 amino acids.
    C. Amino acids in a protein are linked together by peptide bonds.
    D. Enzymes are specialized types of proteins.
    E. Proteins are made from nucleic acid monomers.

29. Which one of the following is most like the process of making proteins?
    A. weaving hair into a braid
    B. sewing a quilt from hundreds of patches of old clothing
    C. making a train by connecting together 20 different types of railroad cars
    D. gathering peas onto a spoon as you prepare to eat them
    E. preparing a stew by cutting up four different vegetables and mixing them with meat

30. The way a protein changes shape when it is heated is most like
    A. tearing sections off a roll of toilet paper.
    B. cutting a piece of string into two.
    C. separating a few cars from a long train.
    D. untangling a garden hose.

31. Examine Figure 3.23 of your text to best understand the levels of protein structure. Which one of the following is most like the secondary level of protein structure?
    A. the sequence of railroad cars on a train
    B. a spool of thread
    C. a tangled ball of string
    D. the links of chain forming a necklace

32. **True or False?** All amino acids have a carboxyl group, an amino group, and a hydrogen atom attached to a central carbon atom. ________________

33. The specific amino acid sequence of a protein is the protein's ________________ structure.

34. If a protein is exposed to a change in temperature or pH, it can lose its normal shape in a process called ________________.

Nucleic Acids

35. Which one of the following statements about nucleic acids is *false*?
    A. The base is the part that varies between nucleotides of DNA.
    B. There are two types of nucleic acids, DNA and RNA.
    C. Nucleic acids are linked into long strands to form nucleotides.
    D. The sequence of amino acids in proteins is determined by the sequence of nucleotides in DNA.
    E. Molecules of DNA form a double helix.

36. The overall shape of a DNA molecule is most similar to the shape of
    A. railroad tracks.
    B. a spiral staircase.
    C. a corkscrew.
    D. a bicycle chain.
    E. the number 8.

37. DNA replication occurs by a process most like
    A. two people getting divorced and each remarrying.
    B. forming two baseball teams from a group of students.
    C. making a photocopy of an important piece of paper.
    D. making a necklace by stringing beads together.
    E. rebuilding an old set of stairs.

38. **True or False?** The genetic material humans and other organisms pass from one generation to the next consists of <u>RNA</u>. ____________________

39. Nucleic acids are made of monomers called ____________________.

Evolution Connection: DNA and Proteins as Evolutionary Tape Measures

40. **True or False?** Scientists expect that the DNA of closely related species will be <u>more</u> similar than the DNA of distantly related species. ____________________

41. The DNA sequences determine the ____________________ sequences in proteins.

Word Roots

di = two (*disaccharide:* two monosaccharides joined together)
glyco = sweet (*glycogen:* a polysaccharide sugar used to store energy in animals)
hydro = water (*hydrolysis:* breaking chemical bonds by adding water)
iso = equal (*isomer:* molecules with similar molecular formulas but different structures)
lyse = break (*hydrolysis:* breaking chemical bonds by adding water)
meros = part (*polymer:* a chain made from smaller organic molecules)
mono = single; **sacchar** = sugar (*monosaccharide:* simplest type of sugar)
philic = loving (*hydrophilic:* water-loving property of a molecule)
phobos = fearing (*hydrophobic:* water-hating property of a molecule)
poly = many (*polysaccharide:* many monosaccharides joined together)

Key Terms

activation energy
active site
amino acid
anabolic steroid
base
carbohydrates
carbon skeleton
cellulose
dehydration synthesis
denaturation
disaccharide
DNA
double helix
fat
feedback regulation
functional group
glycogen
hydrocarbon
hydrolysis
hydrophilic
hydrophobic
induced fit
isomer
lipid
macromolecule
metabolism
monomer
monosaccharide
nucleic acid
nucleotide
organic chemistry
peptide bond
polymer
polypeptide
polypeptide chain
polysaccharide
primary structure
protein
RNA
saturated
starch
steroid
substrate
sugar-phosphate backbone
unsaturated

Crossword Puzzle

Use the Key Terms list from this chapter to fill in the crossword puzzle.

ACROSS

1. an organic molecule composed only of the elements carbon and hydrogen
4. a chemical subunit that serves as a building block of a polymer
5. a complex, extensively branched polysaccharide consisting of many glucose monomers
8. a biological macromolecule made of one or more polypeptides
9. it means "water-fearing"
11. the three letter abbreviation for ribonucleic acid
16. a chain of sugar and phosphate to which the DNA and RNA nitrogenous bases are attached
22. a chemical process in which macromolecules are broken down adding water
24. a large lipid molecule made from glycerol and three fatty acids
25. a sugar molecule consisting of two monosaccharides
26. Organic compounds with the same molecular formula but different structures and properties.
29. the three letter abbreviation for deoxyribonucleic acid
30. a process in which a protein unravels, losing its specific shape and function
31. a type of regulation in which the end product inhibits the pathway
36. the formation of polymers by removing water to link monomers
37. the atoms that form the chemically reactive part of an organic molecule
38. a large molecule consisting of many identical or similar monomers
39. a carbohydrate consisting of hundreds to thousands of monosaccharides
40. a chain of amino acids linked by peptide bonds
41. a reactant on which an enzyme acts
42. the covalent linkage between two amino acids in a polypeptide

DOWN

2. the chain of carbon atoms that forms the structural backbone of an organic molecule
3. the smallest kind of sugar molecule
6. a class of molecules including monosaccharides, disaccharides, and polysaccharides
7. the study of carbon compounds
10. a biological molecule that does not mix with water
12. a polymer consisting of many nucleotide monomers
13. a large polysaccharide composed of many glucose monomers essential to plant cells
14. a synthetic variant of the male hormone testosterone
15. a type of lipid whose carbon skeleton is bent to form four fused rings
17. the amount of energy needed to start a chemical reaction
18. the part of the enzyme where a substrate molecule attaches
19. molecule with a hydrogen atom and carboxyl, amino, and variable side groups bonded to a central carbon atom
20. a substance that decreases the hydrogen concentration in a solution
21. a storage polysaccharide found in the roots of plants
23. the interaction between a substrate molecule and the active site of an enzyme
27. a monomer consisting of a five-carbon sugar, a nitrogenous base, and phosphate group
28. a giant molecule in a living organism
32. pertaining to fats and fatty acids whose hydrocarbon chains contain the maximum number of hydrogens
33. the two complementary strands that form the spiral DNA structure
34. it means "water-loving"
35. the many chemical reactions that occur in organisms

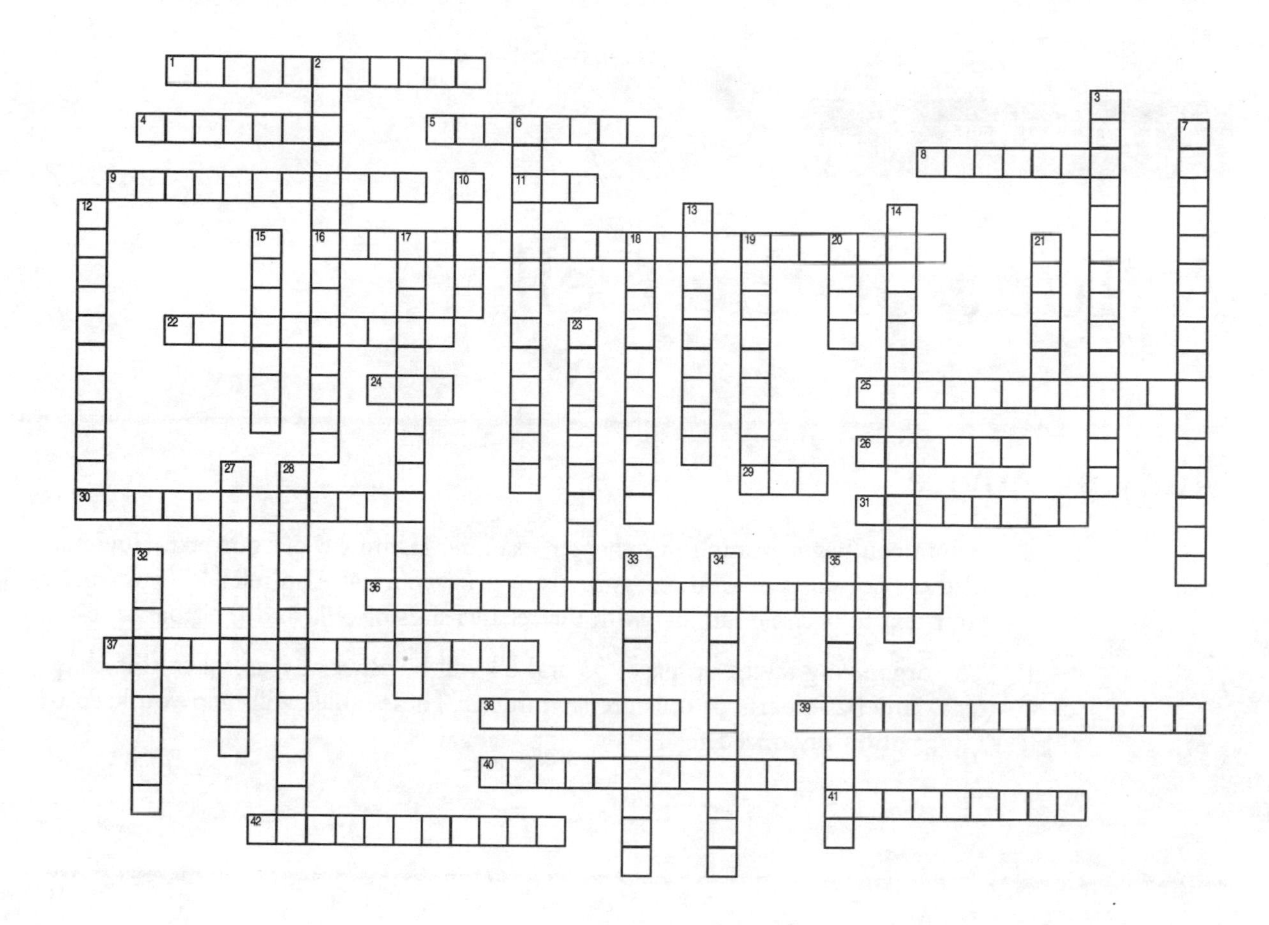

CHAPTER 4

A Tour of the Cell

Studying Advice

a. Before you begin reading this chapter, examine Figure 4.6 of your text. How many of these cell parts do you recognize from prior courses? Also refer to Figure 4.3 of your text to become familiar with the relative sizes of cells and their parts.

b. The organizing tables on pages 33 and 34 will be especially useful in this chapter. Many basic parts of cells are introduced. These tables will help you keep the information organized for review.

Student Media

Biology and Society on the Web

Learn about the downside of antibiotic use.

Activities

4A Metric System Review
4B Prokaryotic Cell Structure and Function
4C Comparing Cells
4D Build an Animal Cell and a Plant Cell
4E Membrane Structure
4F Selective Permeability of Membranes
4G Overview of Protein Synthesis
4H The Endomembrane System
4I Build a Chloroplast and a Mitochondrion
4J Cilia and Flagella
4K Cell Junctions
4L Review: Animal Cell Structure and Function
4M Review: Plant Cell Structure and Function

Case Study in the Process of Science

What Is the Size and Scale of Our World?

Evolution Connection on the Web

Learn what prelife membranes may have been like.

Organizing Tables

Compare the features of prokaryotic and eukaryotic cells. (See Chapter 4 and Figure 4.4 of your text.)

TABLE 4.1

	Prokaryotic Cells	Eukaryotic Cells
Which group has a nucleus bordered by a membrane?		
Which group has organelles?		
How do the sizes compare?		
What are examples of each group?		

Compare the structure, location, and function(s) of the following components of the endomembrane system.

TABLE 4.2

	Structure	Location in the Cytoplasm	Function(s)
Smooth endoplasmic reticulum			
Rough endoplasmic reticulum			
Golgi apparatus			
Lysosome			
Central vacuole			

Compare the features of cilia and flagella.

TABLE 4.3

	Cilia	Flagella
Which are usually longer?		
Which are usually more numerous?		
How does their basic architecture compare?		
Where are they found?		

Content Quiz

Directions: Identify the *one* best answer for the multiple-choice questions. For true/false questions, determine if the statement is true or false. If false, change the underlined word(s) to make the statement true. Finally, add the correct word(s) to the fill-in-the-blank questions to make the statements true.

Biology and Society: Drugs That Target Cells

1. Which one of the following statements about antibiotics is *false*?
   A. Most antibiotics are naturally occurring chemicals derived from other microorganisms.
   B. Many diseases were drastically reduced when antibiotics were introduced.
   C. The goal of treatment with antibiotic drugs is to kill invading bacteria while causing minimal harm to the human host.
   D. Most antibiotics bind only to structures found in human cells.

2. Which one of the following statements about cells is *false*?
   A. Fungi are not made of cells.
   B. Cells are the smallest entity that exhibits all the characteristics of life.
   C. Cells are as fundamental to biology as atoms are to chemistry.
   D. Cells are the building blocks of all life.
   E. All organisms are made of cells.

3. **True or False?** Penicillin, ampicillin, and bacitracin work by disrupting the synthesis of <u>ribosomes</u>. ______________

4. Drugs that disable or kill infectious bacteria are called ______________.

The Microscopic World of Cells

5. Which one of the following statements about cells is *true*?
   A. The magnification of a transmission electron microscope is about 10 times greater than a typical light microscope.
   B. Organelles are small parts of cells with specific functions.
   C. Cell theory states that all living things have a well-defined nucleus bordered by a membranous envelope.
   D. The genetic material of eukaryotic cells is housed in the endoplasmic reticulum.
   E. The cytoplasm consists of various organelles suspended in the nucleus.

6. Examine Figure 4.6 of your text. Which one of the following organelles is physically connected to the nucleus?
   A. mitochondria
   B. Golgi apparatus
   C. rough and smooth endoplasmic reticulum
   D. lysosomes
   E. plasma membrane

7. Which one of the following statements about prokaryotic and eukaryotic cells is *false*?
   A. Only eukaryotic cells have several membrane-enclosed organelles.
   B. Prokaryotic cells are older.
   C. Eukaryotic cells are smaller.
   D. Eukaryotic cells divide the labor of life among many internal compartments.
   E. Only eukaryotic cells have membrane-enclosed organelles.

8. **True or False?** Bacteria and archaea are <u>prokaryotic</u> cells. ______________

9. Cell surfaces are best revealed by a(n) ______________ electron microscope.

10. A nucleus bordered by a membranous envelope is found in ______________ cells, but not in ______________ cells.

Membrane Structure and Function

11. Which one of the following statements about cellular membranes is *false*?
    A. Membrane molecules can move freely past one another.
    B. Most membranes have carbohydrates embedded in the phospholipid bilayer.
    C. Proteins float like icebergs in the phospholipid sea.
    D. Membranes of a cell are composed mostly of lipids and proteins.
    E. Phospholipids in the membrane form a bilayer.

12. **True or False?** The plasma membrane and other membranes of the cell are <u>selectively permeable</u>. ____________________

13. The phospholipids and most of the proteins in a membrane are free to drift about in what is called the ____________________.

The Nucleus and Ribosomes: Genetic Control of the Cell

14. Which one of the following statements about the nucleus and ribosomes is *false*?
    A. Chromosomes are composed of long strands of DNA attached to certain proteins.
    B. Pores in the nuclear envelope allow messenger RNA to move from the nucleus to the cytoplasm.
    C. Ribosomes are constructed in the cytoplasm from parts produced in the nucleolus.
    D. DNA moves from the nucleus to the cytoplasm to direct the production of proteins.
    E. Ribosomes may either work suspended in the cytosol or attached to the endoplasmic reticulum.

15. If we think of the cell as a factory, then the nucleus is its executive boardroom and the top managers are the
    A. genes.
    B. ribosomes.
    C. chromosomes.
    D. mitochondria.
    E. endoplasmic reticulum.

16. **True or False?** The directions to make a protein move from <u>DNA</u> to <u>messenger RNA</u> and then to <u>ribosomes</u>. ____________________

17. Proteins are made on ribosomes in the cytoplasm using the directions from a(n) ____________________ molecule.

The Endomembrane System: Manufacturing and Distributing Cellular Products

18. Which one of the following statements about the components of the endomembrane system is *false*?
    A. Products from the endoplasmic reticulum are modified in the Golgi apparatus.
    B. Lysosomes have several types of digestive functions.
    C. Lipids are typically produced by the rough endoplasmic reticulum.
    D. Central vacuoles contribute to plant growth and may contain pigments and poisons.
    E. The smooth endoplasmic reticulum helps detoxify poisons.

19. Examine Figures 4.11–4.14 of your text. Which one of the following sequences best represents the steps an enzyme would take from initial production to joining with a food vacuole?
    A. rough endoplasmic reticulum, Golgi apparatus, lysosome
    B. Golgi apparatus, rough endoplasmic reticulum, lysosome
    C. lysosome, smooth endoplasmic reticulum, Golgi apparatus
    D. smooth endoplasmic reticulum, Golgi apparatus, lysosome

20. **True or False?** The final destination of proteins in the cell is determined by chemical tags applied by the rough endoplasmic reticulum. ____________________

21. Cellular products produced in the endoplasmic reticulum are next modified and packaged within transport vesicles in the ____________________.

22. Found in animal cells but absent from most plant cells, ____________________ are membrane-bounded sacs of digestive enzymes.

Chloroplasts and Mitochondria: Energy Conversion

23. In a plant cell, where is light energy trapped and converted to chemical energy?
    A. the inner mitochondrial membrane
    B. the outer mitochondrial membrane
    C. the outer chloroplast membrane
    D. the inner chloroplast membrane
    E. the grana in the chloroplast

24. Compare Figures 4.17 and 4.18 of your text. Which of the following pairs of structures are the largest membranes in chloroplasts and mitochondria?
    A. outer chloroplast membrane; outer mitochondrial membrane
    B. inner chloroplast membrane; outer mitochondrial membrane
    C. outer chloroplast membrane; inner mitochondrial membrane
    D. the grana of chloroplasts; the outer mitochondrial membrane
    E. the grana of chloroplasts; the inner mitochondrial membrane

25. **True or False?** Plant cells have chloroplasts and mitochondria.

____

26. Cells use molecules of ________________ as the direct energy source for most of their work.

The Cytoskeleton: Cell Shape and Movement

27. Which one, if any, of the following *is not* a function of the cytoskeleton?
    A. give mechanical support to the cell
    B. anchor organelles
    C. help a cell maintain its shape
    D. help a cell move
    E. All of the above are functions of the cytoskeleton.

28. Which one of the following statements about cilia and flagella is *false*?
    A. Cilia are generally shorter than flagella.
    B. Cilia are more numerous than flagella.
    C. Cilia and flagella have the same basic architecture.
    D. Flagella line the inside of your windpipe.
    E. Human sperm rely on flagella for movement.

29. **True or False?** A specialized arrangement of microfilaments helps move cilia and flagella. ________________

30. The movement of dividing chromosomes is guided by ________________.

Cell Surfaces: Protection, Support, and Cell-Cell Interactions

31. Which one of the following statements about plant cell walls is *false*? Plant cell walls
    A. are generally regarded as the boundary of the living cell.
    B. do not occur around animal cells.
    C. help maintain the shape of plant cells.
    D. provides strength and flexibility to plants.
    E. are composed of cellulose fibers in a matrix of other molecules.

32. What part of a house functions most like plasmodesmata?
    A. an outside window
    B. a door between rooms
    C. the air ducts coming from the furnace
    D. a wall between rooms
    E. the plumbing system

33. Steel-reinforced concrete is most like which one of the following?
    A. a plasma membrane
    B. a plant cell wall
    C. a nuclear membrane
    D. a lysosome
    E. a mitochondrion

34. **True or False?** Plasmodesmata are channels that connect the cytoplasm of adjacent animal cells. ________________

35. Instead of a cell wall, animal cells are surrounded by a(n) ________________.

36. Channels that allow water and other small molecules to flow between neighboring cells are called ________________ junctions.

37. Cells are bound together tightly by ________________ junctions.

38. Cells are attached together with fibers in ________________ junctions, but materials are still allowed to pass along the spaces between the cells.

Evolution Connection: The Origin of Membranes

39. The first cell membranes allowed cells to
    A. regulate their chemical exchanges with the environment.
    B. form an impermeable barrier between the inside and outside of the cell.
    C. use photosynthesis.
    D. reproduce.
    E. make ATP.

40. **True or False?** Phospholipids require no genetic information to assemble into membranes. ________________

41. The main lipids in the earliest cell membranes were probably ________________.

Word Roots

chloro = green (*chloroplast:* the organelle of photosynthesis)

chromo = color (*chromosome:* a threadlike, gene-carrying structure formed from chromatin)

cili = hair (*cilium:* a short, hairlike cellular appendage with a microtubule core)

cyto = cell (*cytoplasm:* cell region between the nucleus and the plasma membrane)

desma = a band or bond (*plasmodesmata:* an open channel in a plant cell wall)

endo = inner (*endocytosis:* the movement of materials into a cell

eu = true (*eukaryote:* cell type with membrane nucleus and other organelles)

exo = outer (*exocytosis:* the movement of materials out of a cell)

extra = outside (*extracellular:* the substance around animal cells)

flagell = whip (*flagellum:* a long, whiplike cellular appendage that moves cells)

karyon = kernel (*prokaryotic:* cell lacking a membrane-enclosed nucleus and other organelles)

micro = small (*microfilament:* the smallest part of the cytoskeleton, made of actin)

plasm = molded (*cytoplasm:* the thin outer membrane of a cell)

pro = before (*prokaryotic:* the first cells, lacking a membrane-enclosed nucleus and other organelles)

reticul = network (*endoplasmic reticulum:* membranous network where proteins are produced in the cytoplasm)

vacu = empty (*vacuole:* sac that buds from the ER, Golgi apparatus, or plasma membrane)

Key Terms

anchoring junctions
cell junctions
cell theory
cell wall
central vacuole
chloroplast
chromatin
chromosome
cilia
communicating junctions
cristae
cytoplasm
cytoskeleton
cytosol
electron microscope (EM)
endomembrane system
endoplasmic reticulum (ER)
eukaryotic cell
extracellular matrix
flagella
fluid mosaic
food vacuoles
gene
Golgi apparatus
grana
light microscope (LM)
lysosomal storage disease
lysosome
magnification
microtubule
mitochondria
nuclear envelope
nucleolus
nucleus
organelles
phospholipid
phospholipid bilayer
plasma membrane
plasmodesmata
prokaryotic cell
resolving power
ribosome
rough ER
scanning electron microscope (SEM)
selectively permeable
smooth ER
stroma
tight junctions
transmission electron microscope (TEM)
transport proteins
transport vesicles
vacuole

Crossword Puzzle

Use the Key Terms list from this chapter to fill in the crossword puzzle.

ACROSS

1. a photosynthetic organelle found in plants and some protists
9. the chromosome-containing organelle of a eukaryotic cell
10. a type of system of membranous organelles dividing eukaryotic cells into functional compartments
12. types of intercellular junctions in animal cells that prevents the leakage of material between cells
15. a type of junction connecting cells within a tissue to one another
17. the membrane at the boundary of every cell that acts as a selective barrier, thereby regulating the cell's chemical composition
19. everything inside a cell between the plasma membrane and the nucleus
20. Initials for an instrument that focuses an electron beam through a specimen or onto its surface
21. a digestive organelle in eukaryotic cells
23. cellular appendages similar to flagella, it moves some protists through water or fluids across cells
24. a tiny membranous sac in a cell's cytoplasm carrying molecules produced by the cell
26. a discrete unit of hereditary information
28. a property of biological membranes that allows some substances to cross more easily than others
31. initials for an optical instrument with lenses that bend visible light to magnify images
33. similar to cilia in structure, these cellular appendages propel protists and sperm
34. a network of fine fibers that provides structural support for a eukaryotic cell
35. types of junctions between cells through which small molecules pass
36. the semifluid medium of a cell's cytoplasm
37. membrane-enclosed sac in a eukaryotic cell with diverse functions
38. a sticky coat secreted by most animal cells

DOWN

2. a bilayer of biological membranes with a polar (hydrophilic) head and a nonpolar (hydrophobic) tail
3. initials for a type of microscope that uses an electron beam to study a specimen's surface
4. a structure with a specialized function within a cell
5. a cell organelle that functions as the site of protein synthesis in the cytoplasm
6. the open channels in a plant cell wall, through which strands of cytoplasm connect to other cells
7. a measure of the clarity of an image
8. a membrane-enclosed sac occupying most of the interior of a mature plant cell
11. the simplest type of digestive cavity, found in protists
13. a type of cell that has a membrane-enclosed nucleus and organelles
14. a threadlike, gene-carrying structure found in the nucleus of a eukaryotic cell
16. the combination of DNA and proteins that constitutes eukaryotic chromosomes
17. a membrane fat with a hydrophilic "head" and two hydrophobic "tails."
18. types of junctions between cells or to an extracellular matrix
22. the thickest of the three main kinds of fibers making up the cytoskeleton of a eukaryotic cell
25. a type of theory that all living things are composed of cells and all cells come from other
27. the place in a eukaryotic cell nucleus where ribosomes are made
29. a protective layer external to the plasma membrane of plant cells
30. initials for a membranous organelle connected to the outer nuclear membrane
32. a network of interconnected membranous sacs studded with ribosomes in a eukaryotic cell

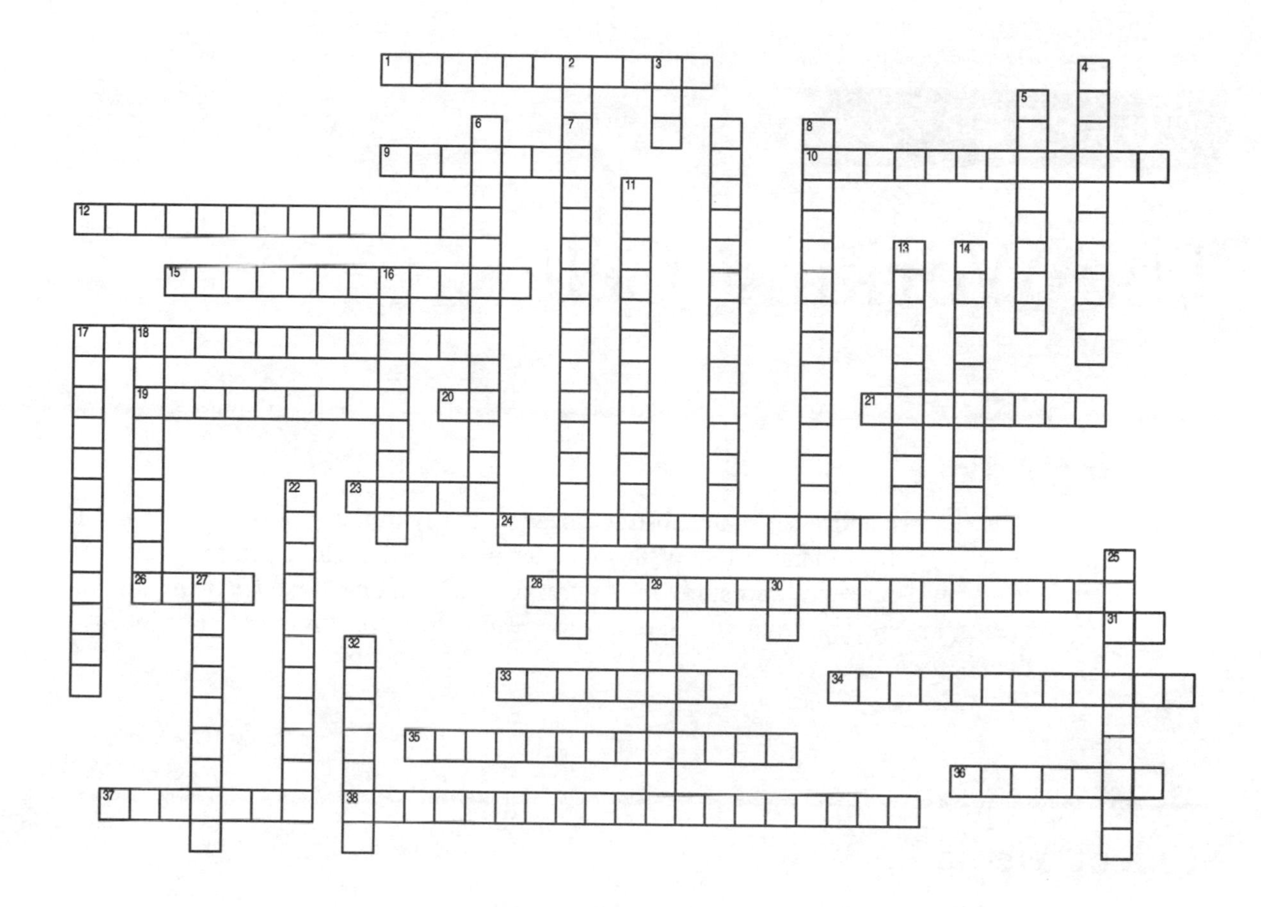
1
2
3
4
5
6
7
8
9
10
11
12
13
14
15
16
17
18
19
20
21
22
23
24
25
26
27
28
29
30
31
32
33
34
35
36
37
38

CHAPTER 5

The Working Cell

Studying Advice

This chapter addresses many abstract ideas that may initially be difficult to grasp. Do not try to understand all of this chapter in a single night. Read the material slowly and carefully, and study the figures as you go along. Take frequent study breaks, and review what you have already read before continuing further into the chapter.

Student Media

Biology and Society on the Web

Learn more about the commercial application of enzymes.

Activities

5A Energy Concepts
5B The Structure of ATP
5C How Enzymes Work
5D Membrane Structure
5E Selective Permeability of Membranes
5F Diffusion
5G Facilitated Diffusion
5H Osmosis and Water Balance in Cells
5I Active Transport
5J Exocytosis and Endocytosis
5K Cell Signaling

Case Studies in the Process of Science

How Is the Rate of Enzyme Catalysis Measured?
How Does Osmosis Affect Cells?
How Do Cells Communicate with Each Other?

Evolution Connection on the Web

Enzymes are very important to life. How did life and enzymes get together?

Organizing Tables

Compare the reactants, products, and efficiencies of cellular respiration and the use of gasoline in an automobile engine by completing the table below.

TABLE 5.1

	Reactants	**Products**	**Efficiency (Percent of Energy Used to Do Work)**
Cellular respiration			
Burning of fuel in an automobile engine			

Compare the reactions of animal and plant cells when placed into isotonic, hypotonic, and hypertonic solutions. Compare your results to Figure 5.13 of your text.

TABLE 5.2

	Isotonic Solution	**Hypotonic Solution**	**Hypertonic Solution**
Animal cell			
Plant cell			

Content Quiz

Directions: Identify the *one* best answer for the multiple-choice questions. For true/false questions, determine if the statement is true or false. If false, change the underlined word(s) to make the statement true. Finally, add the correct word(s) to the fill-in-the-blank questions to make the statements true.

Biology and Society: Stonewashing Without the Stones

1. Which one of the following statements about cellulase is *false*? Cellulase
   A. is a protein.
   B. breaks down cellular membranes.
   C. is a catalyst.
   D. is used by bacteria and fungi to break down plant material.
   E. is used in the process of "biostoning."

2. **True or False?** Using pumice stones is more friendly to the environment than using enzymes to produce stonewashed jeans. ____________

3. The enzyme cellulase breaks down the polysaccharide ____________.

Some Basic Energy Concepts

Conservation of Energy

4. You are riding a bike up and down hills. At which point do you have the greatest potential energy?
   A. at the bottom of the hill
   B. at the top of the hill
   C. riding down the hill at the fastest speed
   D. climbing the hill

5. **True or False?** Energy is defined as the ability to do work. ____________

6. The principle known as ____________ states that it is not possible to destroy or create energy.

Entropy

7. All energy conversions
   A. destroy some energy.
   B. decrease the entropy of the universe.
   C. generate some heat.
   D. produce ATP.

8. **True or False?** The heat produced by an automobile engine is a type of kinetic energy. ________________

9. The term ________________ is used as a measure of disorder or randomness.

Chemical Energy

10. We feel warmer when we exercise because of extra heat produced by
    A. cellular respiration.
    B. breathing faster.
    C. friction of blood flowing through the body.
    D. our movement through the air around us.
    E. sweating.

11. Examine Figure 5.3 of your text. Which one of the following *is not* produced by the engine and cellular respiration?
    A. heat
    B. water
    C. oxygen
    D. carbon dioxide

12. **True or False?** The chemical energy in molecules of glucose and other fuels is a special type of kinetic energy. ________________

13. Molecules of carbohydrates, fats, and gasoline all have structures that make them especially rich in ________________ energy.

Food Calories

14. The energy in a 300 calorie cheeseburger could raise the temperature of
    A. 1 kilogram (kg) of water by 30°C.
    B. 3 kg of water by 100°C.
    C. 3 kg of water by 1°C.
    D. 30 kg of water by 30°C.
    E. 30 kg of water by 100°C.

15. **True or False?** The calories in food are a form of potential energy.

    ____

16. The amount of energy needed to raise 1 gram of water 1°C is a(n) ________________ while 1,000 times this amount of energy is a(n) ________________.

ATP and Cellular Work

The Structure of ATP

17. **True or False?** ATP has <u>less</u> potential energy than ADP. ________________

18. The energy for most cellular work is found in the ________________ of the ATP molecule.

Phosphate Transfer

19. Moving ions across cell membranes is an example of what type of ATP work?
    A. chemical work
    B. transport work
    C. neutral work
    D. mechanical work

The ATP Cycle

20. What happens to the ADP molecule produced when ATP loses a phosphate during an energy transfer?
    A. ADP is used to build parts of cells.
    B. ADP is released from the cells.
    C. ADP is broken down further into separate carbon atoms.
    D. Energy from cellular respiration is used to convert ADP back to ATP.

21. Which one of the following processes is most like energy coupling?
    A. burning coal to produce electricity to run a factory
    B. running a furnace to heat a home
    C. using a car to drive to work
    D. using the energy from the sun to make sugar

22. During the process of energy coupling, ________________ molecules are recycled.

Enzymes

23. The many chemical reactions that occur in organisms are collectively called ________________.

Activation Energy

24. **True or False?** Enzymes <u>raise</u> the activation energy to break the bonds of reactant molecules. ________________

Induced Fit

25. When a substrate molecule slips into an active site, the active site changes shape slightly to embrace the substrate and catalyze the reaction. This interaction is called
    A. substrate locking.
    B. active engagement.
    C. active site shifting.
    D. induced fit.
    E. substrate embracing.

26. The reactant molecule called the ________________ binds at the enzyme's ________________.

Enzyme Inhibitors

27. Examine the role of an inhibitor in Figure 5.10 of your text. The inhibitor functions most like
    A. a teacher giving you instructions for a laboratory exercise.
    B. a friend who helps you do your laundry.
    C. a person who has parked in your parking spot.
    D. a screwdriver used to tighten screws.
    E. a traffic light indicating to a driver when it is safe to cross an intersection.

28. Many antibiotics that kill disease-causing bacteria are also enzyme ________________.

Membrane Transport

Passive Transport: Diffusion Across Membranes

29. Releasing fish into a pond and seeing them swim in all directions is most like the process of
    A. diffusion.
    B. osmosis.
    C. osmoregulation.
    D. active transport.
    E. pinocytosis.

30. Which one of the following most relies upon the process of diffusion?
    A. recognizing a friend eating at another table in a cafeteria
    B. discussing a subject in preparation for an exam
    C. selecting a new perfume or cologne
    D. listening to a song on the radio
    E. checking your pulse after exercise

Osmosis and Water Balance in Cells

31. If you soak your hands in dishwater, you may notice that your skin absorbs water and swells into distinct wrinkles. This is because your skin cells are ________________ to the ________________ dishwater.
    A. hypotonic, hypertonic
    B. hypertonic, hypotonic
    C. hypotonic, hypotonic
    D. isotonic, hypotonic
    E. hypertonic, isotonic

32. You decide to buy a new fish for your freshwater aquarium. When you introduce the fish into its new tank, the fish swells up and dies. You later learn it was a fish from the ocean. Based on what you know of tonicity, the most likely explanation is that unfortunate fish went from a(n) ________________ solution into a(n) ________________ solution.
    A. isotonic, hypotonic
    B. hypertonic, isotonic
    C. hypotonic, hypertonic
    D. hypotonic, isotonic

33. Examine Figure 5.13 of your text. Why does the animal cell but not the plant cell burst when it is in a hypotonic solution?
    A. The plant cell doesn't absorb as much water.
    B. The plant cell is less hypertonic.
    C. The cell wall keeps the cell from bursting.
    D. The water can not easily move through the plant cell wall.

34. When a *Paramecium* uses its contractile vacuole to remove excess water it is engaged in the process called ________________.

35. A solution with a higher solute concentration is ________________ to the solution with a lower solute concentration.

Active Transport: The Pumping of Molecules Across Membranes

36. Moving a molecule across a membrane and against its concentration gradient requires
    A. phospholipids using passive transport.
    B. phospholipids using active transport.
    C. membrane transport proteins using active transport.
    D. membrane transport proteins using passive transport.
    E. membrane transport proteins using receptor-mediated endocytosis.

37. **True or False?** Membrane proteins using <u>passive</u> transport require ATP as a source of energy. ____________

Exocytosis and Endocytosis: Traffic of Large Molecules

38. Which one of the following is a type of endocytosis in which very specific molecules are brought into a cell using specific membrane protein receptors?
    A. osmoregulation
    B. phagocytosis
    C. pinocytosis
    D. receptor-mediated endocytosis
    E. None of the above are correct.

39. In ____________, the cell gulps droplets of fluid by forming tiny vesicles.

The Role of Membranes in Cell Signaling

40. In a cell, the ____________ relays a signal and converts it to chemical forms that work within the cell.
    A. process of exocytosis
    B. process of endocytosis
    C. process of pinocytosis
    D. receptor protein
    E. signal-transduction pathway

41. **True or False?** The three stages of cell signaling are <u>reception</u>, <u>transduction</u>, and <u>response</u>. ____________

Evolution Connection: Evolving Enzymes

42. Which one of the following statements is *false*? Directed evolution
    A. has a specific purpose chosen by the researchers.
    B. may require just a few weeks to produce a new enzyme.
    C. relies upon the natural environment.
    D. can be used to produce enzymes that perform a new function.

43. **True or False?** Natural selection <u>does not</u> have purpose or direction.

    ____

44. Research data suggest that many of our human genes arose through ____________ evolution.

Word Roots

kinet = move (*kinetic:* type of energy, it is the energy of motion)

hyper = excessive (*hypertonic:* in comparing two solutions, it refers to the one with the greater concentration of solutes)

hypo = lower (*hypotonic:* in comparing two solutions, it refers to the one with the lower concentration of solutes)

iso = same (*isotonic:* solutions with equal concentrations of solutes)

phago = eat (*phagocytosis:* cellular eating)

pino = drink (*pinocytosis:* cellular drinking)

tonus = tension (*isotonic:* solutions with equal concentrations of solutes)

Key Terms

activation energy
active site
active transport
ADP
ATP
calorie
cellular respiration
chemical energy
conservation of energy
diffusion
energy
energy coupling
entropy
enzyme
enzyme inhibitors
endocytosis
exocytosis
facilitated diffusion
feedback regulation
heat
hypertonic
hypotonic
induced fit
isotonic
kinetic energy
metabolism
osmoregulation
osmosis
passive transport
phagocytosis
pinocytosis
plasmolysis
potential energy
receptor-mediated endocytosis
signal-transduction pathway
substrate
transport proteins

Crossword Puzzle

Use the Key Terms from this chapter to fill in the crossword puzzle.

ACROSS

1. the passage of a substance across a biological membrane down its concentration gradient
3. a molecule composed of adenosine and two phosphate groups
5. the type of transport that moves a substance across a membrane against its concentration gradient
7. the amount of energy associated with the movement of the atoms and molecules in a body of matter
8. the energy of motion
9. a specific substance (reactant) on which an enzyme acts
10. cellular drinking
12. the capacity to perform work
13. the tendency of molecules to move from high to low concentrations
14. a phenomenon that occurs in plant cells in a hyptertonic environmen
16. the type of energy stored in the chemical bonds of molecules
22. the amount of energy that raises the temperature of 1 gram of water by 1°C
25. the principle that energy can neither be created nor destroyed
27. the transfer of energy from processes that yield energy to those that consume it
28. cellular eating
29. the control of water and solute balance in an organism
30. proteins that help move substances across a cell membrane

DOWN

2. a molecule composed of adenosine and three phosphate groups
3. the part of an enzyme molecule where a substrate molecule attaches
4. the aerobic harvesting of energy from food molecules
5. the type of energy that reactants must absorb before a chemical reaction will start
6. chemicals that interfere with an enzyme's activities
11. of two solutions, the one with the greater concentration of solutes
15. the passive transport of water across a selectively permeable membrane
17. of two solutions, the one with the lesser concentration of solutes
18. the many chemical reactions that occur in organisms
19. the movement of materials into the cytoplasm of a cell via membranous vesicles
20. having the same solute concentration as another solution
21. the movement of materials out of the cytoplasm of a cell via membranous vesicles
23. a measure of disorder, or randomness
24. the interaction between a substrate molecule and the active site of an enzyme
26. a protein that serve as biological catalysts

1
2
3
4
5
6
7
8
9
10
11
12
13
14
15
16
17
18
19
20
21
22
23
24
25
26
27
28
29
30

CHAPTER 6

Cellular Respiration: Harvesting Chemical Energy

Studying Advice

a. The organizing tables should help you understand the basics of cellular respiration. Focus first on the overall process. Then read to gather the details. Table 6.1 should help you organize the definitions of the many pairs of contrasting terms.

b. This chapter addresses many abstract ideas that may initially be difficult to grasp. Do not try to understand all of this chapter in a single night. Read the material slowly and carefully, and study the figures as you go along. Take frequent study breaks, and review what you have already read before continuing further into the chapter.

Student Media

Biology and Society on the Web

Learn about the proper ways to exercise.

Activities

6A Build a Chemical Cycling System
6B Overview of Cellular Respiration
6C Glycolysis
6D The Krebs Cycle
6E Electron Transport
6F Fermentation

Case Study in the Process of Science

How Is the Rate of Cellular Respiration Measured?

Evolution Connection on the Web

Learn about the problems life faced when oxygen first began to accumulate in the atmosphere.

Organizing Tables

Compare the definitions of the following pairs of terms.

TABLE 6.1	
Photosynthesis	Respiration
Autotroph	Heterotroph
Producers	Consumers
Aerobic	Anaerobic
Oxidation	Reduction
Obligate anaerobe	Obligate aerobe

Compare the processes of glycolysis, Krebs cycle, and electron transport chain in the table below. Some cells are already filled in to make the job a little easier!

TABLE 6.2

	Location	Reactants	Products	Energy Yield
Glycolysis			two molecules of pyruvic acid	
Krebs cycle		Acetyl CoA	CO_2	
Electron transport chain		NADH		

Compare the products of each of the following reactions.

TABLE 2.2

Process	Products
Fermentation in human muscles cells	
Fermentation in microorganisms	

Content Quiz

Directions: Identify the *one* best answer for the multiple-choice questions. For true/false questions, determine if the statement is true or false. If false, change the underlined word(s) to make the statement true. Finally, add the correct word(s) to the fill-in-the-blank questions to make the statements true.

Biology and Society: Feeling the "Burn"

1. Which one of the following *does not* occur while you are still within your aerobic capacity?
   A. You use aerobic metabolism.
   B. Oxygen is used to generate ATP.
   C. You reach the maximum rate at which oxygen can be taken in and used by muscle cells.
   D. Your muscles produce lactic acid.

2. **True or False?** Lactic acid is produced during aerobic metabolism.

   ____

3. The burning sensation of an over-exercised muscle is produced by the buildup of ____________________.

Energy Flow and Chemical Cycling in the Biosphere

Producers and Consumers

4. Which one of the following statements about photosynthesis is *false*? Photosynthesis
   A. occurs mainly within the roots of plants.
   B. provides the food for most ecosystems.
   C. provides the energy to power a chemical process that makes organic molecules.
   D. occurs within chloroplasts.

5. **True or False?** Chlorophyll in chloroplasts makes plant leaves appear green.

   ____

6. **True or False?** Plants and other photosynthetic organisms are the consumers in an ecosystem. ____________________

7. Organisms that *cannot* make organic molecules from inorganic ones are called ____________________.

Chemical Cycling Between Photosynthesis and Cellular Respiration

8. Plants use photosynthesis to join together
   A. carbon dioxide and water to make oxygen and glucose.
   B. carbon dioxide and glucose to make water and oxygen.
   C. carbon dioxide and oxygen to make water and glucose.
   D. glucose and oxygen to make water and carbon dioxide.
   E. water and oxygen to make glucose and carbon dioxide.

9. Cellular respiration
   A. occurs in chloroplasts.
   B. is used by plants, but not animals.
   C. is part of photosynthesis.
   D. harvests energy stored in sugars and other organic molecules.
   E. produces glucose and oxygen.

10. Examine Figure 6.3 of your text. In this ecosystem, the carbon dioxide that the trees use in photosynthesis came from
    A. the burning of fossil fuels.
    B. detritovores.
    C. rabbits.
    D. wolves.
    E. Any of the above may be used.

11. **True or False?** Cellular respiration occurs in plant and animal cells in organelles called <u>chloroplasts</u>. ____________________

12. Cellular respiration uses energy extracted from organic fuel to produce another form of cellular energy called ____________________.

13. Photosynthesis combines together ____________________ and ____________________, which are products of cellular respiration.

Cellular Respiration: Aerobic Harvest of Food Energy

The Relationship Between Cellular Respiration and Breathing

14. What do cellular respiration and breathing have in common? Both processes
    A. produce ATP.
    B. produce glucose.
    C. take in carbon dioxide and release oxygen.
    D. take in oxygen and release carbon dioxide.

15. **True or False?** An aerobic cellular process requires carbon dioxide.

____

16. The aerobic harvesting of chemical energy from organic food molecules defines

_______________.

The Overall Equation for Cellular Respiration

17. Which one of the following *is not* produced as a result of the breakdown of a single glucose molecule by cellular respiration?
    A. 6 molecules of carbon dioxide
    B. 6 molecules of water
    C. 6 molecules of oxygen
    D. up to 38 ATP molecules

18. **True or False?** Cellular respiration consists of many steps. _______________

19. Oxygen is a vital part of aerobic respiration because it accepts _______________ atoms from glucose.

The Role of Oxygen in Cellular Respiration

20. Which one of the following occurs during the transfer of hydrogen in cellular respiration?
    A. glucose is reduced, oxygen is oxidized, and energy is released
    B. glucose is oxidized, oxygen is reduced, and energy is released
    C. glucose is reduced, oxygen is oxidized, and energy is consumed
    D. glucose is oxidized, oxygen is reduced, and energy is consumed

21. Which one of the following *does not* occur during the cellular respiration of glucose?
    A. NAD^+ donates electrons to NADH.
    B. NADH transfers electrons from glucose to the top of the electron transport chain.
    C. Electrons cascade down the electron transport chain, giving up a small amount of energy with each transfer.
    D. Oxygen is the final electron acceptor at the bottom of the electron transport chain.

22. Examine Figure 6.6 of your text. The transfer of electrons during cellular respiration is most like
    A. a giant waterfall.
    B. walking down a set of stairs.
    C. climbing a mountain.
    D. running up a set of stairs.
    E. playing catch with a baseball.

23. **True or False?** In the electron transport chain, <u>NAD^+</u> functions like gravity, pulling electrons down the chain.

24. NADH undergoes the process of ________________ when it donates its electrons to the electron transport chain.

The Metabolic Pathway of Cellular Respiration

25. Which one of the following statements about cellular respiration is *false*?
    A. ATP synthase, located within the inner mitochondrial membrane, helps form ATP.
    B. Cellular respiration is a metabolic pathway consisting of more than two dozen chemical reactions, each catalyzed by a specific enzyme.
    C. Inside mitochondria, glycolysis joins a pair of three-carbon pyruvic acid molecules to form a molecule of glucose.
    D. The Krebs cycle breaks down molecules of acetyl CoA to release CO_2 and energy trapped by NADH.

26. The extensive infolding of the inner mitochondrial membrane is likely an adaptation to
    A. make it more difficult for oxygen to pass through.
    B. increase the flexibility of mitochondria.
    C. make it more difficult for glucose to pass through.
    D. increase the surface area of the electron transport system.
    E. increase the surface area for glycolysis.

27. Examine Figure 6.10 of your text. What happens to the two carbons in acetyl CoA during the Krebs cycle? The two carbons are
    A. attached to NADH.
    B. used to make glucose.
    C. released as CO_2.
    D. added to ADP to make ATP.
    E. destroyed in the process.

28. **True or False?** Most of the ATP from cellular respiration is powered by <u>glycolysis</u>. ________________

29. Energy from the electron transport chain pumps ________________ across the inner mitochondrial membrane.

30. Cellular respiration is an example of ________________, the sum of all of the chemical processes that occur in cells.

Fermentation: Anaerobic Harvest of Food Energy

Fermentation in Human Muscle Cells

31. Which one of the following statements about fermentation is *false*?
    A. Your cells can produce ATP when no oxygen is present.
    B. When your muscles use fermentation, glucose is produced.
    C. Fermentation in your cells uses the process of glycolysis.
    D. Fermentation produces about 2 ATP per glucose molecule.
    E. Fermentation occurs when oxygen levels are not sufficient to support cellular respiration.

32. **True or False?** Fermentation is an anaerobic process. ____________________

33. During the process of fermentation, pyruvic acid is converted to

    ____________________.

Fermentation in Microorganisms

34. Yeast used to make beer and bread use fermentation to produce
    A. ATP, carbon dioxide, and oxygen.
    B. ATP, water, and oxygen.
    C. ATP, ethyl alcohol, and carbon dioxide.
    D. ethyl alcohol and oxygen.
    E. carbon dioxide and water.

35. **True or False?** Although our cells behave as facultative anaerobes, our entire body is an obligate aerobe. ____________________

36. Some organisms, called ____________________ anaerobes, are unable to live in the presence of oxygen.

Evolution Connection: Life on an Anaerobic Earth

37. Glycolysis is considered to be an ancient metabolic process because
    A. oxygen was abundant in the early Earth atmosphere.
    B. glycolysis is found only in animals.
    C. glycolysis occurs in the cytosol.
    D. it produces more ATP per glucose molecule than cellular respiration.

38. **True or False?** The process of glycolysis may be more than 3 billion years old.

    ____

39. The early Earth atmosphere had very little ____________________, favoring anaerobic organisms that could use glycolysis.

Word Roots

aero = air (*aerobic:* chemical reaction using oxygen)

auto = self (*autotroph:* Greek word that means "self-feeders")

glyco = sweet (*glycolysis:* process that splits glucose into two molecules)

hetero = other (*heterotroph:* Greek work that means "other feeders")

lysis = split (*glycolysis:* process that splits glucose into two molecules)

photo = light (*photosynthesis:* process using light energy to make organic molecules)

troph = food (*autotroph:* Greek word that means "self-feeders")

Key Terms

aerobic
anaerobic
ATP synthase
autotroph
cellular respiration
consumer
creatine phosphate
electron transport
electron transport chain
facultative anaerobe
fermentation
glycolysis
heterotroph
Krebs cycle
metabolism
NADH
obligate aerobe
obligate anaerobe
oxidation
photosynthesis
producer
redox reaction
reduction

Crossword Puzzle

Use the Key Terms list from this chapter to fill in the crossword puzzle.

ACROSS

1. an organism that cannot survive without oxygen
5. a chemical reaction in which electrons are lost from one substance (oxidation) and added to another (reduction)
9. type of chemical reaction that does not use oxygen
10. cluster of proteins built into inner mitochondrial membrane
12. organism that makes its organic matter from inorganic nutrients
14. type of organism that uses photosynthesis
17. process that harvests energy stored in sugars
19. acceptance of electrons during a redox reaction
20. a redox reaction in which one or more electrons are transferred to carrier molecules
21. abbreviation for type of reaction that transfers electrons
22. an organism that cannot survive in the presence of oxygen

DOWN

1. loss of electrons during a redox reaction
2. process that splits glucose into two molecules
3. heterotroph that eats other heterotrophs or autotrophs
4. anaerobic harvest of food energy
6. a type of chain composed of electron-carrier molecules used to make ATP
7. compound in muscle cells that provides energy
8. a type of anaerobe that can make ATP using aerobic or anaerobic respiration
9. chemical reaction using oxygen
11. general term for all chemical processes in a cell
13. organism that cannot make its own organic food molecules
15. a molecule that carries electrons from glucose and other fuel molecules and deposits them at the top of an electron transport chain
16. type of cycle that continues breakdown of glucose after glycolysis
18. process using light energy to make organic molecules

1
2
3
4
5
6
7
8
9
10
11
12
13
14
15
16
17
18
19
20
21
22

CHAPTER 7

Photosynthesis: Converting Light Energy to Chemical Energy

Studying Advice

a. The organizing table in this chapter should help you understand the two basic steps of photosynthesis and compare these to cellular respiration.

b. Like Chapters 5 and 6, this chapter addresses many abstract ideas that may initially be difficult to grasp. Do not try to understand all of this chapter in a single night. Read the material slowly and carefully, and study the figures as you go along. Take frequent study breaks, and review what you have already read before continuing further into the chapter.

Student Media

Biology and Society on the Web

Learn more about the potential of biomass energy.

Activities

7A Plants in Our Lives
7B The Sites of Photosynthesis
7C Overview of Photosynthesis
7D Light Energy and Pigments
7E The Light Reactions
7F The Calvin Cycle
7G Photosynthesis in Dry Climates
7H The Greenhouse Effect

Case Studies in the Process of Science

How Does Paper Chromatography Separate Plant Pigments?

How Is the Rate of Photosynthesis Measured?

Evolution Connection on the Web

Learn more about the organisms that first released large amounts of oxygen into Earth's atmosphere.

Organizing Table

Compare the reactants, products, and location in the cell for the three reactions listed at the left.

TABLE 7.1

	Reactants	Products	Location In The Cell
Cellular Respiration			
Light Reactions			
Calvin Cycle			

Content Quiz

Directions: Identify the *one* best answer for the multiple-choice questions. For true/false questions, determine if the statement is true or false. If false, change the underlined word(s) to make the statement true. Finally, add the correct word(s) to the fill-in-the-blank questions to make the statements true.

Biology and Society: Plant Power

1. During most of human history, what has been the main source of energy for heat, light, and fuel for cooking?
   A. wood
   B. oil
   C. coal
   D. natural gas
   E. nuclear

2. Burning wood for fuel has many advantages over using fossil fuels. Which one of the following *is not* an advantage of burning wood instead of fossil fuels?
   A. Energy plantations provide wildlife habitat.
   B. Energy plantations reduce soil erosion.
   C. Wood has more sulfur than fossil fuels.
   D. Growing trees for fuel reduces carbon dioxide from the air.

3. **True or False?** All of the food consumed by humans can be traced back to animals. ____________________

4. Over the last century, wood has been largely displaced as an energy source by ____________________.

The Basics of Photosynthesis

Chloroplasts: Sites of Photosynthesis

5. Photosynthesis commonly occurs when sunlight hits chlorophyll pigments in the
   A. thylakoid membrane of a mitochondrion in a cell of the mesophyll layer of a leaf.
   B. stroma of a chloroplast in a cell of the mesophyll layer of a leaf.
   C. thylakoid membrane of a chloroplast in a cell surrounding the stomata of a leaf.
   D. thylakoid membrane of a chloroplast in a cell of the mesophyll layer of a leaf.

6. Examine Figure 7.3 to compare the drawing and photograph of a chloroplast. How is the drawing different from the photograph of an actual chloroplast from a plant leaf?
   A. The grana are not stacked in the drawing.
   B. The grana are not interconnected in the drawing.
   C. There is no stroma in the drawing.
   D. The intermembrane space is much larger in the drawing.

7. **True or False?** Sugars are made in the grana of a chloroplast.

   ____

8. Carbon dioxide enters the leaf, and oxygen exits, through tiny pores called

   ____________________.

The Overall Equation For Photosynthesis

9. Which one of the following statements about photosynthesis is *false*? Photosynthesis
   A. uses the products of respiration.
   B. produces carbon dioxide and oxygen.
   C. uses energy from sunlight to split water and release oxygen into the atmosphere.
   D. boosts the energy in electrons "uphill" to join carbon dioxide to hydrogens from water.

10. **True or False?** Inside chloroplasts, carbon dioxide is split to form hydrogen and oxygen. ____________________

11. The largest molecule produced by photosynthesis is ____________________.

A Photosynthesis Road Map

12. ATP and NADPH, produced by the
    A. light reaction, are used to reduce carbon dioxide to form glucose.
    B. light reaction, are used to reduce glucose to form carbon dioxide.
    C. Calvin cycle, are used to reduce carbon dioxide to form glucose.
    D. Calvin cycle, are used to reduce glucose to form carbon dioxide.

13. Examine the overall equation for photosynthesis. Which one of the following is a product of the Calvin cycle?
    A. ATP
    B. NADPH
    C. glucose
    D. carbon dioxide
    E. water

14. **True or False?** The light reaction does not produce sugar. ________________

15. The Calvin cycle depends on light for a supply of ________________ and ________________.

The Light Reactions: Converting Solar Energy to Chemical Energy

The Nature of Sunlight

16. Which one of the following statements about radiation is *false*?
    A. Waves in the electromagnetic spectrum carry energy.
    B. The electromagnetic spectrum is the full range of radiation.
    C. Energy in wavelengths of visible light are absorbed by components of the Calvin cycle.
    D. The wavelengths of visible light that we see were not absorbed by what we are looking at.
    E. A wavelength is the distance between the crests of two adjacent waves.

17. **True or False?** We can see only a small portion of the entire electromagnetic spectrum. ________________

18. A dark blue shirt will absorb ________________ light energy than a very light blue shirt.

Chloroplast Pigments

19. Chlorophyll *b*
    A. absorbs mainly blue-violet and red light.
    B. participates directly in the light reactions.
    C. absorbs and dissipates excessive light that could damage chlorophyll.
    D. participates indirectly in the light reactions by transferring light energy to chlorophyll *a*.

20. **True or False?** Chloroplast pigments are all located within the outer membrane of the chloroplast. ________________

21. Only the pigment ________________ participates directly in the light reactions.

How Photosystems Harvest Light Energy

22. The photosynthetic pigments work together to function most like
    A. a telephone.
    B. a brain.
    C. an antenna.
    D. an umbrella.
    E. a mirror.

23. Which one of the following is the general sequence of energy transfer during photosynthesis? Light energy in a photon is transferred first to
    A. a primary electron acceptor, then to chlorophyll *a*, and finally to pigment molecules.
    B. pigment molecules, then to a primary electron acceptor, and finally to chlorophyll *a*.
    C. pigment molecules, then to chlorophyll *b*, and finally to a primary electron acceptor.
    D. pigment molecules, then to chlorophyll *a*, and finally to a primary electron acceptor.

24. **True or False?** The reaction center of a photosystem consists of a chlorophyll *a* molecule next to a primary electron acceptor. ________________

25. A fixed quantity of light energy is a(n) ________________.

How the Light Reactions Generate ATP and NADPH

26. Which one of the following parts of a mitochondrion functions most like the thylakoid membrane?
    A. mitochondrial matrix
    B. outer mitochondrial membrane
    C. inner mitochondrial membrane
    D. the intermembrane space

27. The extensive infolding of the inner thylakoid membrane is likely an adaptation to
    A. make it more difficult for oxygen to pass through.
    B. increase the flexibility of chloroplasts.
    C. make it more difficult for hydrogen ions to pass through.
    D. increase the surface area of the electron transport system.
    E. increase the surface area for glycolysis.

28. Examine Figures 7.10, 7.11, and 7.12. Which one of the following steps in Figure 7.10 pumps H^+ across the thylakoid membrane to concentrate H^+ in the thylakoid space?
   A. water-splitting photosystem
   B. electron transport chain
   C. NADPH-producing photosystem
   D. Calvin cycle

29. **True or False?** ATP production in cellular respiration and photosynthesis both have ATP synthases that use the energy stored by the <u>H^+ gradient</u>.

   ____

30. The second photosystem produces ______________.

The Calvin Cycle: Making Sugar from Carbon Dioxide

Water-Saving Adaptations of C_4 and CAM Plants

31. Which one of the following are the products of the Calvin cycle?
   A. $NADP^+$, ADP + P, and G3P
   B. NADPH, ATP, and glucose
   C. $NADP^+$, ADP + P, and glucose
   D. NADPH, ATP, and G3P
   E. G3P and glucose

32. Which one of the following parts of a plant would most likely be very active at night?
   A. mitochondria
   B. the intermembrane space of the chloroplasts
   C. the stroma of the chloroplasts
   D. the thylakoid membranes of the chloroplasts.

33. **True or False?** Plant cells use <u>G3P</u> to make glucose and other organic molecules it needs. ______________

34. Photosynthesis swaps atmospheric ______________ for atmospheric carbon dioxide.

The Environmental Impact of Photosynthesis

How Photosynthesis Moderates the Greenhouse Effect

35. Deforestation contributes to global warming by
    A. decreasing the number of producers and adding CO_2 to the atmosphere.
    B. decreasing the number of consumers and adding CO_2 to the atmosphere.
    C. increasing the number of producers and removing CO_2 from the atmosphere.
    D. increasing the number of consumers and removing CO_2 from the atmosphere.

36. **True or False?** Increasing the total amount of photosynthesis on Earth would increase global warming by removing CO_2 from the atmosphere.

    ____

37. The ________________ is a gradual warming of the Earth caused by increased atmospheric carbon dioxide.

Evolution Connection: The Oxygen Revolution

38. Large amounts of oxygen were first added to the Earth's atmosphere by
    A. land plants.
    B. algae in the oceans.
    C. cyanobacteria.
    D. fungi.
    E. the greenhouse effect.

39. **True or False?** The widespread production of oxygen by cyanobacteria 2.5–3.4 billion years ago permitted the evolution of photosynthetic organisms.

    ____

40. Many anaerobic prokaryotes die if they are exposed to ________________.

Word Roots

chloro = green (*chloroplast:* the organelle of photosynthesis)
meso = middle (*mesophyll:* the green tissue in the middle/inside of a leaf)
phyll = leaf (*chlorophyll:* photosynthetic pigment in chloroplasts)
photo = light (*photosystem:* cluster of pigment molecules)

Key Terms

C_3 plants
C_4 plants
Calvin cycle
CAM plants
chlorophyll *a*
chloroplast
electromagnetic spectrum
global warming
grana
greenhouse effect
light reaction
mesophyll
NADPH
photon
photosystem
primary electron acceptor
reaction center
stomata
stroma
thylakoids
wavelength

Crossword Puzzle

Use the Key Terms list from this chapter to fill in the crossword puzzle.

ACROSS

2. the chlorophyll-a molecule and the primary electron acceptor that trigger the light reactions of photosynthesis
4. membranous sacs suspended in the stroma
6. distance between crests of adjacent light waves
9. the type of plant that uses crassulacean acid metabolism
11. changes in the surface air temperature induced by emission of greenhouse gases
12. reaction converting solar energy to chemical energy
13. pigment that absorbs blue-violet and red light
15. the organelle of photosynthesis
16. the green tissue in the interior of a leaf, it is the main site of photosynthesis
17. the type of spectrum of all types of radiation

DOWN

1. the process of atmospheric carbon dioxide trapping heat
3. a molecule that traps the light-excited electron from the reaction-center chlorophyll in a photosystem
5. thick fluid deep inside a chloroplast
7. stacks of hollow disks inside chloroplasts
8. a fixed quantity of light energy
9. process that makes sugar from carbon dioxide
10. cluster of pigment molecules
14. electron carrier involved in photosynthesis

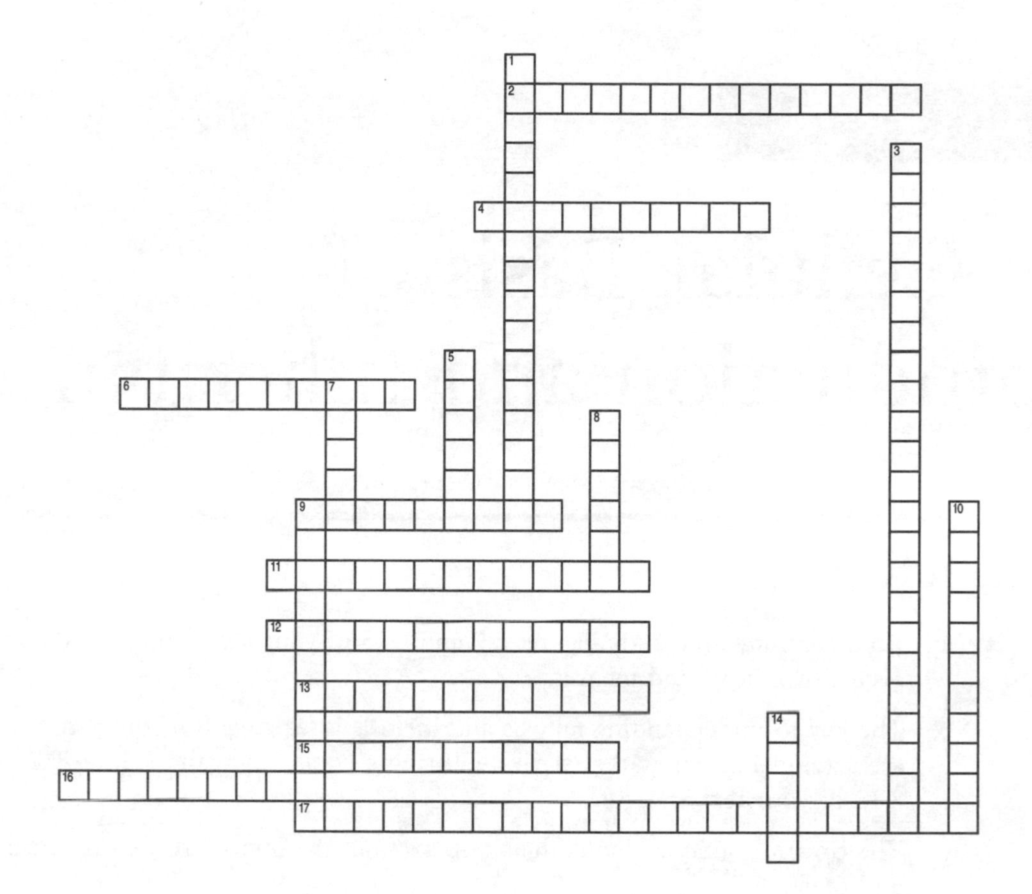
1
2
3
4
5
6
7
8
9
10
11
12
13
14
15
16
17

CHAPTER 8

The Cellular Basis of Reproduction and Inheritance

Studying Advice

a. Read carefully and study Figures 8.7 and 8.15 of your text describing the key events of mitosis and meiosis.

b. The key to understanding mitosis and meiosis is learning how the chromosomes are arranged and how they separate during metaphase. Again, Figures 8.7 and 8.15 of your text are crucial.

c. The organizing tables should help you sort out the important details of mitosis and meiosis.

Student Media

Biology and Society on the Web

Learn more about controversial reproductive technologies.

Activities

8A Asexual and Sexual Reproduction
8B The Cell Cycle
8C Mitosis and Cytokinesis Animation
8D Mitosis and Cytokinesis Video
8E Causes of Cancer
8F Human Life Cycle
8G Meiosis Animation
8H Origins of Genetic Variation
8I Polyploid Plants

Case Studies in the Process of Science

How Much Time Do Cells Spend in Each Phase of Mitosis?

How Can the Frequency of Crossing Over Be Estimated?

Evolution Connection on the Web

Learn more about the role of polyploidy in the evolution of plants.

Organizing Tables

Compare the key events of each stage of the cell cycle.

TABLE 8.1

	Key Events
M Phase	
G_1 Phase	
S Phase	
G_2 Phase	

Compare the following aspects of the G_2 phase of interphase and the stages of mitosis.

TABLE 8.2

Stages	The Arrangement and Behavior of the Chromosomes	The Structure and Functions of the Mitotic Spindle	Other Cellular Details
G_2 interphase: Preparing for mitosis			
Prophase			
Metaphase			
Anaphase			
Telophase			

Compare the following aspects of meiosis I and meiosis II.

TABLE 8.3

Stages of Meiosis	How Are the Chromosomes Arranged During Metaphase? (Draw the arrangement of two tetrads.)	What Separates During Anaphase: Homologous Chromosomes or Sister Chromatids?	Where Does Crossing Over Begin?
Meiosis I			
Meiosis II			

Content Quiz

Directions: Identify the *one* best answer for the multiple-choice questions. For true/false questions, determine if the statement is true or false. If false, change the underlined word(s) to make the statement true. Finally, add the correct word(s) to the fill-in-the-blank questions to make the statements true.

Biology and Society: A $50,000 Egg!

1. Which one of the following does *not* occur in the process of in vitro fertilization (IVF)?
   A. cloning sperm cells in a petri dish
   B. joining sperm and egg in a petri dish
   C. allowing them to grow into an eight-cell embryo
   D. implanting the embryo into the uterus of the mother

2. **True or False?** Infertility affects one in a thousand American couples.

   ____

3. The inability to produce children through normal sexual means after one year of trying defines ________________.

What Cell Reproduction Accomplishes

Passing on the Genes From Cell to Cell

4. Daughter cells formed during typical cell division have
   A. identical sets of chromosomes with identical genes.
   B. different sets of chromosomes with identical genes.
   C. identical sets of chromosomes with different genes.
   D. different sets of chromosomes with different genes.

5. **True or False?** Growth of an organism and the replacement of lost or damaged cells are the main roles of cell division. ________________

6. Before a cell divides, it duplicates its ________________.

The Reproduction of Organisms

7. Sexual reproduction
   A. uses only meiosis.
   B. involves the production of daughter cells with double the genetic material of the parent cell.
   C. uses only ordinary cell division.
   D. requires the fertilization of an egg by a sperm.
   E. is the normal process by which the cells of most organisms divide.

8. **True or False?** A sperm or egg has twice as many chromosomes as its parent cell. ___________________

9. In ___________________ reproduction, the offspring and parent have identical genes.

The Cell Cycle and Mitosis

Eukaryotic Chromosomes

10. Which one of the following statements about chromosomes is *false*?
    A. Sister chromatids remain attached until cell division occurs.
    B. The number of chromosomes in a eukaryotic cell depends upon the species.
    C. Chromosomes are made of a combination of DNA, carbohydrate, and lipid molecules.
    D. Human body cells each typically have 46 chromosomes.
    E. Once separated, sister chromatids go to different cells.

11. **True or False?** The lipids in chromosomes help control the activity of the genes.
    ____

12. Sister chromatids remain attached at a region called the ___________________.

The Cell Cycle

13. What phase of the cell cycle is marked by the presence of sister chromatids and the preparation of the cell for division?
    A. S
    B. M
    C. G_1
    D. G_2
    E. G_3

14. **True or False?** Some highly specialized cells do not undergo a cell cycle.
    ____

15. **True or False?** Prokaryotes do not undergo mitosis. ___________________

16. The M phase of the cell cycle includes ___________________ and ___________________.

17. During the ___________________ phase of the cell cycle, the chromosomes are duplicated.

Mitosis and Cytokinesis

18. Which one of the following statements about mitosis is *false*?
    A. During the G_2 phase of interphase all of the chromosomes are distinct within the nucleus.
    B. During prophase, the sister chromatids become distinct, the nuclear envelope breaks up, and the mitotic spindle begins to form.
    C. During metaphase, the chromosomes line up in the middle of the mitotic spindle.
    D. At the start of anaphase, the sister chromatids are pulled apart and start to move toward opposite spindle poles.
    E. Telophase is like the opposite of prophase: the nuclear envelope reforms, the chromosomes uncoil, and the spindle disappears.

19. Dividing an animal cell in two is most like
    A. building a new wall that divides a room into two.
    B. overtightening the drawstring of sweatpants.
    C. peeling an apple.
    D. cutting a pie into two pieces with a knife.

20. Examine the prophase stages of mitosis in Figure 8.7 of your text. Which one of the following structures invades the nuclear space after the nuclear membrane disintegrates?
    A. chromosomes
    B. centrioles
    C. nucleolus
    D. spindle microtubules

21. Compare the stages in Figure 8.7 of the text to the photograph in Figure 8.3. What stage is indicated by the cell in Figure 8.3 of your text?
    A. interphase
    B. early prophase
    C. late prophase
    D. metaphase
    E. anaphase

22. In animal cells, spindle microtubules grow from ____________________, clouds of cytoplasmic material that contains centrioles.

Cancer Cells: Growing Out of Control

23. The most dangerous property of cancer cells is that they
    A. divide slowly.
    B. spread throughout the body.
    C. form tumors.
    D. do not divide by mitosis.
    E. do not live very long.

24. Which one of the following will *reduce* your risk of developing cancer?
   A. smoking
   B. overexposure to the sun
   C. exercising
   D. eating a low-fiber high-fat diet

25. **True or False?** Radiation and chemotherapy both fight cancer by <u>disrupting the cell cycle</u>. ____________________

26. An abnormal mass of cells that remains at its original site is a ____________________ tumor.

27. The use of drugs to disrupt division of cancerous cells is called ____________________.

Meiosis, the Basis of Sexual Reproduction

Homologous Chromosomes

28. A male human somatic cell has
   A. 46 pairs of chromosomes.
   B. 23 pairs of sex chromosomes.
   C. 23 pairs of autosomes.
   D. 22 pairs of autosomes and 1 pair of sex chromosomes.
   E. 22 pairs of sex chromosomes and 1 pair of autosomes.

29. **True or False?** Most of the chromosomes in humans are <u>sex chromosomes</u>. ____________________

30. The same sequence of genes is found on ____________________ chromosomes.

Gametes and the Life Cycle of a Sexual Organism

31. In the human life cycle, somatic cells are
   A. diploid, while gametes are diploid.
   B. diploid, while gametes are haploid.
   C. haploid, while gametes are diploid.
   D. haploid, while gametes are haploid.

32. **True or False?** If the human reproductive cycle only included diploid cells, the offspring of each generation would have <u>twice as much</u> genetic material in each cell as the parent cells. ____________________

33. If the somatic cells of a species have 25 pairs of homologous chromosomes, the number of chromosomes in haploid gametes would be ____________________.

The Process of Meiosis

34. Which one of the following statements about meiosis is *false*?
    A. Cells dividing by meiosis undergo two consecutive divisions.
    B. Homologous chromosomes may exchange segments before separating from each other.
    C. Homologous chromosomes separate during meiosis I.
    D. Sister chromatids separate during meiosis II.
    E. The chromosomes replicate before meiosis and between meiosis I and meiosis II.

35. In sexually reproducing organisms, cells entering mitosis have
    A. half the amount of DNA as cells entering meiosis.
    B. the same amount of DNA as cells entering meiosis.
    C. twice the amount of DNA as cells entering meiosis.
    D. four times the amount of DNA as cells entering meiosis.

36. **True or False?** Before crossing over occurs, sister chromatids are <u>identical</u>.

    ____

37. The result of meiosis is four (haploid, diploid) ____________________ cells.

Review: Comparing Mitosis and Meiosis

38. Which one of the following is an actual difference between mitosis and meiosis?
    A. A single cell is divided into two cells in mitosis and four cells in meiosis.
    B. Mitosis produces haploid cells, and meiosis produces diploid cells.
    C. Mitosis involves two cellular divisions, and meiosis has only one.
    D. The chromosomes replicate before mitosis and meiosis, but in meiosis they replicate again between the first and second division.

39. **True or False?** The two daughter cells produced by meiosis I <u>do not</u> undergo cytokinesis before meiosis II. ____________________

40. Sister chromatids are separated during mitosis and ____________________.

The Origins of Genetic Variation

41. The random segregation of one member of each homologous pair of chromosomes into gametes defines
    A. random fertilization.
    B. crossing over.
    C. independent assortment.
    D. random assortment.
    E. independent fertilization.

42. **True or False?** Crossing over only occurs in mitosis. ____________________

43. The site of crossing over is a called a(n) ____________________.

When Meiosis Goes Awry

44. In humans,
    A. the absence of a Y chromosome results in maleness.
    B. the presence of only a single X chromosome results in maleness.
    C. a human embryo with an abnormal number of chromosomes is usually normal.
    D. nondisjunction occurs only in males of sexually reproducing, diploid organisms.
    E. the incidence of Down syndrome increases markedly with the age of the mother.

45. Examine the chart in Figure 8.20 of your text. At which of the following maternal age intervals do we find the greatest *increase* in risk of having a child with Down syndrome?
    A. 25–30 years
    B. 30–35 years
    C. 35–40 years
    D. 40–45 years
    E. 45–50 years

46. **True or False?** Klinefelter syndrome and Turner syndrome are results of nondisjunction of autosomes. ____________________

47. A person with trisomy 21 is said to have ____________________.

Evolution Connection: New Species from Errors in Cell Division

48. Polyploid species
    A. are common in mammals.
    B. are more common in animals than plants.
    C. can result if gametes are produced by mitosis.
    D. have two sets of chromosomes in each somatic cell.

49. **True or False?** At least half of all species of flowing plant are polyploid.

    ____

50. The union of a diploid egg and a diploid sperm will produce a

    ____________________ zygote.

Word Roots

a = not or without (*asexual:* type of reproduction not involving fertilization)

auto = self (*autosome:* the chromosomes that do not determine gender)

carcin = an ulcer (*carcinoma:* cancers originating in the external or internal coverings of the body)

centro = the center; **mere** = a part (*centromere:* the centralized region joining two sister chromatids)

chemo = chemical (*chemotherapy:* type of cancer therapy using drugs that disrupt cell division)

chroma = colored (*chromosome:* DNA-containing structure)

cyto = cell (*cytokinesis:* division of the cytoplasm)

di = two (*diploid:* cells that contain two homologous sets of chromosomes)

fertil = fruitful (*fertilization:* process of fusion of sperm and egg cell)

haplo = single (*haploid:* cells that contain only one chromosome of each homologous pair)

homo = like (*homologous:* like chromosomes that form a pair)

inter = between (*interphase:* time when a cell metabolizes and performs its various functions)

karyo = nucleus (*karyotype:* a display of the chromosomes of a cell)

kinet = move (*cytokinesis:* division of the cytoplasm)

poly = many (*polyploid:* having more than two sets of homologous chromosomes in each somatic cell)

sarco = flesh (*sarcoma:* cancers that arise in tissues that support the body)

soma = body (*somatic:* body cells with 46 chromosomes in humans)

Key Terms

anaphase
asexual reproduction
autosome
benign tumor
cancer cells
carcinomas
cell control cycle system
cell cycle
cell division
cell plate
centromere
centrosome
chemotherapy
chiasma
chromatin
chromosome
cleavage furrow
crossing over
cytokinesis
diploid
Down syndrome
fertilization
gamete
genetic recombination
genome
haploid
histones
homologous chromosomes
interphase
karyotype
leukemia
life cycle
lymphoma
malignant tumor
meiosis
metaphase
metastasis
mitosis
mitotic phase
mitotic spindle
nucleosome
nondisjunction
polyploid
prophase
radiation therapy
sarcoma
sexual reproduction
sister chromatids
somatic cell
telophase
trisomy 21
zygote

Crossword Puzzle

Use the Key Terms list from this chapter to fill in the crossword puzzle.

ACROSS

1. exchange of genetic material between homologous chromosomes
4. one of two identical parts of a replicated chromosome
6. division of the cytoplasm
8. the condition which causes Down syndrome (with "21")
10. cells that contain only one chromosome of each homologous pair
11. final mitotic stage when the nuclear envelope reforms
12. the division of a diploid nucleus into four haploid daughter nuclei
15. part of cell cycle when cell is dividing
16. the division of a diploid cell into two diploid cells
17. mitotic stage when sister chromatids separate
19. sequence of events from cell formation to cell division
20. football-shaped structure of microtubules important in mitosis
21. egg or sperm
23. cancers that arise in tissues that support the body
25. cancer of blood forming tissues
30. membranous disk containing cell wall material in plants
31. type of chromosomes that form a pair
33. clouds of cytoplasmic material with centrioles in animal cells
34. a person with trisomy 21
36. cancers originating in the external or internal coverings of the body
37. the sites where crossing-over has occurred
38. production of gene combinations unlike the parent chromosomes
39. the region where two chromatids are joined

DOWN

1. long strands of DNA attached to certain proteins
2. The entire sequence of stages in the life of an organism, from the adults of one generation to the adults of the next.
3. having more than two sets of homologous chromosomes in each somatic cell
5. time when a cell metabolizes and performs its various functions
7. a display of the chromosomes of a cell
9. process of fusion of haploid sperm and haploid egg cell
13. a disease where cells divide excessively and exhibit bizarre behavior
14. type of tumor that stays at its original site of origin
15. type of tumor that migrates away from its site of origin
17. type of reproduction not involving fertilization
18. the chromosomes that do not determine gender
19. indentation at the equator of a cell where it will divide
22. A complete (haploid) set of an organism's genes; an organism's genetic material.
24. the spread of cancer cells beyond their original site
26. cancer of the tissue that forms white blood cells
27. a thread-like, gene-carrying structure formed from chromatin
28. mitotic stage when the nuclear membrane first break ups
29. mitotic stage when the chromosomes are lined up in the cell's middle
32. cells that contain two homologous sets of chromosomes
35. the fertilized egg

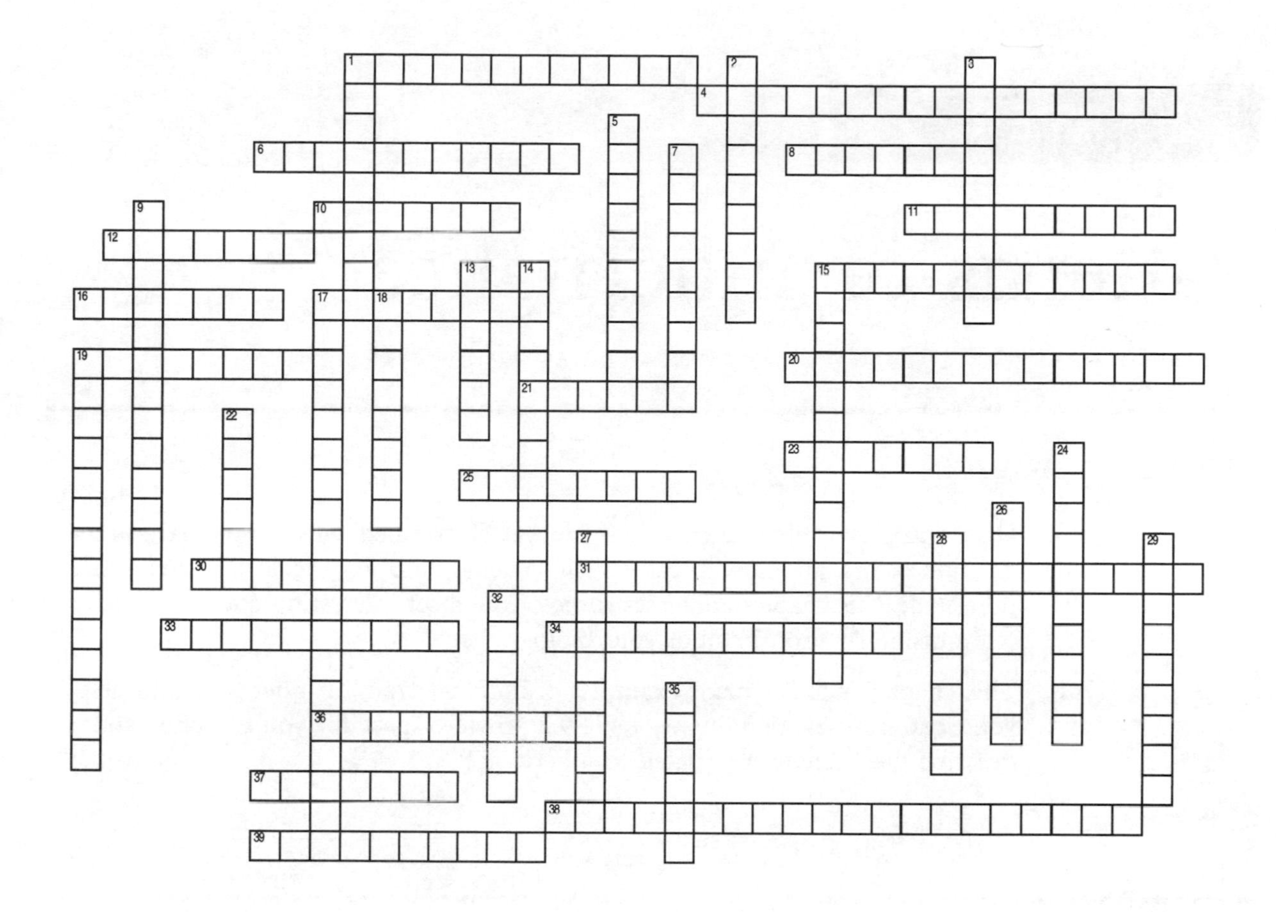

CHAPTER 9

Patterns of Inheritance

Studying Advice

a. Have you ever wondered how your sex was determined, why people have different skin colors, and how diseases such as hemophilia and sickle cell anemia are inherited? This chapter addresses some of the most interesting questions and relevant information in all of your biological studies!

b. This chapter includes many examples of different types of inheritance. To help you organize these definitions, Table 9.2 provides space for you to define and describe these key terms related to inheritance.

Student Media

Biology and Society on the Web

Read about socioeconomic factors related to fetal testing.

Activities

9A Monohybrid Cross

9B Dihybrid Cross

9C Gregor's Garden

9D Incomplete Dominance

9E Linked Genes and Crossing Over

9F Sex-Linked Genes

Case Study in the Process of Science

What Can Fruit Flies Reveal about Inheritance?

Evolution Connection on the Web

Learn more about how the Y chromosome is used to study ancestry.

Organizing Tables

The table below is a Punnett square showing the results of a cross between plants heterozygous for purple flower color. Write the genotypes of the plants in each square as either (a) homozygous dominant, (b) heterozygous, or (c) homozygous recessive. Next, write the phenotypes of the flowers for each square as either (a) purple or (b) white.

TABLE 9.1

Genotype *PP*	**Genotype *Pp***
Genotype in words: ______________ Phenotype: ______________	Genotype in words: ______________ Phenotype: ______________
Genotype *pP*	**Genotype *pp***
Genotype in words: ______________ Phenotype: ______________	Genotype in words: ______________ Phenotype: ______________

Organize and define the key terms related to inheritance.

TABLE 9.2

Term	**Definition**
Recessive disorders	
Dominant disorders	
Incomplete dominance	

continued

TABLE 9.2, *continued*	
Term	**Definition**
Multiple allele inheritance	
Codominance	
Pleiotropy	
Polygenic inheritance	
Linked genes	
Sex-linked genes	

Content Quiz

Directions: Identify the *one* best answer for the multiple-choice questions. For true/false questions, determine if the statement is true or false. If false, change the underlined word(s) to make the statement true. Finally, add the correct word(s) to the fill-in-the-blank questions to make the statements true.

Biology and Society: Testing Your Baby

1. When amniocentesis is performed,
   A. a piece of placenta is removed.
   B. sound waves are used to view a fetus.
   C. the fetus is viewed directly through a thin tube inserted into the uterus.
   D. amniotic fluid is sampled.

2. **True or False?** Amniocentesis or chorionic villus sampling could be used to test for Down syndrome. ____________________

3. Amniocentesis and chorionic villus sampling require the collection of ____________________.

Heritable Variation and Patterns of Inheritance

In an Abbey Garden

4. Crossing members of two different true-breeding varieties of organisms produces
   A. F_1 hybrids.
   B. F_2 hybrids.
   C. P_1 hybrids.
   D. P_2 hybrids.
   E. F_1 true breeders.

5. **True or False?** When Mendel wanted to cross-fertilize his pea plants, he covered the flower with a small bag. ____________________

6. Fertilization between members of the F_1 generation produces members of the ____________________ generation.

Mendel's Principle of Segregation

7. What will be the phenotype of two pea plants with the flower color genotypes *Pp* and *PP*?
   A. Both will be purple.
   B. Both will be white.
   C. One will be white and the other purple.
   D. A Punnett square is needed to figure this out.

8. Which of the following genotypes is homozygous for the dominant allele of purple flower color?
   A. *pp*
   B. *Pp*
   C. *pP*
   D. *PP*

9. Examine Figure 9.9 of your text. Did the three alleles on the top chromosome all come from just one of the homologous chromosomes of a parent?
   A. Yes
   B. No
   C. We can't be certain because crossing over could have occurred.

10. **True or False?** Mendel's principle of segregation states that pairs of alleles segregate during gamete formation and the fusion of gametes at fertilization creates allele pairs again. ________________

11. The physical location of a gene on a chromosome is that gene's ________________.

Mendel's Principle of Independent Assortment

12. If Mendel's principle of independent assortment did *not* apply to the pea shape and color experiment and the two traits in a dihybrid cross *were* inherited together, what would be the expected phenotypic ratio of the F_2 generation?
    A. 9:3:3:1
    B. 1:6:1
    C. 1:2:1
    D. 3:1
    E. 1:1

13. **True or False?** Mendel performed dihybrid crosses involving all seven of his pea characteristics and found 9:3:3:1 ratios in half of them. ________________

14. Mendel's pea shape and color experiment showed that this dihybrid cross was the equivalent of two ________________ crosses occurring simultaneously.

Using a Testcross to Determine an Unknown Genotype

15. A testcross is a mating between
    A. an unknown genotype and a homozygous dominant individual.
    B. an unknown genotype and a heterozygous individual.
    C. an unknown genotype and a homozygous recessive individual.
    D. two heterozygous individuals.
    E. a homozygous dominant and a homozygous recessive individual.

16. **True or False?** Mendel used testcrosses to determine whether he had true-breeding varieties of plants. ________________

17. Testcrosses are used to determine the ________________ of an organism.

The Rules of Probability

18. Using a standard deck of 52 playing cards, the chance of drawing a red card is 1/2 and the chance of drawing a queen is 1/13. To determine the probability of drawing a red queen, we should use the rule of
    A. addition.
    B. subtraction.
    C. multiplication.
    D. division.

19. **True or False?** The probability of producing complex genotypes from parents of known genotypes can be calculated by using the rule of multiplication or <u>a Punnett square</u>. ____________________

Family Pedigrees

20. Consider a trait in which *T* represents the dominant and *t* represents the recessive alleles. Which of the following genotypes will exhibit the recessive phenotype?
    A. *TT*
    B. *Tt*
    C. *tT*
    D. *tt*
    E. More than one of the above are correct.
    F. None of the above are correct.

21. Examine the patterns of inheritance of deafness in Figures 9.15 and 9.16 of your text. Does it appear that this inheritance pattern is sex-linked?
    A. Yes
    B. No
    C. We can't be certain because crossing over could have occurred.

22. **True or False?** <u>Dominant phenotypes</u> are more common than <u>recessive phenotypes</u>. ____________________

23. **True or False?** A family <u>pedigree</u> is used to assemble information about the occurrence of heritable characters in parents and their offspring across several generations. ____________________

24. People who have one copy of an allele for a recessive disorder but do not show symptoms are ____________________ of the disorder.

Human Disorders Controlled By a Single Gene

25. Dominant lethal alleles
    A. are more common than lethal recessive alleles.
    B. are commonly carried by heterozygotes without affecting them.
    C. are never inherited from a heterozygote.
    D. cause most human genetic disorders.
    E. cause achondroplasia and Huntington disease.

26. **True or False?** Inbreeding is <u>less</u> likely to produce offspring that are homozygous for a harmful recessive trait. ____________________

27. Most people born with recessive disorders are born to parents who are both ____________________, or carriers for the recessive allele.

28. The illness called ________________ is a degeneration of the nervous system that usually does not begin until middle age.

29. The most common lethal genetic disease in the United States is ________________.

Beyond Mendel

Incomplete Dominance in Plants and People

30. If a mating occurs between two parent plants that are both heterozygous for a trait with incomplete dominance, the expected ratio of phenotypes will be
    A. 1:3.
    B. 1:2:1.
    C. 1:1.
    D. 9:3:3:1.
    E. 2:1.

31. **True or False?** Mendel's work involved plants that <u>showed</u> incomplete dominance. ________________

32. Humans with the disease ________________ are homozygous or heterozygous for an allele that causes dangerously high cholesterol levels in the blood.

Multiple Alleles and Blood Type

33. What is the expected <u>phenotypic</u> ratio of children from parents with blood genotypes $I^A I^A$ and $I^B i$?
    A. 1/2 AB and 1/2 A
    B. 1/2 AB and 1/2 B
    C. 1/4 A, 1/2 AB, and 1/4 B
    D. 1/2 AB and 1/2 O
    E. 1/4 B, 1/2 A, and 1/4 O

34. Examine the blood test results in Figure 9.20 of your text. Why aren't any of the blood cells clumped in the test of AB blood?
    A. There must have been a mistake.
    B. AB blood has antibodies to types A and B carbohydrates.
    C. AB blood does not have antibodies to types A or B carbohydrates.
    D. AB blood has antibodies to type A carbohydrates.
    E. AB blood has antibodies to type B carbohydrates.

35. **True or False?** When a trait exhibits codominance, both <u>recessive</u> phenotypes are expressed in heterozygotes with two dominant alleles. ________________

36. People with type ________________ blood show codominance because both alleles are expressed.

Pleiotropy and Sickle-Cell Disease

37. In tropical Africa, resistance to malaria occurs in people who are
    A. homozygous for the sickle-cell allele.
    B. heterozygous for the sickle-cell allele.
    C. homozygous for the nonsickle-cell allele.
    D. born with an extra nonsickle-cell containing chromosome.

38. **True or False?** As many as one in every ten African Americans are <u>heterozygotes</u> for the sickle-cell trait. ________________

39. Sickle-cell anemia is an example of ________________, when a single gene impacts more than one characteristic.

Polygenic Inheritance

40. In the example of polygenic inheritance described in the text, regarding human skin color, which of the following genotypes would produce the darkest skin?
    A. *AAbbcc*
    B. *AaBbCc*
    C. *aabbcc*
    D. *AABbcc*
    E. *AaBbCC*

41. **True or False?** In polygenic inheritance of skin color, only the <u>dominant</u> alleles contribute to darker skin color. ________________

42. The opposite of pleiotropy, ________________ involves two or more genes affecting a single characteristic.

The Role of Environment

43. If we examine a real human population for the skin color phenotype, we would see more shades than just seven because
    A. of the effects of environmental factors.
    B. some genes fade over time, producing light effects.
    C. there appear to be more genes involved in skin tone than have been described.
    D. skin color has no genetic component.

44. **True or False?** Height clearly <u>has</u> a large environmental component.

    ____

45. Although organisms result from a combination of genetic and environmental factors, only ________________ factors are passed on to the next generation.

The Chromosomal Basis of Inheritance

Gene Linkage

46. Linked genes tend to be inherited together because
    A. they affect the same characteristic.
    B. they determine different aspects of the same trait.
    C. they have similar structures.
    D. they are on the same chromosome.
    E. they have the same alleles.

47. Which one of the following processes can result in the separate inheritance of linked genes?
    A. polygenic inheritance
    B. crossing over
    C. incomplete dominance
    D. pleiotropy
    E. independent assortment

48. **True or False?** Linked genes <u>do not</u> follow the typical patterns of inheritance.

    ____

49. The ____________________ states that genes are located on chromosomes and that chromosomal behavior during meiosis and fertilization accounts for inheritance patterns.

50. The probability of crossover between two linked genes is greatest when the genes are
    A. located on separate chromosomes.
    B. close together on the same chromosome.
    C. farthest apart on the same chromosome.
    D. pleiotropic.
    E. dominant.

Genetic Recombination: Crossing Over and Linkage Maps

51. **True or False?** The assumption that crossover is equally likely at all points on a chromosome was later shown to be <u>correct</u>. ____________________

52. Researchers frequently use ____________________ to study the relationship between chromosome behavior and inheritance.

Sex Chromosomes and Sex-Linked Genes

Sex Determination in Humans and Fruit Flies

53. Which one of the following statements about sex chromosomes is *false*?
    A. Humans have 44 autosomes and 2 sex chromosomes.
    B. In both humans and fruit flies, a male inherits an X and Y sex chromosome.
    C. A gene called *SRY,* found on the Y chromosome, triggers testicular development.
    D. Each human gamete contains both sex chromosomes.
    E. Most sex-linked genes unrelated to sex determination are found on the X chromosome.

Sex Linked Genes

54. **True or False?** Any gene located on a sex chromosome is a <u>sex-linked gene</u>.

    ____

55. The ____________________ chromosomes of a human female are XX.

Sex-Linked Disorders in Humans

56. For a sex-linked recessive allele to be expressed, a man would have to inherit
    A. only one recessive allele, but a woman would have to inherit two.
    B. two recessive alleles, but a woman would have to inherit only one allele.
    C. two recessive alleles, just like a woman.
    D. only one recessive allele, just like a woman.

57. **True or False?** Sex-linked human diseases <u>do not occur</u> in females.

    ____

58. **True or False?** Red-green color blindness is a common sex-linked disorder involving <u>one</u> X-linked gene(s). ____________________

59. Factors involved in blood clotting are missing in people suffering from ____________________.

60. A sex-linked recessive disorder called ____________________ is a condition characterized by a progressive weakening and loss of muscle tissue.

Evolution Connection: The Telltale Y Chromosome

61. Which one of the following statements about the human Y chromosome is *false*? The human Y chromosome
    A. is only about 1/3 the size of the X chromosome.
    B. carries only 1/100 as many genes as the X chromosome.
    C. carries most of the genes that code for maleness and male fertility.
    D. does not engage in crossover with the X chromosome during meiosis.
    E. may have evolved from an X chromosome about 300 million years ago.

62. **True or False?** Most of the DNA of the Y chromosome passes intact from father to son. ____________________

63. Studies of the ____________________ chromosome have been used to trace the evolution of humans.

Word Roots

co = together (*codominance:* phenotype in which both dominant alleles in a heterozygous individual are expressed)

di = two (*dihybrid:* a type of cross that mates varieties differing in two characteristics)

geno = offspring (*genotype:* an organism's genetic makeup)

hetero = different (*heterozygous:* when an organism has different alleles for a gene)

homo = alike (*homozygous:* when an organism has the same alleles for a gene)

hyper = excessive (*hypercholesterolemia:* a condition of incomplete dominance resulting in elevated blood cholesterol levels)

mono = one (*monohybrid:* a type of cross between organisms that differ in only one trait)

pheno = appear (*phenotype:* an organism's physical traits)

pleio = more (*pleiotropy:* when single gene impacts more than one characteristic)

poly = many; **gen** = produce (*polygenic:* type of inheritance in which two or more genes affect a single trait)

ultra = beyond (*ultrasound:* the procedure that uses sound waves to view a fetus)

Key Terms

- ABO blood groups
- achondroplasia
- alleles
- autosome
- carrier
- chromosome theory of inheritance
- codominance
- cross
- cross-fertilization
- dihybrid cross
- dominant allele
- Duchenne muscular dystrophy
- F_1 generation
- F_2 generation
- genotype
- hemophilia
- heterozygous
- homozygous
- Huntington disease
- hybrid
- hypercholesterolemia
- inbreeding
- incomplete dominance
- linkage map
- linked genes
- loci
- Mendel's principle of independent assortment
- Mendel's principle of segregation
- monohybrid cross
- pedigree
- P generation
- phenotype
- pleiotropy
- polygenic inheritance
- Punnett square
- recessive allele
- recombination frequency
- red-green color blindness
- rule of multiplication
- self-fertilize
- sex chromosome
- sex-linked gene
- sickle-cell disease
- testcross
- true-breeding
- wild-type

Crossword Puzzle

Use the Key Terms list from this chapter to fill in the crossword puzzle.

ACROSS

3. type of genes that are located close together on the same chromosome
5. in a heterozygote, the type of allele that determines the phenotype
11. an organism's physical traits
13. the three letters representing the three types of humans blood alleles
16. a thread-like, gene-carrying structure formed from chromatin
18. fusion of sperm and egg from different organisms
21. phenotype in which both dominant alleles in a heterozygous individual are expressed
24. alternate forms of genes
25. a hybridization
27. an organism that has one allele for a recessive disorder but shows no symptoms
28. a type of map based on the frequencies of recombinations during crossover
29. when a single gene impacts more than one characteristic
30. production of offspring with inherited trait(s) identical to the parents
31. a type of cross between organisms that differ in only one trait

DOWN

1. the parental organisms
2. family tree describing the occurrence of heritable characters in parents and offspring
4. a form of muscular dystrophy characterized by a progressive weakening and loss of muscle tissue
6. the chromosomes that do not determine gender
7. a type of colorblindness commonly sex-linked in humans
8. type of inheritance in which two or more genes affect a single trait
9. a condition of incomplete dominance resulting in elevated blood cholesterol levels
10. a sex linked recessive trait that causes excessive bleeding
11. a device for predicting the results of a genetic cross
12. in a heterozygote, the type of allele that has no noticeable affect on the phenotype
13. a form of dwarfism caused by a dominant allele
14. a type of degenerative disease of the nervous system caused by a dominant allele
15. particular sites where a gene is found on a chromosome
17. when an organism has the same alleles for a gene
19. results of a cross of close relatives
20. an organism's genetic makeup
22. the offspring of two different true-breeding varieties
23. when an organism has different alleles for a gene
26. a type of cross mating varieties that differ in two characteristics

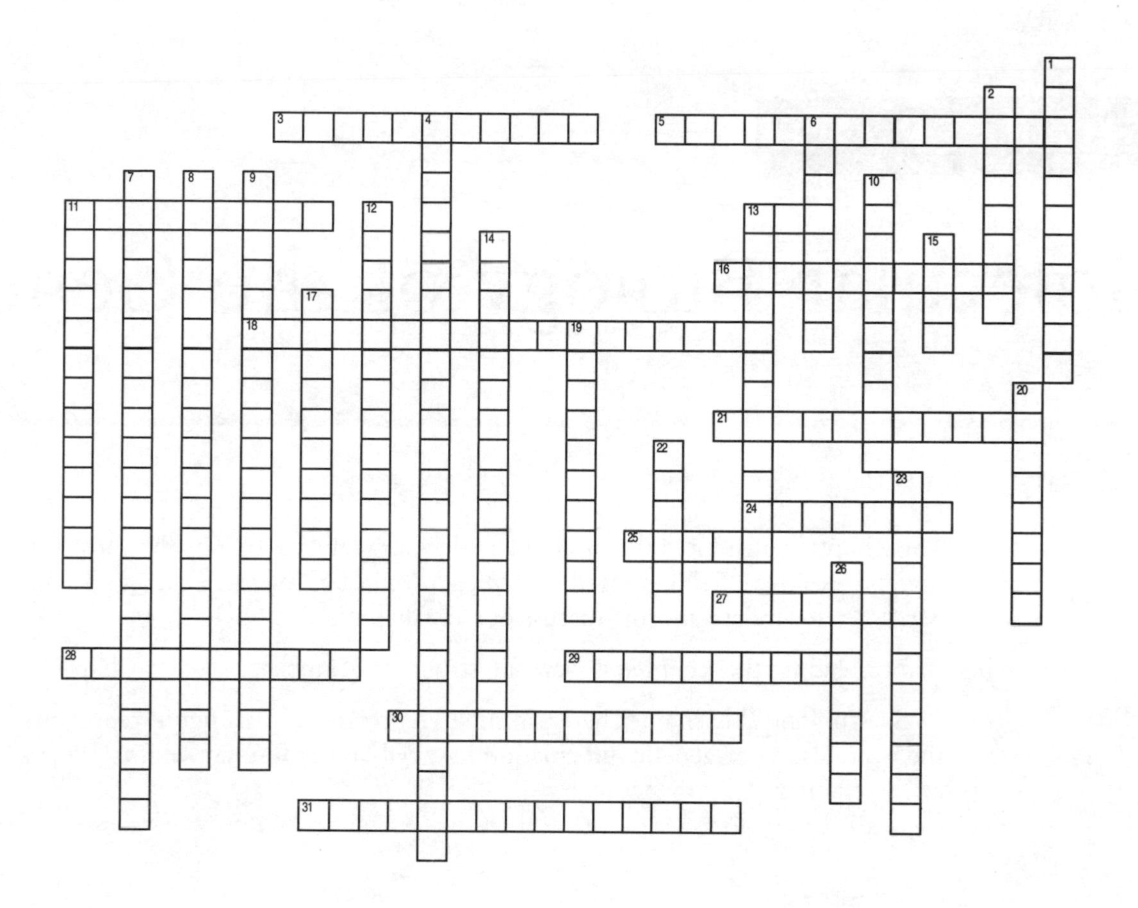

1
2
3
4
5
6
7
8
9
10
11
12
13
14
15
16
17
18
19
20
21
22
23
24
25
26
27
28
29
30
31

CHAPTER 10

Molecular Biology of the Gene

Studying Advice

a. This chapter introduces the basics of molecular genetics, one of the most exciting and explosive fields of modern biology. A mastery of this chapter's content is necessary to understand the discussions in Chapters 11 and 12.

b. Before reading this chapter, review the structure of nucleic acids in Chapter 3.

c. Begin studying this chapter by examining Figure 10.14. This figure represents the overall flow of genetic information in a cell and is fundamental to the rest of the chapter.

Student Media

Biology and Society on the Web

Learn about other drugs that combat HIV.

Activities

10A The Hershey-Chase Experiment
10B DNA and RNA Structure
10C DNA Double Helix
10D DNA Replication
10E Overview of Protein Synthesis
10F Transcription and RNA Processing
10G Translation
10H Causes of Cancer
10I Simplified Reproductive Cycle of a Virus
10J Phage Lytic Cycle
10K Phage Lysogenic and Lytic Cycles
10L HIV Reproductive Cycle

Case Studies in the Process of Science

What Is the Correct Model for DNA Replication?

How Is a Metabolic Pathway Analyzed?

How Do You Diagnose a Genetic Disorder?

What Causes Infections in AIDS Patients?

Why Do AIDS Rates Differ Across the United States?

Evolution Connection on the Web

Discover why you can get the flu year after year after year . . .

Organizing Tables

Compare the structure of DNA and mRNA in the table that follows.

TABLE 10.1

	Type of Sugar	Types of Bases	Overall Shape of the Molecule
DNA			
mRNA			

Compare the structures and functions of the different types of RNA molecules in the table below.

TABLE 10.2

	Structure	Function	Location in the Cell
mRNA			

continued

TABLE 10.2, *continued*			
	Structure	**Function**	**Location in the Cell**
tRNA			
rRNA			

Content Quiz

Directions: Identify the *one* best answer for the multiple-choice questions. For true/false questions, determine if the statement is true or false. If false, change the underlined word(s) to make the statement true. Finally, add the correct word(s) to the fill-in-the-blank questions to make the statements true.

Biology and Society: Sabotaging HIV

1. Which one of the following statements is *false*?
   A. AIDS is one of the most significant health challenges facing the world today.
   B. The letters in AIDS stand for Acquired ImmunoDeficiency Syndrome.
   C. AIDS is spread through the exchange of bodily fluids.
   D. An HIV-infected mother can pass the virus to her baby during childbirth.
   E. In the United States, a half-million HIV-infected babies are born each year.

2. A molecule of AZT has a shape very similar to
   A. ATP, an important molecule in energy transport inside cells.
   B. thymine, one of the four nucleotides that comprise DNA.
   C. water.
   D. ribosomal RNA.
   E. glucose, a common type of sugar found in cells.

3. **True or False?** AZT prevents the synthesis of HIV <u>RNA</u>. ____________________

The Structure and Replication of DNA

4. In the 1950s, scientists understood the functions of DNA to be all of the following *except:*
   A. the capacity to *store* genetic information.
   B. the capacity to *create* new genetic information.
   C. the capacity to *copy* genetic information.
   D. the capacity to *pass along* genetic information from generation to generation.

5. **True or False?** Mendel worked on inheritance patterns without knowing about DNA's role in heredity. ________________

6. By the 1950s, a race was on to understand the ________________ of DNA.

DNA and RNA: Polymers of Nucleotides

7. The backbone of DNA and RNA polynucleotides consists of a repeating pattern of
   A. sugar, base, sugar, base.
   B. phosphate, base, phosphate, base.
   C. sugar, phosphate, sugar, phosphate.
   D. sugar, base, phosphate, sugar, base, phosphate.

8. **True or False?** The "D" in DNA comes from deoxyribose because the sugar in DNA has an extra oxygen atom. ________________

9. The sugars of DNA and RNA are different. DNA has the sugar ________________, and RNA has the sugar ________________.

10. RNA uses the nitrogenous base ________________, which DNA does not use.

11. The two types of bases in DNA are the single-ring bases, ________________ and ________________, and the double-ring structures, ________________ and ________________.

12. In DNA and RNA, the polymers are ________________ and the monomers are ________________.

13. The shape of a DNA molecule is most like the shape of
    A. railroad tracks.
    B. a spiral staircase.
    C. the strings on a tennis racket.
    D. the letter X.

14. If adenine paired with guanine and cytosine paired with thymine in DNA, then
    A. DNA would have irregular widths along its length.
    B. the DNA molecule would be much longer.
    C. the DNA molecule would be much shorter.
    D. the sequential information would be lost.
    E. the DNA molecule would be circular.

15. **True or False?** Watson and Crick discovered that the backbone of DNA was located on the <u>inside</u> of the molecule. ________________

16. If one side of a DNA molecule has the bases CGAT, the opposite side would have the bases ________________.

DNA Replication

17. When a DNA molecule is copied, how much of the original DNA is included in the new copy?
    A. none
    B. 25%
    C. 50%
    D. 75%
    E. 100%

18. DNA replication is most like
    A. picking students to make two baseball teams.
    B. two people getting divorced followed by each person remarrying.
    C. splitting a large plant into two and planting each half.
    D. mixing vinegar and oil to make salad dressing.

19. **True or False?** The DNA molecule of a eukaryotic chromosome has <u>a single replication origin</u>. ________________

20. Covalent bonds between the nucleotides of a new DNA strand are made by the enzyme ________________.

21. DNA repair is accomplished by the enzyme ________________ and some of the proteins associated with DNA replication.

The Flow of Genetic Information from DNA to RNA to Protein

How an Organism's DNA Genotype Produces Its Phenotype

22. The Beadle and Tatum hypothesis about the function of genes is best stated as one gene–
    A. one protein.
    B. one enzyme.
    C. one DNA.
    D. one polypeptide.
    E. one monomer.

23. Which one of the following best represents the flow of genetic information in a cell?
    A. RNA → transcription → DNA → translation → PROTEIN
    B. DNA → transcription → RNA → translation → PROTEIN
    C. DNA → transcription → PROTEIN → translation → RNA
    D. DNA → translation → RNA → transcription → PROTEIN

24. **True or False?** The molecular basis of the phenotype lies in an organism's <u>DNA</u>.

    ____

25. An organism's ____________________ is its genetic makeup. An organism's specific traits are its ____________________.

26. Cell's produce RNA by the process of ____________________ and proteins by the process of ____________________.

From Nucleotide Sequence to Amino Acid Sequence: An Overview

27. If a gene consisted of 60 bases, the protein made from the gene would be
    A. 180 amino acids long.
    B. 60 amino acids long.
    C. 20 amino acids long.
    D. 10 amino acids long.

28. **True or False?** A DNA molecule may contain <u>thousands</u> of genes.

    ____

29. **True or False?** A gene may consist of <u>thousands</u> of nucleotides.

    ____

30. A codon consists of ______________ base(s) in a DNA or RNA molecule.

31. The sequence of nucleotides of the RNA molecule dictates the sequence of ______________ of the polypeptide.

The Genetic Code

32. Of the 64 possible codons,
    A. 61 code for an amino acid and 3 are stop codons.
    B. all of them code for at least one amino acid.
    C. 20 code for 2 amino acids and the rest code for a single amino acid.
    D. 2 are start codons, 3 are stop codons, and 59 code for a single amino acid.
    E. 1 is a start codon, 1 is a stop codon, and the remaining 62 code for a single amino acid.

33. Examine the sets of codons in Figure 10.11. How are the codons in each bracket similar to each other?
    A. The first letter in the codon is most likely to be different.
    B. The second letter in the codon is most likely to be different.
    C. The third letter in the codon is most likely to be different.
    D. The second and third letters in the codon are always the same.
    E. The first and third letters in the codon are always the same.

34. **True or False?** There are gaps between the codons of RNA.

    ____

35. **True or False?** Different codons may code for the same amino acid.

    ____

36. The set of rules relating nucleotide sequence to amino acid sequence is the ______________.

Transcription: From DNA to RNA

37. Which one of the following statements about transcription is *false*?
    A. In RNA, U, rather than T, pairs with A.
    B. The RNA molecule is built one nucleotide at a time.
    C. Both DNA strands serves as the template for one RNA.
    D. Transcription begins when RNA polymerase attaches to the promoter.
    E. As the RNA molecule is produced, it peels away from its DNA template.

38. **True or False?** The terminator sequence indicates the end of a gene.

    ____

39. The ________________ dictates which of the two DNA strands is to be transcribed.

40. The RNA nucleotides are linked by the enzyme ________________.

The Processing of Eukaryotic RNA

41. In eukaryotic cells, RNA transcribed from DNA undergoes processing before leaving the nucleus. In this processing,
    A. a cap and tail are added, introns are edited out, and exons are joined together to make mRNA.
    B. a cap and tail are removed, introns are edited out, and exons are joined together to make mRNA.
    C. a cap and tail are added, exons are edited out, and introns are joined together to make mRNA.
    D. a cap is removed, a tail is added, exons are edited out, and introns are joined together to make mRNA.

42. **True or False?** The cap and tail on an mRNA molecule protects mRNA from cellular enzymes and helps ribosomes recognize RNA as mRNA.

    ____

43. RNA splicing involves the removal of ________________ and the joining of ________________ to produce mRNA.

Translation: The Players

44. The actual translation of codons into amino acids is the job of
    A. mRNA.
    B. tRNA.
    C. rRNA.
    D. ribosomes.
    E. RNA polymerase.

45. Which one of the following statements about tRNA is *false*?
    A. An anticodon recognizes a particular mRNA codon by using base-pairing rules.
    B. There is a slightly different version of tRNA for each amino acid.
    C. Some parts of the tRNA molecule twist, fold around, and base-pair with itself.
    D. Each tRNA molecule must pick up the appropriate amino acid.
    E. A tRNA molecule is made of a double strand of RNA about 800 nucleotides long.

46. A ribosome consists of
   A. one subunit made up of proteins and DNA.
   B. two subunits, each made up of tRNA and rRNA.
   C. two subunits, each made up of proteins and mRNA.
   D. two subunits, each made up of proteins and rRNA.
   E. three subunits, each made up of proteins, mRNA, and tRNA.

47. **True or False?** The part of tRNA molecule that binds to a codon is a parallel codon. ________________

48. A fully assembled ribosome has a binding site for ________________ on its small subunit and a binding site for ________________ on its large subunit.

Translation: The Process

49. During the initiation stage of translation,
   A. a ribosome assembles with the mRNA and the initiator tRNA bearing the first amino acid.
   B. additional amino acids are brought in, one at a time, as a polynucleotide forms.
   C. the ribosomal subunits form by the combination of rRNA and proteins.
   D. the mRNA molecule is edited further, beginning with the removal of the cap and tail.

50. **True or False?** An incoming tRNA molecule first binds with the mRNA codon at the A site. ________________

51. After a new amino acid is brought in, the polypeptide leaves the tRNA in the P site and attaches to the ________________ on the tRNA in the A site.

52. Once a new amino acid is added to a polypeptide, the tRNA and mRNA move together from the ________________ site to the ________________ site. This process is known as ________________.

53. Elongation continues until a(n) ________________ reaches the ribosome's A site.

Review: DNA → RNA → Protein

54. Which one of the following is the correct molecular sequence in the combined processes of transcription and translation?
   A. DNA → mRNA → polypeptide → protein
   B. mRNA → DNA → polypeptide → protein
   C. DNA → polypeptide → mRNA → protein
   D. DNA → mRNA → protein → polypeptide
   E. mRNA → polypeptide → DNA → protein

Mutations

55. Compare the two sentences below.
    1) The cat hit the red toy pig.
    2) The caz thi tth ere dto ypi.

    The second sentence has been changed in a way that is most like a mutation caused by

    A. a base substitution.
    B. a single base deletion.
    C. a single base addition.
    D. a multiple base deletion.
    E. a multiple base addition.

56. In the mutation above, the second sentence no longer reads properly because
    A. of a shift in the reading frame.
    B. most of the letters are different.
    C. several key letters were lost.
    D. of word substitutions.
    E. of word deletions.

57. **True or False?** Alleles frequently differ by only a single base pair difference.

    ____

58. **True or False?** A base substitution shifts the reading frame.

    ____

59. **True or False?** Mutations are usually beneficial. ____________________

60. Any change in the nucleotide sequence of DNA is a(n) ____________________.

61. Chemicals and X-rays that cause mutations are called ____________________.

Viruses: Genes in Packages

Bacteriophages

62. In the lytic cycle of a bacteriophage,
    A. the DNA inserts by genetic recombination into the bacterial chromosome.
    B. the phage genes remain inactive.
    C. the viral DNA can be passed along through many generations of infected bacteria.
    D. the DNA immediately turns the cell into a virus-producing factory.
    E. the phage DNA is referred to as a prophage.

63. Which one of the following is most like the way that phage DNA is replicated during a lysogenic cycle?
   A. Having a friend copy your two page essay while he is copying a 50-page article.
   B. Having your friend wash a shirt for you while she washes her clothes.
   C. Asking a friend to help you while you change the tires on your truck.
   D. Inviting friends over to your house to share a good homecooked meal.
   E. Working with a group of friends to design an experiment for a science class.

64. **True or False?** The lysogenic cycle always leads to the lysis of the host bacterial cell. ____________________

65. **True or False?** Once a prophage forms, it can not leave its chromosome.

____

66. **True or False?** Prophage DNA is replicated along with the host cell's DNA.

____

67. Phage DNA that is incorporated into the bacterial chromosome is referred to as a(n) ____________________.

Plant Viruses

68. Which one of the following statements about plant viruses is *false*?
   A. Most plant viruses discovered to date have RNA instead of DNA as their genetic material.
   B. Most plant viruses consist of RNA surrounded by proteins.
   C. Plant viruses can not easily penetrate a healthy plant epidermis.
   D. Plant viruses can spread within a plant by moving through plasmodesmata.
   E. Unlike animals, there are many simple cures for viral diseases of plants.

Animal Viruses

69. Which one of the following does *not* occur during the reproductive cycle of an enveloped RNA virus?
   A. A protein-coated RNA enters the host cell. Once inside the host cell, the protein coat around the virus is removed.
   B. A viral enzyme starts making complementary strands of the viral RNA.
   C. Some of the new RNA serve as mRNA for the synthesis of new viral proteins.
   D. The new viral proteins assemble around new viral RNA.
   E. The new viruses leave the cell by causing the cell to rupture.

70. **True or False?** Antibiotics are effective for the treatment of viral infections.

____

71. **True or False?** We usually recover from colds by replacing cells damaged by the virus. ____________________

72. Much like a prophage, the herpes virus DNA may insert itself into the host cell's DNA as a(n) ____________________.

73. Most RNA viruses get their outer membranes from the ____________________ of the host cell. But DNA viruses like the herpes virus get their outer membranes from the ____________________ of the host cell.

HIV, the AIDS Virus

74. Which one of the following statements about HIV is *false*?
    A. The virus that is transmitted into cells contains RNA.
    B. Once in a host cell, the viral RNA synthesizes more RNA from the host's DNA.
    C. Double-stranded DNA produced by reverse transcription is inserted into the host cell's chromosomal DNA as a provirus.
    D. The provirus is transcribed and translated into viral proteins.
    E. New HIV leaves without killing the host cell.

75. The outer envelope of HIV is derived from the ____________________ of a previous host cell.

76. HIV is an example of a(n) ____________________, a virus that reproduces by means of a host DNA molecule.

77. HIV uses the enzyme ____________________ to catalyze reverse transcription.

Evolution Connection: Emerging Viruses

78. RNA viruses mutate more quickly than our DNA because RNA viruses
    A. come into contact with many mutagens.
    B. are made of more fragile amino acids.
    C. are smaller.
    D. lack proofreading steps after replication.
    E. have weaker covalent bonds between their bases.

79. **True or False?** Annual flu vaccines are necessary because last year's flu virus may be mutated enough that you have little immunity against it.

    ____

80. **True or False?** New viral diseases may result when an old virus is <u>introduced to a new host</u>. ____________________

81. **True or False?** The first viruses may have evolved from <u>fragments of cellular nucleic acid</u> that moved from one cell to another. ____________________

Word Roots

anti = opposite (*anticodon:* the triplet that is complimentary to a codon triplet on mRNA)

muta = change; **gen** = producing (*mutagen:* a physical or chemical agent that causes mutations)

phage = eat (*bacteriophages:* viruses that attack bacteria)

poly = many (*polynucleotide:* a polymer of many nucleotides)

pro = before (*prophage:* phage DNA inserted into the bacterial chromosome before viral replication)

retro = backward (*retrovirus:* an RNA virus that reproduces by first transcribing its RNA into DNA then inserting the DNA molecule into a host DNA.)

trans = across; **script** = write (*transcription:* the transfer of genetic information from DNA into an RNA molecule)

Key Terms

adenine (A)
AIDS
anticodon
bacterial chromosome
bacteriophages
cap
codon recognition
codon
cytosine (C)
DNA polymerase
double helix
exons
genetic code
guanine (G)
HIV
intron
lysogenic cycle
lytic cycle
messenger RNA
mutagen
mutation
nucleotide
peptide bond formation
phages
polynucleotide
promoter
prophage
provirus
reading frame
retrovirus
reverse transcriptase
ribosomal RNA (rRNA)
RNA polymerase
RNA splicing
stop codon
sugar-phosphate backbone
tail
terminator
thymine (T)
transcription
transfer RNA (tRNA)
translation
translocation
uracil (U)

Crossword Puzzle

Use the Key Terms list from this chapter to fill in the crossword puzzle.

ACROSS

3. the nitrogenous base that pairs with cytosine
6. viral DNA that inserts into a host genome
7. a type of viral replication cycle that releases new phages by the death of the host cell
10. phage DNA inserted into the bacterial chromosome
11. a type of viral replication cycle in which viral DNA replication occurs without phage production
14. a physical or chemical agent that causes mutations
15. the triplet grouping of a mRNA molecule
20. the abbreviation for acquired immuno deficiency syndrome
24. type of bond between adjacent amino acids
26. the triplet that is complimentary to a codon triplet on mRNA
28. the type of chromosome found in bacteria
32. a virus that infects bacteria
33. a nucleotide sequence that signals to start transcribing
34. the selective editing out of introns
35. a change in the nucleotide sequence of DNA
36. the type of structural backbone of polynucleotides
37. a monomer of a nucleic acid

DOWN

1. viruses that attack bacteria
2. a triplet that signals for translation to stop
4. the nitrogenous base that pairs with adenine
5. the abbreviation for the type of polynucleotide that serves as a molecular interpreter
6. a polymer of nucleotides
8. the nitrogenous base found only in RNA
9. the alternating chain of sugar and phosphate to which the DNA and RNA nitrogenous bases are attached
12. an enzyme that catalyzes reverse transcription
13. the transcription enzyme used to link RNA nucleotides
16. an RNA virus that reproduces by mean of a DNA molecule
17. the set of rules relating nucleotide sequence to amino acid sequence
18. a nucleotide sequence that signals the end of a gene
19. the enzyme that makes the covalent bonds during DNA replication
20. the nitrogenous base that pairs with thymine
21. the three base word used during translation
22. the abbreviation for the human immunodeficiency virus
23. the general shape of a DNA molecule
25. the abbreviation for the type of polynucleotide that helps to form ribosomes
27. the abbreviation for the type of polynucleotide transcribed from a gene
29. the nitrogenous base that pairs with guanine
30. noncoding stretch of nucleotides in RNA transcripts
31. coding stretches of nucleotides in RNA transcripts

CHAPTER 11

Gene Regulation

Studying Advice

a. The content of this chapter provides great depth and understanding to big questions in biology. For example: (1) What causes cancer and *what can you do* to reduce your chances of getting cancer? (2) How do our cells specialize and arrange themselves as we develop? and (3) Are there common principles to animal development? Although the material is challenging, it is a chance to understand the cutting edge of research in these important areas of biology and medicine.

b. Use the two organizing tables to keep track of the details in the chapter. The tables can serve as important reference points as you read further and review for exams.

Student Media

Biology and Society on the Web

Learn about the use of umbilical cord blood in the treatment of disease.

Activities

11A The *lac* Operon in *E. coli*
11B Gene Regulation in Eukaryotes
11C Review: Gene Regulation in Eukaryotes
11D Signal-Transduction Pathway
11E Causes of Cancer

Case Study in the Process of Science

How Do You Design a Gene Expression System?

Evolution Connection on the Web

You may be more like a fruit fly than you think.

Organizing Tables

Using the following table, describe the main functions of each component of the lactose (*lac*) operon and the regulatory genes and repressor proteins that affect the operon. Some of the cells of the table are already filled in to make this task a bit easier.

TABLE 11.1

	Location	Function	What Controls Them
Promoter			
Operator			
Three enzyme genes	All three enzyme genes are located together next to the operator.		
Regulatory gene		Produces the repressor protein.	Nothing regulates it. It is always on!
Repressor protein	The repressor proteins are free floating in the cytoplasm.		Lactose, when present, binds and prevents the repressor from attaching to the promoter.

Use the following table to compare the genes and process of gene regulation in prokaryote and eukaryotes. Some of the cells of the table are already filled in to make this task a bit easier.

TABLE 11.2

	Prokaryotes	Eukaryotes
Which group(s) uses regulatory proteins that attach to DNA?		
Which group uses genes with their own promoter and control sequences?		
Which is used more often, activators or repressors?		
Are the operators (prokaryotes) or enhancers (eukaryotes) far away or close to the genes they help regulate?		
Which group(s) modifies mRNA before it is translated?	Prokaryotes do not usually modify mRNA.	
How long do the mRNA molecules last in the cell?	Prokaryotic mRNA are degraded by enzymes after only a few minutes.	

Content Quiz

Directions: Identify the *one* best answer for the multiple-choice questions. For true/false questions, determine if the statement is true or false. If false, change the underlined word(s) to make the statement true. Finally, add the correct word(s) to the fill-in-the-blank questions to make the statements true.

Biology and Society: Baby's First Bank Account

1. Which one of the following statements is *false*?
   A. Umbilical cord blood is stored frozen.
   B. The American Academy of Pediatrics recommends cord blood banking for all newborn babies.
   C. Umbilical cord blood is rich in stem cells.
   D. Stem cells can give rise to many types of body cells.
   E. Most cells of the adult body lack the ability to give rise to many types of body cells.

2. **True or False?** Most cells of an adult lack the ability to develop into a wide variety of different body cells. ________________

3. The power of ________________ cells lies in their ability to develop into a wide variety of different body cells.

From Egg to Organism: How and Why Genes Are Regulated

4. Genes determine the nucleotide sequences of
   A. DNA.
   B. proteins.
   C. lipids.
   D. amino acids.
   E. mRNA.

5. **True or False?** The genes for specialized proteins are expressed in all cells.
   ____

6. Individual cells must undergo ________________, that is, they must become specialized in structure and function.

Patterns of Gene Expression in Differentiated Cells; DNA Microarrays: Visualizing Gene Expression

7. When using a microarray, a researcher begins by collecting ____________ from a particular type of cell.
   A. mRNA
   B. DNA
   C. proteins
   D. lipids
   E. ATP

8. **True or False?** A DNA microarray consists of a glass slide containing thousands of different single stranded mRNA fragments arranged in a grid.

    ____

9. Researchers can use microarrays to learn what ____________________ are active in different tissues.

The Genetic Potential of Cells

10. The process of cloning shows that a mature plant cell can lose its differentiation and then ______________ to give rise to all the specialized cells of a new plant.
    A. dedifferentiate
    B. redifferentiate
    C. multiply
    D. reproduce
    E. mutate

11. **True or False?** Hundreds or thousands of genetically identical clones can be produced from the somatic cells of a single plant. ____________________

12. **True or False?** The process of cloning shows that differentiation does not involve irreversible changes in the DNA. ____________________

13. Salamanders are capable of ____________________, the regrowth of lost body parts.

Reproductive Cloning of Animals

14. In the process of nuclear transplantation, the nucleus from a donor cell is transplanted into
    A. an egg in which the nucleus has been removed.
    B. a normal egg, allowing the two nuclei to fuse.
    C. the nucleus of another adult cell.
    D. another adult cell in which the nucleus has been removed.
    E. a sperm that is then used to fertilize an egg.

15. Reproductive cloning is used to
    A. produce herds of farm animals with specific sets of desirable traits.
    B. help researchers identify the roles of specific genes.
    C. produce pigs for organ donation that lack a gene coding for a protein that can cause immune system rejection in humans.
    D. restock populations of endangered animals.
    E. All of the above.

16. **True or False?** Dolly, the first mammal to be cloned, genetically resembled the egg donor. ____________

17. **True or False?** Animal cloning was first performed in the 1990s.

____

Therapeutic Cloning and Stem Cells

18. Which one of the following statements is *false*?
    A. When grown in laboratory culture, embryonic stem cells can divide indefinitely.
    B. Adult stem cells are much more difficult than embryonic stem cells to grow in culture.
    C. If the right conditions are used, scientists can induce changes in gene expression that cause differentiation of embryonic stem cells into a particular cell type.
    D. In the future, embryos may be created using a cell nucleus from a patient so that embryonic stem cells can be harvested and induced to develop into whole organs.
    E. Embryonic stem cells are partway along the road to differentiation and usually give rise to only a few related types of specialized cells.

19. **True or False?** Embryonic stem cells generate replacements for nondividing differentiated cells in adults. ____________

20. The purpose of ____________ cloning is to produce embryonic stem cells.

The Regulation of Gene Expression

Gene Regulation in Bacteria

21. In bacteria, gene expression is mainly controlled by
    A. deleting certain genes from chromosomes.
    B. moving DNA into special capsules.
    C. limiting DNA replication.
    D. turning transcription on and off.
    E. making extra copies of chromosomes.

22. Examine Figure 11.10 in the text. A gene mutation would produce the greatest effects
    A. when changes are made to the polypeptide in the cytoplasm.
    B. during translation.
    C. during processing of RNA in the nucleus.
    D. during transcription.

23. Which one of the following statements about the *lac* operon is *false*?
    A. Enzymes that help absorb and process lactose are produced by *E. coli* when lactose is absent.
    B. RNA polymerase attaches to the promoter and initiates transcription.
    C. The operator determines whether RNA polymerase can attach to the promoter.
    D. The repressor protein binds to the operator and blocks the attachment of RNA polymerase to the promoter.
    E. When lactose is present it interferes with the attachment of the *lac* repressor to the promoter by binding to the repressor and changing its shape.

24. Examine the functions of the *lac* operon in Figure 11.11. The way that lactose works in the *lac* operon is most like
    A. adding milk and sugar to coffee to improve its flavor.
    B. a boy distracting his mom while his brother takes some cookies.
    C. cooking a meal and serving it to guests.
    D. putting an ATM card into an ATM to get some money.
    E. advertising a restaurant to attract customers.

25. **True or False?** A gene that is turned on is being transcribed into <u>protein</u>.

    ____

26. A cluster of genes with related functions, along with the control sequences, is called a(n) ____________________.

Gene Regulation in the Nucleus of Eukaryotic Cells

27. The extra X chromosome in females
    A. is expressed at about the same level as the other X chromosome.
    B. is eliminated from the cell early in embryonic development.
    C. is highly compacted and inactivated.
    D. is less folded and more frequently expressed.

28. Eukaryotes usually
    A. have operons.
    B. have a promoter and other control sequences for each gene.
    C. have regulatory proteins that bind to DNA.
    D. have more repressor genes than activators.

29. The process in Figure 11.14 is most like
    A. baking a pie and a cake and cutting them up to serve to guests.
    B. clipping news and sports articles out of newspapers to make two scrap books.
    C. sorting out beads to make two different necklaces.
    D. editing eight hours of film in different ways to produce two different movies.

30. In eukaryotes, the most important stage for regulating gene expression is the
    A. unpacking of chromosomal DNA.
    B. breakdown of mRNA.
    C. removal of introns from RNA.
    D. initiation of transcription.
    E. transport of mRNA from the nucleus to the cytoplasm.

Regulation in the Cytoplasm

31. Which one of the following is *not* a mechanism used to regulate gene expression after eukaryotic mRNA is transported to the cytoplasm?
    A. The mRNA molecule is typically broken down within minutes.
    B. Different mRNA molecules combine in the cytoplasm to form new mRNA molecules.
    C. Proteins in the cytoplasm regulate translation.
    D. Proteins are edited after translation.
    E. The final protein products often last only a few minutes or hours.

32. **True or False?** DNA packing tends to promote gene expression.

    ____

33. **True or False?** Eukaryotic cells have more elaborate mechanisms than bacteria do for regulating the expression of their genes. ____________________

34. **True or False?** Both prokaryotes and eukaryotes regulate transcription by using regulatory proteins that bind to mRNA. ____________________

35. Eukaryotic genes may be turned on when activators bind to DNA sequences called ____________________, which are usually located far away from the gene they help regulate.

36. Turning on a eukaryotic gene involves regulatory proteins called ____________________, in addition to RNA polymerase.

37. Repressor proteins bind to DNA sequences called ____________________ to inhibit the start of transcription.

Cell Signaling

38. **True or False?** Cell-to-cell signaling is a key mechanism in development and in the coordination of cellular activities throughout an organism's life.

    ____

39. In cell-to-cell signaling, a signal molecule usually acts by binding to a receptor protein in the plasma membrane of the target cell and initiating a ________________ ________________ pathway.

The Genetic Basis of Cancer

Genes That Cause Cancer

40. Which one of the following does *not* typically promote cancer?
    A. viruses transmitting tumor-suppressor genes
    B. mutations in proto-oncogenes that code for growth factors
    C. the inactivation of tumor-suppressor genes that inhibit cellular growth
    D. a mutation that causes the ras protein to be hyperactive

41. Cancers usually take a long time to develop because
    A. oncogenes are rare.
    B. only old cells can become cancerous.
    C. several specific mutations must occur.
    D. cancer cells usually grow very slowly.

42. Finding a single cure for all cancer is unlikely because
    A. we've made so little progress in recent years.
    B. we just don't understand the genetics of the disease.
    C. the rapidly dividing cells can not be killed.
    D. cancer is caused by many different factors.
    E. cancer cells migrate throughout the body.

43. **True or False?** Some viruses carry <u>oncogenes</u>. ________________

44. **True or False?** For a proto-oncogene to become an oncogene, a cell's <u>RNA</u> must become mutated. ________________

45. **True or False?** Cancer <u>always</u> results from changes in DNA.

    ____

46. **True or False?** Most cases of breast cancer are <u>caused</u> by inherited mutations.

    ____

47. Many proto-oncogenes code for ________________, proteins that stimulate cell division.

The Effects of Lifestyle on Cancer Risk

48. The one substance known to cause many types of cancer is
    A. alcohol.
    B. asbestos.
    C. X-rays.
    D. table salt.
    E. tobacco.

49. **True or False?** Cancers can be caused by factors that promote cell division.

    ____

50. **True or False?** People can reduce their risks of developing cancer by avoiding tobacco products, alcohol, animal fat, and time spent in the sun.

    ____

51. Cancer-causing compounds called ______________ include ultraviolet light and tobacco smoke.

Evolution Connection: Homeotic Genes

52. Homeoboxes, found in homeotic genes,
    A. are types of proto-oncogenes that are a common cause of breast cancer.
    B. promote cancer by increasing the rate of cellular division.
    C. are characteristic of only mammals and birds.
    D. appear to be of recent evolutionary origin, coding for many animal traits that have only recently evolved.
    E. are very similar in many diverse organisms, suggesting a common evolutionary heritage.

53. **True or False?** Homeoboxes containing homeotic genes have different developmental roles in mice and fruit flies. ______________

54. Researchers studying homeotic genes from different species found a common sequence of 180 ______________.

Word Roots

homeo = alike (*homeoboxes:* a 180-nucleotide sequence within a homeotic gene)

onkos = tumor (*oncogene:* a gene that causes cancer)

proto = first (*proto-oncogene:* a normal gene with the potential to become an oncogene)

Key Terms

activators
adult stem cells
alternative RNA splicing
carcinogens
cellular differentiation
clones
DNA microarray
embryonic stem cells (ES cells)
enhancers
gene expression
growth factors
homeoboxes
homeotic genes
nuclear transplantation
oncogene
operator
operon
promoter
proto-oncogene
regeneration
repressor
reproductive cloning
silencers
therapeutic cloning
transcription factors
tumor-suppressor genes
X chromosome inactivation

Crossword Puzzle

Use the Key Terms list from this chapter to fill in the crossword puzzle.

ACROSS

1. the type of gene that codes for proteins that normally help prevent uncontrolled cell growth
5. a glass slide containing thousands of kinds of single-stranded DNA fragments arranged in an array
6. the type of cloning that scientists use to help patients with irreversibly damaged tissues
7. a normal gene with the potential to become an oncogene
13. a protein secreted by certain body cells that stimulate other cells to divide
15. type of adult cell that generates replacements for nondividing differentiated cells
16. the process of cells becoming specialized in structure and function
19. a way for an organism to generate more than one polypeptide from a single gene
22. cancer-causing agents

DOWN

2. the overall process by which genetic information flows from genes to proteins
3. DNA sequences where eukaryotic activators bind
4. 180-nucleotide sequences within a homeotic gene
8. undifferentiated cells in an embryo which undergo unlimited division and produce several different types of cells
9. genetically identical organisms
10. eukaryotic gene regulatory proteins
11. a site where the transcription enzyme RNA polymerase attaches and initiates transcription
12. using a somatic cell to make one or more genetically identical individuals
14. master control genes that regulate batteries of other genes
15. proteins that switch on a gene or group of genes
17. a type of switch between the promoter and the enzyme genes
18. a cancer-causing gene
20. the replacement of body parts
21. a molecule that can turn off transcription
23. a cluster of genes with related functions along with a promoter and an operator
24. DNA sequences where eukaryotic repressor proteins bind

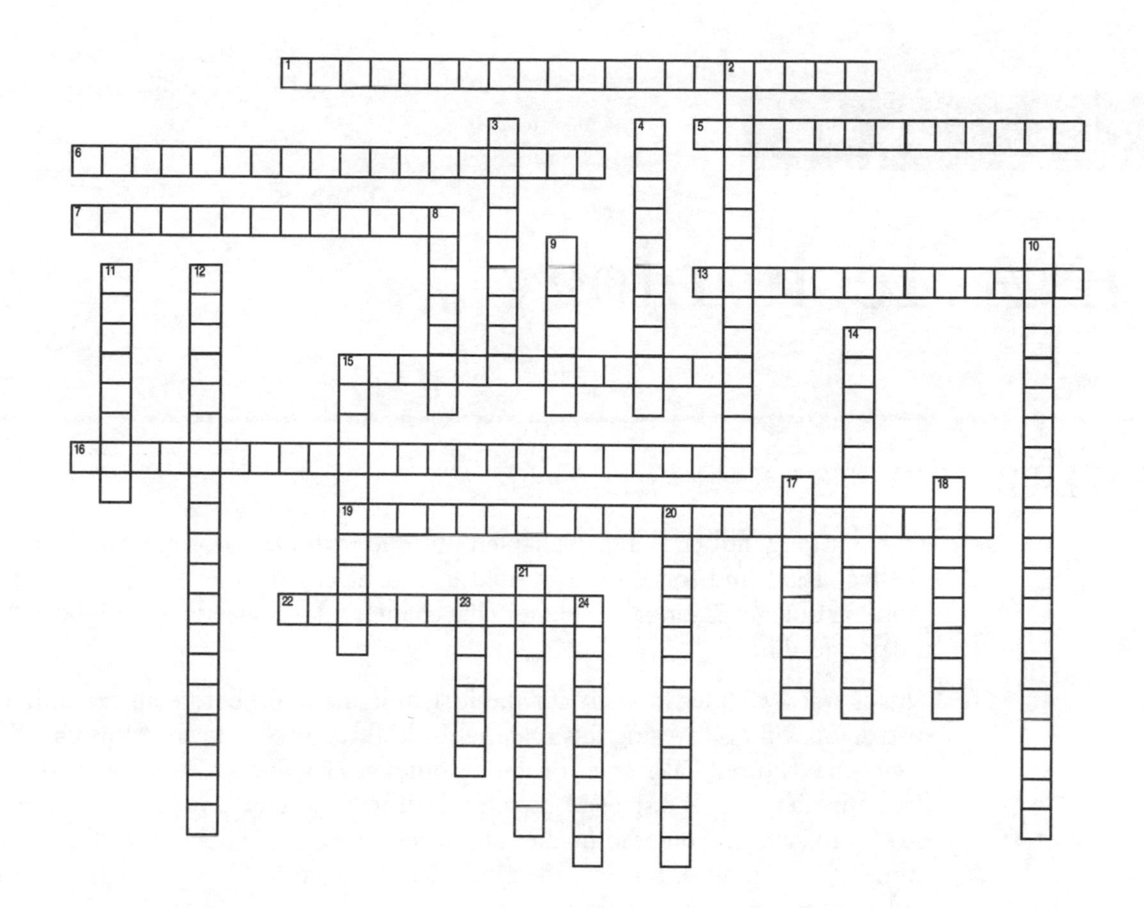

CHAPTER 12

DNA Technology

Studying Advice

a. This is a long and challenging chapter, but one with plenty of rewards. The medical, legal, and social issues related to genetic engineering are now a common part of our national dialogue. This chapter addresses issues straight out of today's headlines.

b. This is not a chapter to study for the first time the night before an exam. If your instructor will be covering this material in lecture, be sure to read this chapter *before* the lectures. Take notes on the techniques as you read. Make your own diagrams detailing the steps of these procedures, and create your own miniglossary to refer to as you read further and listen in lecture. This is one of the hottest fields in all of science, affecting what food we eat, how we fight disease, and the way we view ourselves.

Student Media

Biology and Society on the Web

Learn more about the Human Genome Project.

Activities

12A Applications of DNA Technology
12B DNA Technology and Golden Rice
12C Cloning a Gene in Bacteria
12D Restriction Enzymes
12E DNA Fingerprinting
12F Gel Electrophoresis
12G Analyzing DNA Fragments Using Gel Electrophoresis
12H The Human Genome Project: Human Chromosome 17

Case Studies in the Process of Science

How Are Plasmids Introduced into Bacterial Cells?
How Can Gel Electrophoresis Be Used to Analyze DNA?

Evolution Connection on the Web

Investigate relationships among different life forms by examining genes.

Organizing Tables

In Table 12.1, indicate the starting materials and products of each of these procedures described in the text. To help you in this task, the procedure cells in the table are already completed.

TABLE 12.1

Process	Starting Materials	Procedure	Product
Cloning genes in a recombinant plasmid		1. Use the restriction enzyme to cleave the plasmid in only one DNA sequence. 2. Use the same restriction enzyme to cleave the foreign DNA into many pieces.	
Making pure genes using reverse transcriptase		1. Use the reverse transcriptase to make a pure gene. 2. Insert the pure gene into a bacterium.	

Describe the function of each of the enzymes listed in the table below.

TABLE 12.2

Enzyme	Enzyme Function
DNA ligase	

continued

TABLE 12.2, *continued*	
Enzyme	**Enzyme Function**
DNA polymerase	
Restriction enzymes	
Reverse transcriptase	

Content Quiz

Directions: Identify the *one* best answer for the multiple-choice questions. For true/false questions, determine if the statement is true or false. If false, change the underlined word(s) to make the statement true. Finally, add the correct word(s) to the fill-in-the-blank questions to make the statements true.

Biology and Society: Hunting for Genes

1. Which one of the following is *not* yet a product of DNA technology?
   A. Corn has been genetically modified to produce its own insecticide.
   B. DNA fingerprints have solved crimes.
   C. DNA fingerprints have been used to study the origins of ancient people.
   D. Many fatal genetic diseases have been cured.
   E. Hundreds of human disease-associated genes have been identified.

2. Which one, if any, of the following is *not* a way that DNA technology plays a significant role in our lives?
   A. the comparison of whole genomes as an investigative tool
   B. the use of recombinant DNA to produce useful products
   C. the use of human gene therapy for the treatment of diseases
   D. the application of DNA fingerprinting in forensic science
   E. All of the above are ways that DNA technology plays a significant role in our lives.

3. **True or False?** Using data from the Human Genome Project, scientists have recently learned that some cases of Parkinson disease are linked to a specific environmental factor. ____________________

4. DNA technology is a set of methods for studying and manipulating ____________________.

Recombinant DNA Technology

5. Recombinant DNA technology combines together
   A. genes from more than one source.
   B. the nucleus of one cell with the cytoplasm of another.
   C. all of the genetic material of two cells.
   D. proteins from one cell and DNA from another.
   E. proteins from two different cells.

6. **True or False?** A host that carries recombinant DNA is called a transgenic or genetically modified organism. ____________________

7. **True or False?** Biotechnology was first used thousands of years ago to make bread and wine. ____________________

8. Research on *E. coli* led to the development of ____________________, a set of laboratory techniques for combining genes from different sources into a single DNA molecule.

9. The use of organisms to perform practical tasks defines ____________________.

From Humulin to Genetically Modified Foods

10. Which one of the following, if any, is *not* produced by recombinant DNA technology?
   A. Humulin
   B. human growth hormone
   C. vaccines
   D. insect-resistant plants
   E. All of the above are produced by recombinant DNA technology.

11. **True or False?** In the year 2002, about half of all American corn and soybean crops were genetically engineered. ____________________

12. Genetic engineering has produced rice that can help prevent ____________________ deficiency, a disease that often leads to vision impairment and increases susceptibility to disease.

13. Genetic engineering has now produced __________________ that synthesize and secrete Humulin.

14. Genetic engineering has produced potatoes that may help provide immunity against the disease __________________.

Recombinant DNA Techniques

Sequencing: Number the following seven steps in the order that they occur in the process of cloning genes in recombinant plasmids.

_____ 15. Recombinant DNA plasmids are now formed, most without the desired gene, but one with the desired gene.

_____ 16. Sticky ends of the human DNA fragments base-pair with the complementary sticky ends of the plasmids.

_____ 17. A molecular biologist finds the one bacterial clone that contains the desired gene.

_____ 18. The plasmid and human DNA containing the desired gene are treated with the same restriction enzyme.

_____ 19. The genetically engineered bacteria are cloned.

_____ 20. DNA ligase joins the two DNA molecules by helping form covalent bonds.

_____ 21. The recombinant plasmids are then mixed with bacteria.

22. When biologists want to customize bacteria to produce a specific protein, the gene for that protein is inserted into
   A. the chromosome of another bacterium.
   B. the coat of a phage.
   C. the DNA of a phage.
   D. a plasmid.
   E. the chromosome of the bacterium.

23. **True or False?** The DNA of a plasmid is <u>part of</u> the bacterial chromosome.

____

24. **True or False?** A nucleic acid probe <u>can be used</u> when only a part of the nucleotide sequence of a gene is already known. __________________

25. **True or False?** In a genomic library, each cell within a clone carries <u>a different</u> recombinant plasmid. __________________

26. **True or False?** A <u>nucleic acid probe</u> can be used to identify a bacterial clone carrying a particular gene of interest amongst the thousands of clones produced by shotgun cloning. __________________

27. A(n) ________________ is a short, single-stranded molecule of radioactively labeled DNA whose nucleotide sequence is complementary to part of the gene or other DNA of interest.

28. The cutting tools for making recombinant DNA are bacterial enzymes called ________________.

29. The entire collection of cloned DNA fragments from a shotgun experiment, in which the starting material is bulk DNA from whole cells, is called a(n) ________________.

30. When plasmids function as DNA carriers, moving genes from one cell to another, they are acting as ________________.

31. The places where DNA is cut by restriction enzymes are called ________________.

32. A recombinant DNA molecule is finally produced when DNA pieces are connected into a continuous strand by ________________, which forms covalent bonds between adjacent nucleotides.

DNA Fingerprinting and Forensic Science

Murder, Paternity, and Ancient DNA

33. Which one of the following, if any, is not a typical step in the DNA fingerprinting process?
    A. DNA samples are collected from different sources.
    B. When the genetic material is insufficient to analyze, DNA in the sample is amplified.
    C. Proteins are produced from the DNA.
    D. The DNA samples are compared to each other.
    E. All of the above steps are used in a typical DNA fingerprinting process.

34. **True or False?** DNA from semen can be compared to DNA <u>from blood</u> to help solve crimes. ________________

35. Since its introduction in 1986, ________________ has become a standard part of law enforcement.

DNA Fingerprinting Techniques

36. Which one of the following techniques is used to sort macromolecules, primarily on the basis of their electrical charge and length?
    A. gel electrophoresis
    B. RFLP analysis
    C. recombinant DNA technology
    D. gene cloning
    E. polymerase chain reaction

37. The ________________ is a technique by which any segment of DNA can be cloned.

38. The goal of DNA fingerprinting by ________________ is to determine whether samples of DNA contain identical markers or not.

39. The key to PCR is an unusual ________________ that can withstand the heat needed to separate DNA strands.

40. The science of crime scene investigation is called ________________.

Genomics

The Human Genome Project

41. The first targets of genomics research were
    A. algae.
    B. human cells.
    C. eukaryotic disease-causing microbes.
    D. pathogenic bacteria.
    E. viruses.

42. Which one of the following is *false*?
    A. The human genome carries between two and three thousand genes.
    B. The human genome contains approximately 3.2 billion nucleotide pairs.
    C. Much of the DNA between genes consists of repetitive DNA.
    D. As of 2003, the genomes of over 100 organisms have been sequenced
    E. About 97% of the entire human genome consists of DNA that does not code for proteins.

43. **True or False?** Repetitive DNA sequences at the ends of chromosome are called endomeres. ________________

44. The science of studying whole genomes is called ________________.

45. Much of the DNA between genes consists of nucleotide sequences present in many copies, called ________________.

Tracking the Anthrax Killer

46. When the genomes of the anthrax spores used in the 2001 bioterrorist attacks were compared, it was determined that the mailed spores
    A. were not identical.
    B. came from different laboratories.
    C. were all a harmless veterinary vaccine strain.
    D. were all from the Ames strain.
    E. all came from one particular laboratory.

47. **True or False?** Comparative genomics has revealed that humans and chimpanzees share about 65% of their DNA. ________________

Genome Mapping Techniques

48. The stage of the Human Genome Project that uses restriction enzymes to break the DNA of each chromosome into a number of identifiable fragments, which are then cloned, is the
    A. genetic mapping stage.
    B. DNA sequencing stage.
    C. physical mapping stage.
    D. the whole-genome shotgun method.
    E. assembly stage.

49. Which one of the following did Celera Genomics pioneer as part of the Human Genome Project?
    A. genetic mapping stage
    B. DNA sequencing stage
    C. physical mapping stage
    D. the whole-genome shotgun method
    E. assembly stage

50. **True or False?** The functions of all human genes have been determined for the human genome. ________________

51. Parts of the Human Genome Project required that sections of DNA were ________________ so that the exact order of A, T, C, and G nucleotides were determined.

Human Gene Therapy

Treating Severe Combined Immunodeficiency

52. Which one of the following statements is *false*?
    A. The goal of gene therapy is to replace a mutant version of a gene with a properly functioning one within a living person.
    B. It was not until April 2000 that the first scientifically strong evidence of effective gene therapy was reported.
    C. Most human gene therapy experiments to date have been preliminary.
    D. Severe combined immunodeficiency (SCID) is a fatal inherited disease caused by a single defective gene.
    E. Ideally, the nonmutant version of a gene would be inserted into cells that will soon die.

53. **True or False?** Human gene therapy is a recombinant DNA procedure that alters a living genome by introducing natural genetic material. ________________

54. One of the prime targets for gene therapy are ________________ cells that give rise to all the cells of the blood and immune system.

Safety and Ethical Issues

The Controversy Over Genetically Modified Foods

55. Which one of the following statements is *false*?
    A. The European Union suspended the introduction of new GM crops.
    B. GM strains account for a significant percentage of several agricultural crops.
    C. The labeling of GM foods is now being debated but has not yet become law.
    D. Lawn and crop grasses commonly exchange genes with wild relatives via pollen transfer.
    E. Today, most public concern about possible hazards centers not on GM foods, but on recombinant microbes.

56. **True or False?** The U.S. National Academy of Sciences released a study finding no scientific evidence that transgenic crops pose any special health or environmental risks. ________________

Ethical Questions Raised by DNA Technology

57. With respect to the ethical issues raised by genetic engineering, the authors argue that
    A. there really isn't much to be concerned about.
    B. the issues are so troubling that current research should be stopped.
    C. there are serious societal issues that need to be addressed.
    D. the scientific community will find answers to these concerns.
    E. scientists have not been concerned about these issues.

58. **True or False?** Genetic engineering of gametes (sperm or ova) and zygotes in humans has not been attempted. ________________

Evolution Connection: Genomes Hold Clues to Evolution

59. Research on the genetics of organisms at all levels of biological organization suggest that
    - A. organisms are not as interrelated as we think.
    - B. life does not have unifying principles.
    - C. life is more interrelated than we knew.
    - D. the genetics of yeast and humans are fundamentally different systems.
    - E. although primitive forms of life are similar, they share little with multicellular eukaryotes.

60. **True or False?** The DNA sequences determined to date do not confirm the evolutionary connections between distantly related organisms. ________________

61. **True or False?** Yeast has a number of genes close enough to the human versions that they can substitute for them in a human cell. ________________

Word Roots

liga = tied (*DNA ligase:* the enzyme that permanently "pastes" together DNA fragments)

telo = an end (*telomeres:* the repetitive DNA at chromosome ends)

Key Terms

biotechnology
DNA fingerprinting
DNA ligase
forensics
gel electrophoresis
gene cloning
genetically modified (GM) organism
genetic marker
genomic library
genomics
human gene therapy
nucleic acid probe
plasmid
polymerase chain reaction (PCR)
recombinant DNA
recombinant DNA technology
repetitive DNA
restriction enzyme
restriction sites
RFLP analysis
sequenced
telomere
transgenic organism
vaccine
vector

Crossword Puzzle

Use the Key Terms list from this chapter to fill in the crossword puzzle.

ACROSS

6. another term for a genetically modified organism
12. any DNA segment that varies from person to person
13. a specific sequence on a DNA strand that is recognized as a "cut site" by a restriction enzyme
14. much of the DNA between genes
15. initials for a technique used to obtain many copies of a DNA molecule
16. a small ring of DNA separate from the chromosome(s)
17. a repetitive DNA at each end of a eukaryotic chromosome
18. the cutting tools for making recombinant DNA
19. a type of DNA molecule carrying genes derived from two or more sources
20. in DNA technology, a labeled single-stranded nucleic acid molecule used to find a specific gene
21. the study of whole sets of genes and their interactions

DOWN

1. the entire collection of cloned DNA fragments from a shotgun experiment
2. the alternation of the genes of a person afflicted with a genetic disease
3. the process of detecting an individual's unique collection of DNA restriction fragments through electrophoresis and nucleic acid probes
4. the use of organisms to perform practical tasks
5. the enzyme that permanently "pastes" together DNA fragments
7. techniques for combining genes from different sources into a single DNA molecule and then transferring the new molecule into cells, where it can be replicated and expressed
8. a method for sorting macromolecules primarily on the basis of their size and electric charge
9. the production of multiple copies of a gene
10 a harmless variant or derivative of a pathogen used to prevent disease
11 the role of a plasmid when it carries extra genes to another cell

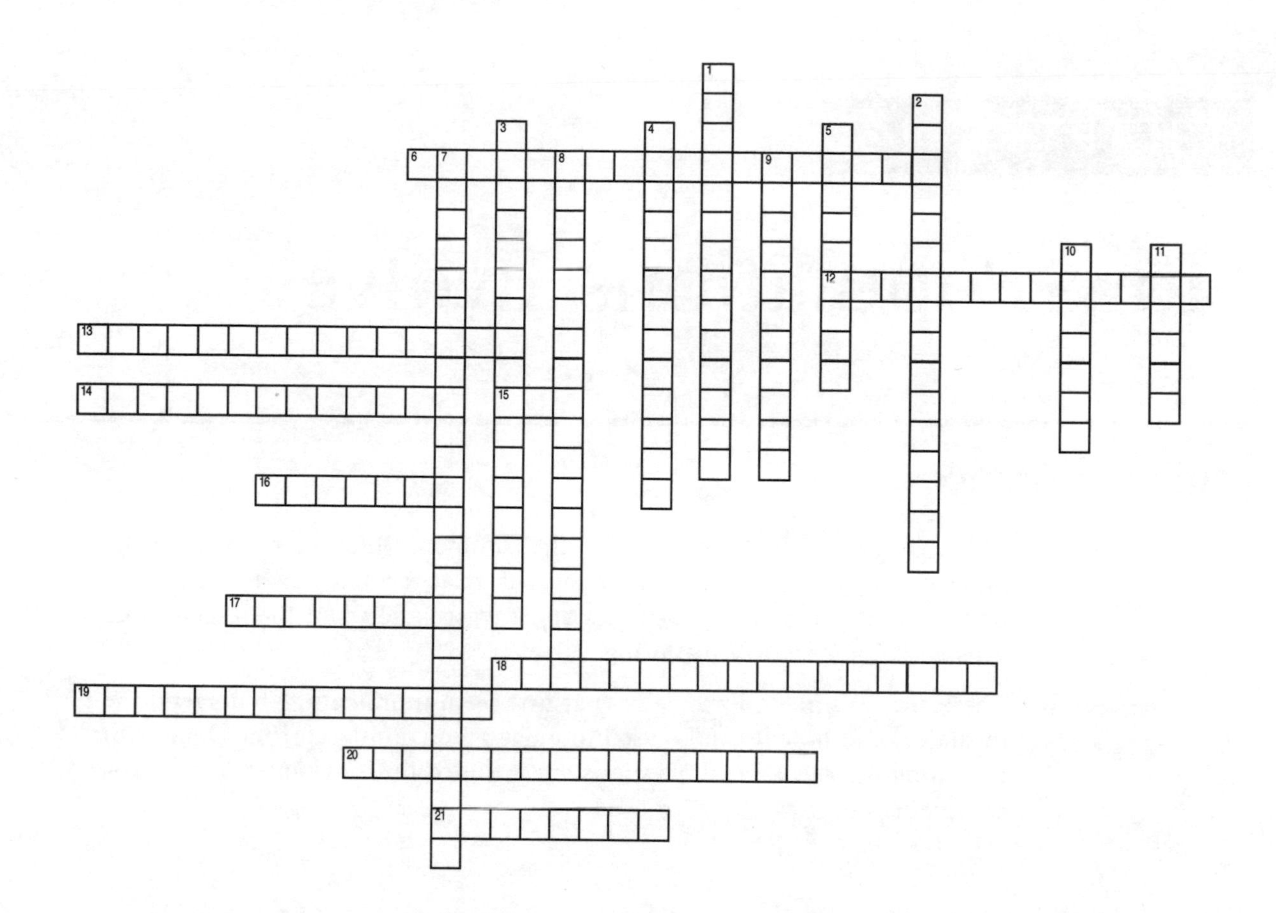
1
2
3
4
5
6
7
8
9
10
11
12
13
14
15
16
17
18
19
20
21

CHAPTER 13

How Populations Evolve

Studying Advice

a. This chapter lays the foundation for the common evolutionary theme of the other chapters. It is largely conceptual with relatively few new terms. A thorough understanding of evolution is necessary to best understand the rest of biology. This is a chapter worth mastering.

b. The Hardy-Weinberg formula may at first seem intimidating, but spend the time to understand how it can be used to make predictions about the chances of inheriting disease. Several questions are included in this chapter to quiz your comprehension.

Student Media

Biology and Society on the Web

Learn about the evolution of bacteria that are resistant to antibiotics.

Activities

13A The Voyage of the *Beagle:* Darwin's Trip Around the World

13B Darwin and the Galápagos Islands

13C Reconstructing Forelimbs

13D Genetic Variation from Sexual Recombination

13E Causes of Microevolution

Case Studies in the Process of Science

What Are the Patterns of Antibiotic Resistance?

How Do Environmental Changes Affect a Population of Leafhoppers?

How Can Frequency of Alleles Be Calculated?

Evolution Connection on the Web

Learn more about sickle-cell disease.

Organizing Tables

Define and compare the pairs of concepts in the following.

TABLE 13.1

	Genetic Drift	**Gene Flow**
Mechanism: What is happening?		
Consequence: How does this affect the populations?		
	Founder Effect	**Bottleneck Effect**
Mechanism: What is happening?		
Consequence: How does this affect the populations?		

Compare the three types of selection in the following aspects.

TABLE 13.2

	Directional Selection	**Diversifying Selection**	**Stabilizing Selection**
What members of the population are being favored?			
How does the population curve change as a result of this selective force?			

Content Quiz

Directions: Identify the *one* best answer for the multiple choice questions. For true/false questions, determine if the statement is true or false. If false, change the underlined word(s) to make the statement true. Finally, add the correct word(s) to the fill-in-the-blank questions to make the statements true.

Biology and Society: Persistent Pests

1. The World Health Organization's effort to spray mosquitoes' habitat with dichlorodiphenyltrichloroethane (DDT) in tropical regions of the world ended because
   A. it was very successful and no longer needed.
   B. it killed the mosquitoes but also the animals that feed on them.
   C. the DDT accumulated in the ecosystem and caused too many other animals to die.
   D. it encouraged the development of new strains of mosquitoes resistant to DDT.
   E. None of the above.

2. Pesticide resistance quickly evolves in insects because
   A. the pesticide creates new individuals that can resist it.
   B. new species evolve that eat the pesticide.
   C. those that survive produce offspring who inherit the genes for pesticide resistance.
   D. the pesticide changes the DNA and creates resistant individuals.
   E. the insects just get used to the presence of the insecticide.

3. **True or False?** After one application of a strong pesticide, the percentage of the population that is resistant to the pesticide will <u>decrease</u>. ____________________

4. **True or False?** Natural selection <u>creates</u> organisms that can resist pesticides.

   ____

5. All life is united by ____________________ from the first microbes that populated the primordial planet.

Charles Darwin and *The Origin of Species*

6. Which one of the following people helped set the stage for Darwin by proposing that adaptations evolve as a result of interactions between organisms and their environments?
   A. Buffon
   B. Lyell
   C. Lamarck
   D. Wallace
   E. Anaximander

7. In his book, *The Origin of Species,* Charles Darwin developed two main points. These were that
   A. the world is very old and that evolution occurs by natural selection.
   B. evolution occurs slowly and that new species form by spontaneous generation.
   C. new species evolve by mutations and that new species don't reproduce with old species.
   D. modern species descended from ancestral species and that organisms evolve by natural selection.

8. Which one of the following statements about Darwin's voyage on the *H.M.S. Beagle* is *false*?
   A. During the trip, Darwin read and was strongly influenced by the book *Principles of Geology* by Charles Lyell.
   B. The concept of natural selection came quickly to Darwin when he discovered 13 species of finches on the Galápagos.
   C. Darwin thought that most of the animals of the Galápagos resembled species living on the South American mainland.
   D. Darwin collected many South American species of plants and animals while on the trip.

9. Which one of the following also suggested the idea of natural selection?
   A. Buffon
   B. Lyell
   C. Lamarck
   D. Wallace
   E. Anaximander

10. In his book *Principles of Geology,* Lyell argued that
    A. the Earth was about 6,000 years old.
    B. the Earth was ancient, sculpted by gradual geological processes that continue today.
    C. fossils could be explained by a single worldwide flood that occurred in the last 10,000 years.
    D. erosion, earthquakes, and other geological forces are new geological events and did not occur regularly in the past.

11. **True or False?** Darwin <u>was</u> the first to suggest that life evolves.

    ____

12. **True or False?** The Greek philosopher <u>Aristotle</u> suggested that simpler forms of life preceded more complex ones. ____________________

13. **True or False?** Lamarck proposed that by using or not using its body parts, an individual develops certain characteristics, which it passes on to its offspring.

____

14. **True or False?** Lamarck explained evolution as a process of adaptation.

____

Evidence of Evolution

The Fossil Record

15. When we examine the fossil record, we find that
    A. older fossils are in strata below younger fossils.
    B. amphibians appeared before the first fishes.
    C. transitional fossil forms are missing.
    D. the sediments of each stratum all come from a common geological region.
    E. there are major inconsistencies with molecular and cellular evidence.

16. **True or False?** The oldest known fossils are single-celled eukaryotes.

____

17. A preserved remnant or impression of an organism is a(n) _______________.

Biogeography; Comparative Anatomy

18. Which one of the following types of evidence first suggested to Darwin that modern organisms evolved from ancestral forms?
    A. comparative embryology
    B. the fossil record
    C. comparative anatomy
    D. biogeography
    E. molecular biology

19. Which one of the following is most like the concept of homology?
    A. mass-producing the Model T car
    B. repairing a car that has been in an accident
    C. automakers each producing their own version of a pickup truck
    D. replacing worn tires, changing the oil, and tuning up the engine of a car

20. Which one of the following is most like the way that evolution works?
    A. remodeling your snow skis into water skis now that you've moved from Alaska to Florida
    B. building a house by gathering together a group of architects and engineers to develop a new design
    C. mixing together chemicals to find a solution to remove graffiti from walls
    D. hiring a chemist to discover the secret formula of Coca-Cola

21. **True or False?** Biogeography <u>supports</u> the concept that species were individually placed into suitable environments. ________________

Comparative Embryology; Molecular Biology

22. The embryos of vertebrates
    A. appear most similar at the earliest stages of development.
    B. appear most different at the earliest stages of development.
    C. develop similar structures from different embryonic parts.
    D. are dramatically different throughout development.

23. Evolution predicts that two closely related species, such as humans and chimpanzees, would have
    A. many similar genes and proteins.
    B. similar anatomical patterns.
    C. common embryological stages.
    D. a common fossil ancestor.
    E. All of the above.

24. **True or False?** Molecular biology <u>confirmed</u> the fossil record and other evidence supporting Darwin's view of the interrelatedness of all life. ________________

25. Similar genes in two species suggest that the genes have been copied from a(n) ________________.

Natural Selection and Adaptive Evolution

Darwin's Theory of Natural Selection

26. Which one of the following is *not* an assumption of natural selection?
    A. Populations typically produce more offspring than can survive.
    B. Resources to support a population are unlimited.
    C. Individuals in a population vary.
    D. Many individual traits are inherited.

27. Darwin suggested that the animals on the Galápagos Islands had become new species because they
    A. were all much larger than their ancestors.
    B. were reproducing very quickly.
    C. ate similar foods as their ancestors.
    D. were adapting to new, local environments.

28. **True or False?** The different beaks of the many Galápagos finch species <u>are homologous</u>, because they are variations on the ancestral body plan.

    ____

29. **True or False?** The most fit individuals tend to leave the <u>fewest</u> fertile offspring.

____

30. Darwin thought that the Galápagos Islands were first colonized by animals from

____________________.

31. According to natural selection, a population's inherent variability is screened by the ____________________.

Natural Selection in Action

32. Instead of being a creative mechanism, natural selection is really more a process of
    A. editing.
    B. remixing.
    C. integrating.
    D. converging.
    E. directing.

33. **True or False?** An adaptation in one environment may <u>be useless or harmful</u> in different circumstances. ____________________

34. **True or False?** Doctors have documented an <u>increase</u> in drug-resistant strains of HIV. ____________________

The Modern Synthesis: Darwinism Meets Genetics

Populations as the Units of Evolution

35. Which one of the following is the smallest unit of evolution?
    A. a cell
    B. an individual organism
    C. a population
    D. a species
    E. an ecosystem

36. **True or False?** Members of separate populations of a species <u>can not</u> interbreed.

____

37. **True or False?** <u>Individuals</u> evolve. ____________________

38. The field of ____________________ emphasizes the genetic variation within populations and tracks the genetic composition of populations over time.

39. The fusion of genetics with evolutionary biology has become known as the

____________________.

Genetic Variation in Populations

40. Which one of the following processes is *not* very significant in the evolution of most animal and plant species?
    A. mutation
    B. sexual recombination
    C. meiosis
    D. random fertilization

41. **True or False?** The phenotype of an organism results from the combined influence of genotype and the environment. ________________

42. The ABO blood groups of humans are an example of a(n) ________________ characteristic in which several forms, or morphs, are common in a population.

Analyzing Gene Pools; Population Genetics and Health Science

Matching: Match each part of the Hardy-Weinberg formula, $p^2 + 2pq + q^2 = 1$, to what it represents.

____ 43.	p^2	A. the frequency of the homozygous recessive genotype
____ 44.	$2pq$	B. the total of the frequency of all genotypes
____ 45.	q^2	C. the frequency of the homozygous dominant genotype
____ 46.	1	D. the frequency of the heterozygous genotype

Matching: Match the term or phrase on the left to its best description on the right.

____ 47.	population	A. when two or more morphs are noticeably present
____ 48.	population genetics	B. the smallest biological unit that evolves
____ 49.	modern synthesis	C. all the alleles in all individuals in a population
____ 50.	gene pool	D. a field that studies genetic variation in a population
____ 51.	polymorphic	E. the combination of Darwin's and Mendel's work

52. If 1% of a population is homozygous recessive for a trait with two alleles, then
    A. 63% are homozygous dominant.
    B. 49% are homozygous dominant.
    C. 18% are heterozygous.
    D. 25% are heterozygous.
    E. 81% are heterozygous.

Microevolution as Change in a Gene Pool

53. In a nonevolving population, from generation to generation the frequency of alleles
    - A. changes, but the frequency of the genotypes stays the same.
    - B. and the frequency of the genotypes stays the same.
    - C. and the frequency of the genotypes changes.
    - D. stays the same, but the frequency of the genotypes changes.

54. The genetic equilibrium of a nonevolving population is also called the ____________________ equilibrium.

55. A generation-to-generation change in a population's frequencies of alleles defines ____________________.

Mechanisms of Microevolution

Gene Drift; Gene Flow; Mutations

56. Imagine that a global disease spread quickly across the Earth killing every human except the people in your biology class. Although the future human population might some day recover, future generations would not represent the current diversity of all humans living today. This loss of diversity and change in the gene pool would be an example of
    - A. the founder effect.
    - B. sexual recombination.
    - C. gene flow.
    - D. the bottleneck effect.
    - E. Hardy-Weinberg equilibrium.

57. Another disaster strikes. Your teacher takes your class on a spectacular field trip to study marine mammals, but your ship encounters a terrific storm. Your class and ship's crew land on a remote and uninhabited Pacific island where they live and prosper, never again making contact with other humans. As future generations come and go, the appearance of your island population takes on its own characteristics, representing the process of
    - A. the founder effect.
    - B. sexual recombination.
    - C. gene flow.
    - D. the bottleneck effect.
    - E. Hardy-Weinberg equilibrium

58. Which one of the following statements about gene flow is *false*?
   A. Modern transportation and the frequent relocation of people has increased gene flow in the human population.
   B. Gene flow occurs when fertile individuals move between populations.
   C. Populations may gain or lose alleles through gene flow.
   D. Gene flow tends to increase genetic differences between populations.

59. Which one of the following would *not* increase gene flow?
   A. very light seeds blown several miles in high storm winds
   B. a river flooding over into nearby ponds
   C. a storm blowing a flock of birds out to sea, where they land on an uninhabited island
   D. a giant redwood tree dropping its seeds all around the base of its grand trunk

60. **True or False?** Genetic bottlenecks usually increase the genetic variability in a population. ____________________

61. **True or False?** Some human diseases are more abundant in small populations due to genetic drift. ____________________

62. **True or False?** Genetic drift and gene flow cause microevolution.

    ____

63. **True or False?** Any one gene locus mutation alone has a considerable quantitative effect on a large population in a single generation because it is a rare event.

    ____

64. The founder and bottleneck effects are examples of ____________________, an evolutionary mechanism by which the gene pool of a small population changes due to chance.

65. A(n) ____________________ is a change in an organism's DNA.

Natural Selection: A Closer Look

66. Which one of the following terms best represents the process of evolution?
   A. deliberate
   B. intentional
   C. fixed
   D. directional
   E. responsive

67. Which one of the following organisms has the greatest Darwinian fitness?
   A. a mother robin that finds more food than any other female robin but lays no eggs
   B. a mother robin that has six chicks hatch, but only one lives to reproduce
   C. a mother robin that has two chicks that both live to reproduce
   D. the largest mother robin in the entire population, who lays one large egg
   E. a mother robin that lays the most eggs, but none of them hatch

Matching: Match the types of selection on the left to descriptions on the right. One of the items on the left matches two items on the right.

____ 68. directional selection	A. selection favoring two or more contrasting morphs
____ 69. stabilizing selection	B. selection favoring one extreme phenotype
____ 70. diversifying selection	C. selection favoring a particular trait within a narrow range
	D. the most common type of selection

71. **True or False?** The Hardy-Weinberg equilibrium demands that all individuals in a population <u>be equal</u> in their ability to survive and reproduce.

   ____

72. Human birth weights are kept in the range of 3–4 kilograms by ____________________ selection.

73. The fossil record indicates that when a population is challenged by a new environmental problem, the most common result is ____________________.

Evolution Connection: Population Genetics of the Sickle-Cell Allele

74. Which one of the following statements about sickle-cell anemia is *false*?
   A. The sickle-cell allele is most common in Africa where the malarial parasite is most common.
   B. People heterozygous for the sickle-cell allele are relatively resistant to malaria.
   C. People heterozygous for the sickle-cell allele are less common than people who are homozygous for the sickle-cell allele.
   D. Only people homozygous for the sickle-cell allele develop the disease.

75. The Hardy-Weinberg formula indicates that in a population with a 20% sickle-cell allele frequency
   A. the sickle-cell allele benefits more of the population than it hurts by causing sickle-cell disease.
   B. the sickle-cell allele benefits less of the population than it hurts by causing sickle-cell disease.
   C. the sickle-cell allele benefits about the same amount of the population that it hurts by causing sickle-cell disease.
   D. nearly everyone will develop sickle-cell disease, and no one will benefit.
   E. everyone will benefit, and nobody will develop sickle-cell disease.

76. **True or False?** Heterozygous individuals are relatively resistant to malaria.

   ____

77. People who are sickle-cell disease ____________________ have a single copy of the sickle-cell allele.

Word Roots

homo = alike (*homology:* traits that appear similar due to common ancestry)

micro = small (*microevolution:* evolution at its smallest scales)

poly = many; **morph** = form (*polymorphic:* a characteristic of a population in which two or more forms are clearly present)

Key Terms

biogeography
bottleneck effect
comparative anatomy
comparative embryology
Darwinian fitness
Darwinian medicine
directional selection
diversifying selection
evolution
evolutionary adaption
fossil record
fossil
founder effect
gene flow
gene pool
genetic drift
Hardy-Weinberg equilibrium
Hardy-Weinberg formula
homology
microevolution
modern synthesis
mutation
natural selection
polymorphic
population
population genetics
stabilizing selection

Crossword Puzzle

Use the Key Terms list from this chapter to fill in the crossword puzzle.

ACROSS

2. evolution at its smallest scale
4. genetic drift due to a drastic reduction in population size
7. the name of the formula and type of equilibrium that describes a population's gene pool
8. the comparison of body structures between different species
12. the comparison of structures that appear during the development of different organisms
16. an individual's contribution to the next generation's gene pool relative to the contribution of other individuals
19. all of the alleles in all of the individuals in a population
20. a trait that appears similar due to common ancestry
22. genetic drift in a new colony
23. the study of genetic variation within a population and over time

DOWN

1. the type of selection that can lead to a balance of two or more different morphs
3. a group of interacting individuals belonging to one species and living in the same geographic area
4. the study of the geographic distribution of species
5. a population in which two or more morphs are clearly present
6. the mechanism for descent with modification
9. genetic change in a population or species over generations
10. the study of health problems in an evolutionary context
11. a change in a gene pool of a small population due to chance
13. the type of selection that shifts the phenotypic curve of a population in favor of some extreme phenotype
14. genetic exchange with another population
15. the fusion of genetics with evolutionary biology
17. a preserved remnant or impression of an organism that lived in the past
18. the type of selection that maintains variation for a particular trait within a narrow range
21. a change in the nucleotide sequence of DNA

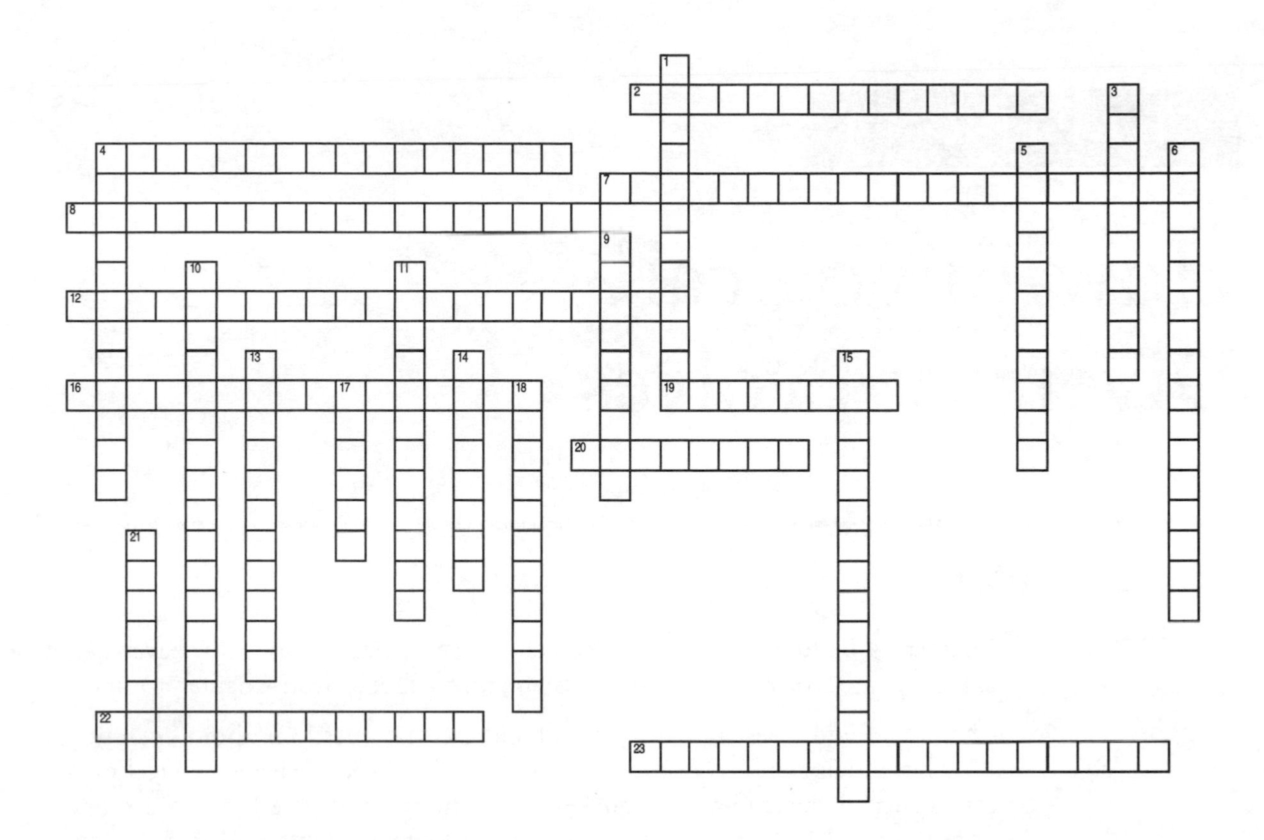

1
2
3
4
5
6
7
8
9
10
11
12
13
14
15
16
17
18
19
20
21
22
23

CHAPTER 14

How Biological Diversity Evolves

Studying Advice

a. Chapter 14 builds heavily upon Chapter 13. If for some reason, you haven't studied Chapter 13 before reaching this chapter, you should read Chapter 13 first.

b. Chapter 14 addresses some very broad and exciting questions about evolution: Why have major groups of organisms gone extinct? How have geological processes such as volcanoes, earthquakes, and continental drift affected the evolution of animals? How do new species evolve? Many students find this chapter to be one of the most interesting in the book. The new terminology is limited. So enjoy the ideas as you reflect on the major forces that have influenced the evolution of life on Earth!

Student Media

Biology and Society on the Web

Learn about the evidence for an asteroid impact 250 million years ago that may have contributed to mass extinctions at that time.

Activities

14A Mechanisms of Macroevolution
14B Exploring Speciation on Islands
14C Polyploid Plants
14D Paedomorphosis: Morphing Chimps and Humans
14E The Geologic Time Scale
14F Classification Schemes

Case Studies in the Process of Science

Investigate how new species arise by genetic isolation.

Analyze proteins to build phylogenetic trees.

Evolution Connection on the Web

Explore a website that explains common misconceptions about evolution.

Organizing Table

Compare the five Kingdoms by completing the table below. Some table cells have already been filled in to help you.

TABLE 14.1

Kingdom	Examples	Prokaryotic or Eukaryotic Cells?	Single or Multicellular?	Domain (in the three-domain system)
Monera				Bacteria *and* Archaea
Protista			Single *and* multicellular	
Plantae				
Fungi				
Animalia				

Content Quiz

Directions: Identify the *one* best answer for the multiple-choice questions. For true/false questions, determine if the statement is true or false. If false, change the underlined word(s) to make the statement true. Finally, add the correct word(s) to the fill-in-the-blank questions to make the statements true.

Biology and Society: The Impact of Asteroids

1. Which one of the following statements is *false*? During the period of time when dinosaurs went extinct,
   A. the climate cooled.
   B. the shallow seas were receding from continental lowlands.
   C. plants required by plant-eating dinosaurs died out first.
   D. the rock deposited at the time contained a thin layer of clay that is rich in iridium.
   E. mammals of all sizes were abundant.

2. **True or False?** Mass extinctions like the one associated with the death of the dinosaurs have happened before. ____________________

3. Many paleontologists conclude that a layer of ____________________ is the result of fallout from a huge cloud of dust that billowed into the atmosphere when a large meteorite or asteroid hit Earth.

Macroevolution and the Diversity of Life

4. Which one of the following is *not* included in the field of macroevolution?
   A. the multiplication of species
   B. the origin of evolutionary novelty
   C. the explosive diversification following some evolutionary breakthrough
   D. mass extinctions
   E. All of the above are included in macroevolution.

5. **True or False?** Biological diversity is generated by nonbranching evolution.

   ____

6. Darwin visited the volcanic ____________________, named for its giant tortoises.

7. Nonbranching and branching are two types of ____________________, the origin of new species.

The Origin of Species

What Is a Species?

8. According to the biological species concept, what keeps species separate?
   A. time
   B. differences in diet
   C. natural selection
   D. reproductive barriers
   E. appearance

9. **True or False?** All humans belong to the same species. ________________

10. **True or False?** Members of a biological species can not interbreed with members of other species. ________________

11. The biological species concept was the original idea of ________________, who developed the idea while studying birds in New Guinea.

Reproductive Barriers Between Species

12. Which one of the following is a post-zygotic reproductive barrier?
    A. temporal isolation
    B. habitat isolation
    C. behavioral isolation
    D. hybrid inviability
    E. gamete incompatibility
    F. anatomical incompatibility

13. Examine the two birds in Figure 14.4 and read the figure legend. Which one of the following reproductive barriers is keeping these species separate?
    A. behavioral isolation
    B. temporal isolation
    C. mechanical isolation
    D. habitat isolation
    E. gametic isolation

14. **True or False?** The evolution of reproductive barriers is the key to the origin of new species. ________________

15. A mule, formed by the hybridization of a female horse and a male donkey, is an example of ________________, a form of reproductive isolation.

Mechanisms of Speciation

16. Sympatric speciation
    A. typically occurs over millions of years.
    B. occurs when a population becomes geographically isolated.
    C. is widespread among animals and rare in plants.
    D. can occur in a single generation.
    E. is the mechanism by which most Galápagos species have evolved.

17. In North America, some rare salamanders are known to be triploid, possessing three copies of every chromosome. (They are an all-female species that reproduces by producing triploid eggs that develop without the addition of a sperm nucleus.) It is believed that these salamanders formed when the sperm from one species fertilized a diploid egg of another species. This evolutionary scenario is an example of
    A. hybrid sterility.
    B. hybrid inviability.
    C. allopatric speciation.
    D. sympatric speciation.

18. **True or False?** Most polyploid species arise from <u>one</u> parent species.

    ____

19. Many of the plant species we grow for food are the result of

    ____________________ speciation.

In the blanks for questions 20–24, write A if the example is of allopatric speciation or S if the example is of sympatric speciation.

_____ 20. A flock of birds is blown by a hurricane onto a remote Pacific island, where the birds remain isolated and become a new species.

_____ 21. A plant seed fails to undergo meiosis, fertilizes itself, and doubles the number of chromosomes in the next generation. These new plants do not interbreed with the parent species.

_____ 22. The long peninsula becomes a string of islands, separating the ant populations and evolving new ant species.

_____ 23. A glacier advances through the midwestern United States, splitting the populations of rabbits into eastern and western species.

_____ 24. Two species hybridize, producing polyploid individuals that remain reproductively isolated and forming a new polyploid species.

What Is the Tempo of Speciation?

25. The concept of punctuated equilibrium suggests that a species
    A. evolves quickly when it first forms and then changes very little for a long time.
    B. stays very similar when it first forms but then undergoes drastic change just before becoming another species.
    C. stays the same when it first evolves, then cycles change repeatedly between a lot and then not much at all until it becomes another species.
    D. changes a lot when it first evolves, then goes extinct.

26. The somewhat sudden switch in the music industry from vinyl records to compact discs is most like
    A. sympatric speciation.
    B. allopatric speciation.
    C. punctuated equilibrium.
    D. microevolution.

27. **True or False?** A "sudden" geological appearance of a species may actually be 50,000–100,000 years. ________________

28. **True or False?** Once a species forms, species in a stable environment may experience much noticeable change. ________________

The Evolution of Biological Novelty

Adaptation of Old Structures for New Functions

29. Which one of the following terms is closest in meaning to the term *exaptation*?
    A. building
    B. remodeling
    C. destroying
    D. repairing
    E. anticipating

30. **True or False?** Lightweight bones could have evolved in the ancestors of birds in anticipation of the evolution of flight in birds. ________________

31. A structure that evolved in one context but which conveyed advantages for other functions is a(n) ________________.

"Evo-Devo": Development and Evolutionary Novelty

32. Which one of the following is *not* a paedomorphic trait of humans?
    A. large brain
    B. rounded skull
    C. large teeth
    D. flat face
    E. small jaws

33. The evolution of dramatically different species by the process of paedomorphosis is most like
    A. ordering a birthday cake without frosting.
    B. blending a new flavor of wine by mixing wine from red and white grapes.
    C. adding extra luxuries to an automobile to make a special edition.
    D. repairing a truck after a collision.
    E. creating a tetraploid plant by hybridizing two diploid species.

34. Examine the transformation grids in Figure 14.17. Which one of the following shows *the greatest* change in the development of the chimpanzee into an adult?
    A. top of the skull
    B. ear region
    C. rear of the skull
    D. region around the mouth (upper and lower jaw)
    E. bottom rear of the skull

35. **True or False?** Slight genetic changes in key genes controlling development can dramatically change the appearance of an animal. ________________

36. The retention of juvenile traits in the adult, or ________________, has occurred in salamanders and humans.

Earth History and Macroevolution

Geological Time and the Fossil Record

37. Pulling away many layers of wallpaper as a person remodels an old home is most like
    A. the way that the continents formed.
    B. the distribution of fossils within a sedimentary layer.
    C. exploring layers of sedimentary rock.
    D. the effects of earthquakes on the continents.
    E. radiometric dating.

38. Which one of the following commonly marks the boundaries of geological eras?
    A. evidence of extensive earthquake activity
    B. sudden changes in ocean levels
    C. little change in biological diversity
    D. mass extinctions
    E. alignment of certain stars

39. Which one of the following is the correct sequence of the four major eras?
    A. Paleozoic, Mesozoic, Precambrian, Cenozoic
    B. Precambrian, Paleozoic, Mesozoic, Cenozoic
    C. Cenozoic, Mesozoic, Paleozoic, Precambrian
    D. Cenozoic, Precambrian, Paleozoic, Mesozoic
    E. Precambrian, Mesozoic, Paleozoic, Cenozoic

40. Examine Figure 14.19 and read the figure legend to understand the half-life of an element. If an element has a half-life of 20,000 years, and we start out with 1 kilogram (kg) of the element, how much of the element will be present after 40,000 years?
   A. 4 kg
   B. 2 kg
   C. 0.5 kg
   D. 0.25 kg
   E. 0.10 kg

41. **True or False?** Examining the fossils in layers of sedimentary rock reveals the <u>absolute</u> age of the layers. ________________

42. The rate of radioactive decay of specific isotopes can be used to date fossils in a process called ________________.

Continental Drift and Macroevolution

43. Which one of the following did *not* occur as a result of the formation of Pangaea?
   A. ocean levels were lowered
   B. shallow marine communities were flooded
   C. the total amount of shoreline was reduced
   D. populations of organisms that had evolved in isolation were brought together
   E. the continental interior increased in size

44. If we compared the fossils of vertebrates found in Africa and South America, we would expect that fossils found about 200 million years ago would be
   A. similar, but those from animals 50 million years ago would be quite different.
   B. different, but those from animals 50 million years ago would be quite similar.
   C. similar, just like fossils from animals 50 million years ago.
   D. different, just like fossils from animals 50 million years ago.

45. **True or False?** The breakup of Pangaea probably caused <u>sympatric</u> speciation in many animals. ________________

46. When Pangaea broke up, the continents moved apart because of

   ________________.

Mass Extinctions and Explosive Diversification of Life

47. Which of the following is considered to be a contributing factor to the demise of the dinosaurs 65 million years ago?
   A. increased volcanic activity
   B. a cooled global climate
   C. an asteroid or comet striking the Earth
   D. All of the above are possible factors.

48. Throughout the history of life on Earth, mass extinctions have been followed by periods of
    A. little life.
    B. recovery of many of the same species.
    C. no life at all.
    D. an explosion of new diversity.

49. **True or False?** Over the last 600 million years, there have been <u>two</u> distinct periods of mass extinctions. ________________

Classifying the Diversity of Life

Some Basics of Taxonomy

50. Which one of the following is the correct sequence of taxonomic groups moving from most general to most specific?
    A. kingdom, phylum, class, order, family, genus, and species
    B. kingdom, order, class, family, phylum, genus, and species
    C. phylum, kingdom, class, order, genus, family, and species
    D. kingdom, phylum, order, class, family, genus, and species
    E. class, kingdom, phylum, order, genus, family, and species

51. Which one of the following is the correct way to write the scientific name for humans?
    A. *Homo Sapiens*
    B. homo Sapiens
    C. *homo sapiens*
    D. Homo sapiens
    E. *Homo sapiens*

52. The hierarchical system used to classify life is most like
    A. the many ranks in the military.
    B. the many names for colleges in the United States.
    C. the different types of fruits and vegetables in a grocery store.
    D. the first, last, and middle names that people use.

53. **True or False?** The first part of the scientific name of several species in the same genus will <u>all be the same</u>. ________________

54. Carolus Linnaeus suggested a system that gives each species a two part name, or ________________.

Classification and Phylogeny

55. Homologous structures are related to each other in the same way as
    A. two different animal figures are to each other when both are carved from bars of soap.
    B. two buildings are to each other when one is made from brick and the other from steel.
    C. pens are to pencils.
    D. the brain is to the stomach.
    E. water is to fish.

56. Organisms that are very closely related are more likely to have
    A. different genes and few homologous structures.
    B. different genes and many homologous structures.
    C. similar genes but few homologous structures.
    D. similar genes and many homologous structures.

57. Examine Figure 14.23 showing the classification of four carnivores. Which one of the following pairs of genera is most closely related to each other?
    A. *Canis* and *Mephitis*
    B. *Canis* and *Lutra*
    C. *Panthera* and *Mephitis*
    D. *Panthera* and *Canis*
    E. *Mephitis* and *Lutra*

58. Examine the evolutionary relationships in Figure 14.25. The common ancestor of the group that included just the red kangaroo and North American beaver had these characteristics:
    A. vertebral column but no hair, no mammary glands, and no gestation
    B. vertebral column, hair, no mammary glands, and no gestation
    C. vertebral column, hair, mammary glands, but no gestation
    D. vertebral column, hair, mammary glands, and gestation
    E. vertebral column, hair, mammary glands, and long gestation

59. **True or False?** Convergent evolution often produces structures that are <u>homologous</u> but not <u>analogous</u>. ____________________

60. Homologous structures have evolved from ____________________ in a common ancestor.

61. Structures that are ____________________ to each other are similar in function.

62. Phylogenetic trees are based upon ____________________ relationships.

63. Very complex systems are less likely to be the result of ____________________ evolution and more likely to be the result of ____________________.

64. One of the most widely used methods in systematics, ________________, uses computers to analyze data to identify clades with unique homologies.

Arranging Life into Kingdoms: A Work in Progress

65. In the three-domain system, the new domains Bacteria and Archaea, are
    A. subdivisions of the kingdom Protista.
    B. subdivisions of the kingdom Fungi.
    C. subdivisions of the kingdom Plantae.
    D. groups that were not discovered until the 1970s.
    E. both prokaryotic cells.

66. **True or False?** Although seaweeds are multicellular, they are commonly classified into the typically unicellular kingdom Plantae. ________________

67. The revision of the five-kingdom system to the three-domain system was largely based upon molecular studies and ________________.

Evolution Connection: Just a Theory?

68. In science, a *theory:*
    A. is about the same as a hypothesis.
    B. accounts for many facts and attempts to explain many phenomena.
    C. is a strict rule that has no exceptions and is not to be questioned.
    D. is really a type of law, a rule that is widely applied in all fields of science.
    E. is an idea in its earliest stages of development, awaiting experimental evidence to test it.

69. **True or False?** Among scientists, debates over evolution are like debates over gravity as scientists seek to understand the mechanisms of obvious phenomena. ________________

Word Roots

allo = other; **patric** = country (*allopatric:* the type of speciation in which a new species forms by geographical isolation)

bi = two; **nom** = name (*binomial:* a two-part name used to identify a species)

con = together (*convergent evolution:* the type of evolution in which unrelated organisms evolve structures with similar functions)

macro = large (*macroevolution:* major biological changes evident in the fossil record)

sym = together (*sympatric:* the type of speciation in which a new species forms in the same geographical region)

paedos = child; **morphosis** = shaping (*paedomorphosis:* the retention of juvenile traits into the adult)

post = after (*post-zygotic barriers:* the type of barriers that prevent development of a zygote that is a hybrid between species)

pre = before (*pre-zygotic barriers:* the type of barriers that prevent the fertilization of the eggs of different species)

Key Terms

allopatric speciation
analogy
binomial
biological species concept
cladistic analysis
class
convergent evolution
domain
exaptation
family
five-kingdom system
genus
geologic time scale
kingdoms
orders
paedomorphosis
phyla
phylogenetic tree
phylogeny
post-zygotic barriers
pre-zygotic barriers
punctuated equilibrium
radiometric dating
speciation
species
sympatric speciation
systematics
taxonomy
three-domain system

Crossword Puzzle

Use the Key Terms list from this chapter to fill in the crossword puzzle.

ACROSS

9. anatomical similarity due to convergent evolution
10. the type of systematics that uses computers and homologous structures
11. a group of orders
16. a two-part, latinized name of a species; for example, *Homo sapiens*
17. the system of classification that divides all life into five groups
19. the evolution of a structure in one context that becomes adapted for other functions
21. the type of concept that defines a species
22. a table of historical periods grouped into four eras
23. the type of evolution in which unrelated organisms
24. the process by which new species form in spurts of rapid change followed by periods of slow speciation

DOWN

1. the system of classification that divides all life into three groups
2. types of barriers that prevent development of a zygote that is a hybrid between species
3. a taxonomic category above the kingdom level
4. groups of phyla
5. a group of genera
6. the origin of new species
7. the retention of juvenile body features in the adult stage
8. in classification, the taxonomic category just below genus
9. the type of speciation in which a new species forms by geographical isolation
12. a groups of classes
13. a branching diagram that represents a hypothesis about evolutionary relationships among organisms
14. the evolutionary history of a species
15. barriers that prevent the fertilization of an egg of a different species
18. a groups of families
20. the first part of a binomial

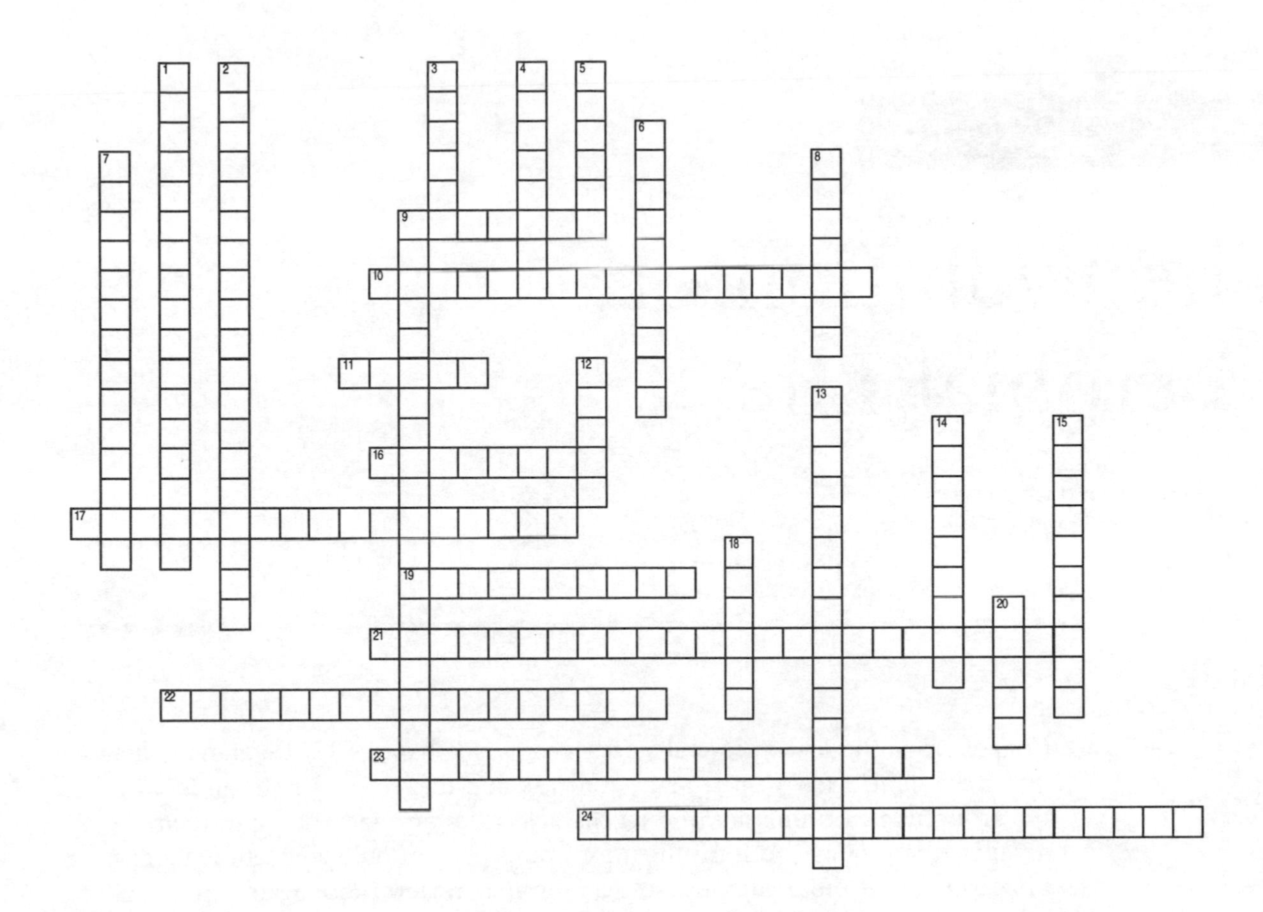

1
2
3
4
5
6
7
8
9
10
11
12
13
14
15
16
17
18
19
20
21
22
23
24

CHAPTER 15

The Evolution of Microbial Life

Studying Advice

a. Chapter 15 is the first of several survey chapters (Chapters 15, 16, and 17) that introduce quite a few names. Many students find that making note cards, with the group name on one side and its characteristics on the back, helps them organize the groups and learn them for exams. As you quiz yourself, remember to keep track of those you miss so that you may review those again.

b. The five organizing tables will help you organize some of the chapter information.

Student Media

Biology and Society on the Web

Learn more about the history of biowarfare.

Activities

15A The History of Life

15B Prokaryotic Cell Structure and Function

15C Diversity of Prokaryotes

Case Studies in the Process of Science

How Did Life Begin on Early Earth?

What Are the Modes of Nutrition in Prokaryotes?

What Kinds of Protists Are Found in Various Habitats?

Evolution Connection on the Web

Learn about what may be the oldest fossils of multicellular organisms.

Organizing Tables

In Table 15.1, identify the key events that occurred during the periods in Earth's history.

TABLE 15.1

Billions of Years Ago (BYA)	**Key Events in Earth's History and the Evolution of Life**
4.5	
4.0	
by 3.5	
3.5–2.5	
2.5	
1.7	
About 1.0	
0.57 (570 Million Years Ago [MYA])	
0.475 (475 MYA)	

Describe the four main stages, in order, of the four-stage hypothesis for the evolution of life on Earth.

TABLE 15.2

	Key Events of Stage
Stage 1	
Stage 2	
Stage 3	
Stage 4	

Use Table 15.3 to compare and contrast the prokaryotes and eukaryotes.

TABLE 15.3

	Prokaryotes	Eukaryotes
Domains		
Kingdoms		
Examples of each kingdom		
Is a true nucleus present?	❑ Yes or ❑ No	❑ Yes or ❑ No
Are other membrane-enclosed organelles present?	❑ Yes or ❑ No	❑ Yes or ❑ No
Are cell walls present? If yes, are they made of cellulose?	❑ Yes or ❑ No ❑ Yes or ❑ No	❑ Yes or ❑ No ❑ Yes or ❑ No

Use Table 15.4 to compare and contrast the different types of bacteria.

TABLE 15.4		
Bacterial Cell Shape and Organization	**Draw the Shape**	**Provide an Example from the Text**
Cocci		
Staphylococci		
Streptococci		
Bacilli		
Spirochetes		

Use Table 15.5 to compare and contrast the types of prokaryotic nutritional strategies.

TABLE 15.5			
Prokaryotic Group	**Use Light Energy to Synthesize Organic Compounds?**	**Require at Least One Organic Nutrient as a Source of Carbon?**	**Examples Given in Text**
Photoautotrophs	❑ Yes or ❑ No	❑ Yes or ❑ No	
Chemoautotrophs	❑ Yes or ❑ No	❑ Yes or ❑ No	
Photoheterotrophs	❑ Yes or ❑ No	❑ Yes or ❑ No	
Chemoheterotrophs	❑ Yes or ❑ No	❑ Yes or ❑ No	

Content Quiz

Directions: Identify the *one* best answer for the multiple-choice questions. For true/false questions, determine if the statement is true or false. If false, change the underlined word(s) to make the statement true. Finally, add the correct word(s) to the fill-in-the-blank questions to make the statements true.

Biology and Society: Bioterrorism

1. Which one of the following types of organisms has not been used for bioterrorism?
   A. animals
   B. plants
   C. fungi
   D. viruses
   E. All of the above have been used for bioterrorism.

2. **True or False?** The United States once had a <u>bioweapons</u> program.

   ____

3. During the Middle Ages, the bacterium that causes ________________ was used as a biowarfare agent.

Major Episodes in the History of Life

4. Which one of the following is the *correct* sequence of the evolution of early life?
   A. prokaryotic cells, photosynthetic prokaryotes, prokaryotes using cellular respiration, eukaryotic cells, multicellular eukaryotes
   B. photosynthetic prokaryotes, prokaryotic cells, eukaryotic cells, prokaryotes using cellular respiration, multicellular eukaryotes
   C. eukaryotic cells, photosynthetic eukaryotes, prokaryotes using cellular respiration, prokaryotic cells, multicellular prokaryotes
   D. multicellular eukaryotes, eukaryotic cells, prokaryotic cells, photosynthetic prokaryotes, prokaryotes using cellular respiration,
   E. prokaryotic cells, prokaryotes using cellular respiration, photosynthetic prokaryotes, eukaryotic cells, multicellular eukaryotes

5. Which one of the following is *false*?
   A. Multicellularity evolved about 1 billion years ago.
   B. The oldest known eukaryotes are about 3 billion years old.
   C. Life has been confined to water for nearly 85% of its existence on Earth.
   D. Plants, along with fungi, were the first to colonize land.
   E. Reptiles are the common ancestors of birds and mammals.

6. Examine the evolutionary relationships in Figure 15.2. Which one of the following pairs is most closely interrelated?
   A. archaea and plants
   B. archaea and protists
   C. bacteria and protists
   D. fungi and animals
   E. animals and protists

7. **True or False?** For almost 2 billion years, eukaryotes lived alone.

   ____

8. **True or False?** Mitochondria and chloroplasts are descendants of prokaryotes.

   ____

9. The addition of great amounts of oxygen into our atmosphere about 2.5 billion years ago favored organisms that could use the process of ____________________.

The Origin of Life

Resolving the Biogenesis Paradox

10. Biogenesis is the idea that
    A. life emerges by spontaneous generation.
    B. the first cells ever to evolve arose by spontaneous generation.
    C. it takes a male and female organism to produce new life.
    D. life is cellular.
    E. life gives rise to life.

11. **True or False?** The conditions on Earth when life first arose were very similar to conditions today. ____________________

12. The idea that life can come from nonliving matter, called ____________________, was commonly accepted before modern science demonstrated otherwise.

A Four-Stage Hypothesis for the Origin of Life; From Chemical Evolution to Darwinian Evolution

13. Which one of the following is the *correct* sequence of the four stages of the commonly accepted hypothesis for the origin of the first cells?
    A. joining of monomers into polymers, abiotic synthesis of small organic molecules, packaging into pre-cells, and origin of self-replicating molecules
    B. origin of self-replicating molecules, joining of monomers into polymers, abiotic synthesis of small organic molecules, and packaging into pre-cells
    C. origin of self-replicating molecules, abiotic synthesis of small organic molecules, joining of monomers into polymers, and packaging into pre-cells
    D. abiotic synthesis of small organic molecules, joining of monomers into polymers, origin of self-replicating molecules, and packaging into pre-cells

14. Which of the following molecules were produced by experiments simulating conditions on primitive Earth?
    A. monomers of DNA and RNA
    B. sugars and lipids
    C. ATP
    D. all 20 amino acids
    E. All of the above have been produced experimentally.

15. **True or False?** There is evidence that some of the early organic molecules could have come from space. ________________

16. **True or False?** Natural selection would not have operated on the first pre-cells to have formed. ________________

17. The early replication of RNA may have been catalyzed by ________________.

Prokaryotes

They're Everywhere!

18. Which one of the following statements about prokaryotes is *false*?
    A. Some prokaryotes decompose organic matter and thus help recycle essential nutrients.
    B. Prokaryotes live in many places on Earth where eukaryotes can not survive.
    C. Some prokaryotes cause serious disease.
    D. Some prokaryotes provide us with vitamins.
    E. Prokaryotes have lived with eukaryotes for about 3.5 billion years.
    F. Some prokaryotes prevent fungal infections in our body.

19. **True or False?** Without prokaryotes, eukaryotic life would be dead.

____

20. **True or False?** Without eukaryotes, prokaryotic life would be dead.

____

21. Tuberculosis, cholera, food poisoning, and many sexually transmissible infections are caused by _________________.

The Two Main Branches of Prokaryotic Evolution: Bacteria and Archaea

22. Bacteria and the archaea both
    A. live in extreme environments.
    B. occur in about equal numbers in most ecosystems.
    C. have a prokaryotic cellular organization.
    D. produce methane as a waste product of metabolism.
    E. aid digestion in cattle.

23. **True or False?** Some archaea live in aerobic environments and give off methane as a by-product. _________________

24. The _________________ are the prokaryotic group most closely related to the eukaryotes.

The Structure, Function, and Reproduction of Prokaryotes

Matching: Match the word on the left to its best description on the right.

____ 25. spirochete	A. chains of cocci
____ 26. bacilli	B. clusters of cocci
____ 27. cocci	C. rod-shaped bacteria
____ 28. streptococci	D. dormant bacterial cells
____ 29. staphylococci	E. spherical bacteria
____ 30. endospores	F. spiral-shaped bacteria

31. Which one of the following statements about prokaryotes is *false*?
    A. Very few prokaryotic species are motile.
    B. Some prokaryotic species exhibit simple multicellular organization.
    C. Some endospores can remain dormant for centuries and survive boiling water.
    D. Many motile prokaryotes use flagella to move about.
    E. If sufficient resources are available, prokaryotic populations can double every 20 minutes.

32. **True or False?** Most bacteria have a cell wall exterior to their plasma membrane that is made of <u>the same</u> materials as plant cells. ____________________

33. Prokaryotes reproduce by the process of ____________________.

The Nutritional Diversity of Prokaryotes

Matching: Match the word on the left to its best description on the right.

_____ 34.	photoheterotroph	A. energy source is sunlight; carbon source is CO_2
_____ 35.	chemoautotroph	B. energy source is sunlight; carbon source is organic compounds
_____ 36.	chemoheterotroph	C. energy source is inorganic compounds; carbon source is CO_2
_____ 37.	photoautotroph	D. energy source and carbon source are both organic compounds

38. Which of the following nutritional strategies are unique to prokaryotes?
    A. photoautotrophs
    B. chemoheterotrophs
    C. photoautotrophs and chemoheterotrophs
    D. chemoautotrophs and photoheterotrophs

39. **True or False?** All plants and algae are <u>chemoautotrophs</u>. ____________________

40. All fungi and animals are ____________________.

The Ecological Impact of Prokaryotes

41. Most pathogenic bacteria cause disease by
    A. eating our living flesh.
    B. consuming so much oxygen that our tissues die.
    C. producing nitrogen-containing compounds.
    D. producing exotoxins and endotoxins.
    E. living in our cells and disrupting normal cellular processes.

42. The nitrogen that plants use to make proteins and nucleic acids comes from
    A. photoautotrophs that live in the air.
    B. prokaryotic metabolism in the soil.
    C. bioremediation.
    D. pathogenic bacteria that live inside leaves.
    E. genetically engineered bacteria that we add to the soil.

43. **True or False?** Most bacteria <u>are</u> pathogenic. ____________________

44. Bacteria are used to break down sewage and to clean up oil spills and other toxic environments through the process of ________________.

Protists

The Origin of Eukaryotic Cells

45. Eukaryotic cells evolved by
   A. mitosis and meiosis.
   —B. inward folds of the plasma membrane and endosymbiosis.
   C. inward folds of the plasma membrane and bioremediation.
   D. endosymbiosis and bioremediation.
   E. mitosis and endosymbiosis.

46. Which one of the following statements about chloroplasts and mitochondria is *false*? Both chloroplasts and mitochondria
   A. possess DNA.
   B. make some of their enzymes.
   C. reproduce by binary fission.
   D. appear to have evolved by endosymbiosis.
   E. are found in all eukaryotic cells.

47. Examine Figure 15.18. According to the endosymbiosis theory, the inner mitochondrial membrane corresponds to the
   A. outer membrane of a chloroplast.
   B. nuclear membrane.
   C. original aerobic bacterial membrane.
   D. original membrane of a photosynthetic bacterium.
   E. endoplasmic reticulum.

48. Examine Figure 15.18 again. According to the endosymbiosis theory, the outer chloroplast membrane corresponds to the
   A. inner mitochondrial membrane.
   B. nuclear membrane.
   C. plasma membrane of the cell that engulfed it.
   D. endoplasmic reticulum.
   E. the outer membrane of an aerobic bacterium.

49. **True or False?** The ancestors of mitochondria were probably <u>anaerobic</u> bacteria.

____

50. Almost all eukaryotes have ________________, but only some have chloroplasts.

51. The first eukaryotic cells to evolve were ________________.

The Diversity of Protists

Matching: Match the group on the left to its best description on the right.

_____ 52. diatoms	A. large, multicellular marine algae
_____ 53. flagellates	B. protozoans covered by cilia
_____ 54. forams	C. grass-green chloroplasts which includes *Volvox*
_____ 55. dinoflagellates	D. protozoans with great flexibility and no locomotory organelles
_____ 56. amoebas	E. all parasitic protozoans with an apex for penetrating hosts
_____ 57. ciliates	F. fungal life cycle with a plasmodium feeding stage
_____ 58. plasmodial slime molds	G. protozoans such as amoebas, but without pseudopods
_____ 59. apicomplexans	H. silica cell wall that consists of two pieces
_____ 60. cellular slime molds	I. external plates of cellular and two flagella
_____ 61. seaweeds	J. fungal life cycle with a solitary amoeboid feeding stage
_____ 62. green algae	K. protozoans with flagella

Matching: Match the protist group on the left to its characteristic on the right.

_____ 63. flagellates	A. mined and used as filtering material or an abrasive
_____ 64. plasmodial slime molds	B. their gel-forming substances are used in ice cream and pudding
E 65. ciliates	C. these are responsible for red tides and massive fish kills
_____ 66. seaweeds	D. can be found among leaf litter on a forest floor
_____ 67. apicomplexans	E. *Paramecium* is an example
_____ 68. dinoflagellates	F. includes the organism that causes malaria
_____ 69. diatoms	G. includes *Giardia* and the trypanosomes that cause sleeping sickness

70. Which one of the following is found in nearly all protists?
    A. unicellularity
    B. flagella
    C. chloroplasts
    D. mitochondria
    E. pseudopods
    F. cilia

71. Which one of the following groups consists of organisms that are decomposers?
   A. protozoans
   B. slime molds
   C. unicellular algae
   D. seaweeds

72. Which one of the following groups consists of multicellular photosynthetic protists?
   A. protozoans
   B. slime molds
   C. unicellular algae
   D. seaweeds

73. The cilia on the surface of ciliates move these protists in a way most similar to
   A. a paddleboat.
   B. a Viking ship with many sets of oars.
   C. a sailboat.
   D. a tugboat pushing a ship out to sea.
   E. a submarine using a nuclear engine to turn its propeller.

74. **True or False?** The closest relatives of seaweeds are <u>unicellular algae</u>.

   ____

75. The ____________________ are photosynthetic protists that are most closely related to plants.

76. Seaweeds are classified into three groups based upon the type of ____________________ present along with chlorophyll *a*.

Evolution Connection: The Origin of Multicellular Life

77. A major advantage of multicellularity is
   A. the ability of cells to multiply.
   B. the smaller size of the organism.
   C. faster reproductive cycles.
   D. the ability for cells to specialize.

78. **True or False?** Multicellular organisms are <u>fundamentally different from</u> unicellular organisms. ____________________

79. The organism, ____________________, is a colonial green algae that shows some of the traits of early multicellular life.

Word Roots

api = the tip (*apicomplexans:* all parasitic protists named for an apparatus at their apex that is specialized for penetrating host cells)

archae = ancient (*archaea:* prokaryotic group most closely related to eukaryotes)

bacill = a little stick (*bacilli:* rod-shaped prokaryotes)

bio = life; **genesis** = origin (*biogenesis:* the principle that life gives rise to life)

chemo = chemical; **auto** = self (*chemoautotrophs:* organisms that need only carbon dioxide as a carbon source and extract energy from inorganic substances)

dinos = whirling (*dinoflagellates:* protists with two flagella that cause them to spin)

endo = inner (*endospores:* bacterial resting cells)

exo = outside (*exotoxins:* toxic proteins secreted by bacterial cells)

flagell = a whip (*flagellates:* protozoa that move by means of one of more flagella)

hetero = different (*chemoheterotrophs:* organisms that must consume organic molecules for both energy and carbon)

patho = disease (*pathogens:* organisms that cause disease)

photo = light (*photoautotrophs:* photosynthetic organisms, including the cyanobacteria)

planktos = wandering (*plankton:* communities of mostly microscopic organisms that drift or swim near the water surface)

protos = first (*protists:* the first eukaryotic cells to evolve)

pseudo = false; **pod** = foot (*pseudopodia:* temporary extensions of a cell used for locomotion and/or feeding)

spiro = spiral (*spirochetes:* spiral-shaped bacteria)

sym = together (*endosymbiosis:* the process whereby mitochondria and chloroplasts exist as associations between prokaryotic cells living within larger prokaryotic cells)

zoan = animal (*protozoan:* protists that live primarily by ingesting food)

Key Terms

algae
amoebas
apicomplexans
archaea
bacilli
bacteria
binary fission
biogenesis
bioremediation
cellular slime molds
chemoautotrophs
chemoheterotrophs
ciliates
cocci
diatoms
dinoflagellates
endospores
endosymbiosis
endotoxins
exotoxins
flagellates
forams
green algae
pathogens
photoautotrophs
photoheterotrophs
plankton
plasmodial slime mold
protists
protozoans
pseudopodia
ribozymes
seaweeds
spirochetes
spontaneous generation

Crossword Puzzle

Use the Key Terms list from this chapter to fill in the crossword puzzle.

ACROSS

2. large spiral-shaped prokaryotic cells
5. organisms that obtain energy from sunlight and carbon from organic sources
8. photosynthetic organisms including the cyanobacteria
10. enzymatic RNA molecules that catalyzes reactions during RNA splicing
11. photosynthetic, plantlike protists
14. type of cell division in which each daughter cell receives a copy of the single parental chromosome
15. one of two prokaryotic domains, the other being the bacteria
16. relatively simple eukaryotes
17. one of two prokaryotic domains, the other being the archaea
18. disease-causing organisms
19. communities of mostly microscopic organisms that drift or swim near the water surface
20. the principle that life gives rise to life
21. protozoans that move by means of one or more flagella
22. rod-shaped prokaryotic cells
24. protists including Volvox with grass-green chloroplasts
26. organisms with solitary amoeboid cells that can form a sluglike colony
27. toxic proteins secreted by bacterial cells
28. cellular extensions of amoeboid cells used in moving and feeding
29. protists with two flagella
30. the process that generated mitochondria and chloroplasts
31. large, multicellular marine algae

DOWN

1. organisms that must consume organic molecules for both energy and carbon
3. organisms that need only carbon dioxide as a carbon source and extract energy from inorganic substances
4. bacterial resting cells
6. protists with a glassy silica containing cell wall
7. algae and other organisms, mostly microscopic, that drift passively in ponds, lakes and oceans
9. type of protozoan that moves by means of cilia
11. type of protists characterized by great flexibility and the presence of pseudopodia
12. group of parasitic protozoans, some of which cause human diseases
13. protist that has a plasmodium feeding stage
17. the use of organisms to remove pollutants from water, air, and soil
23. toxic chemical components in the cell walls of certain bacteria
25. protozoans without pseudopodia

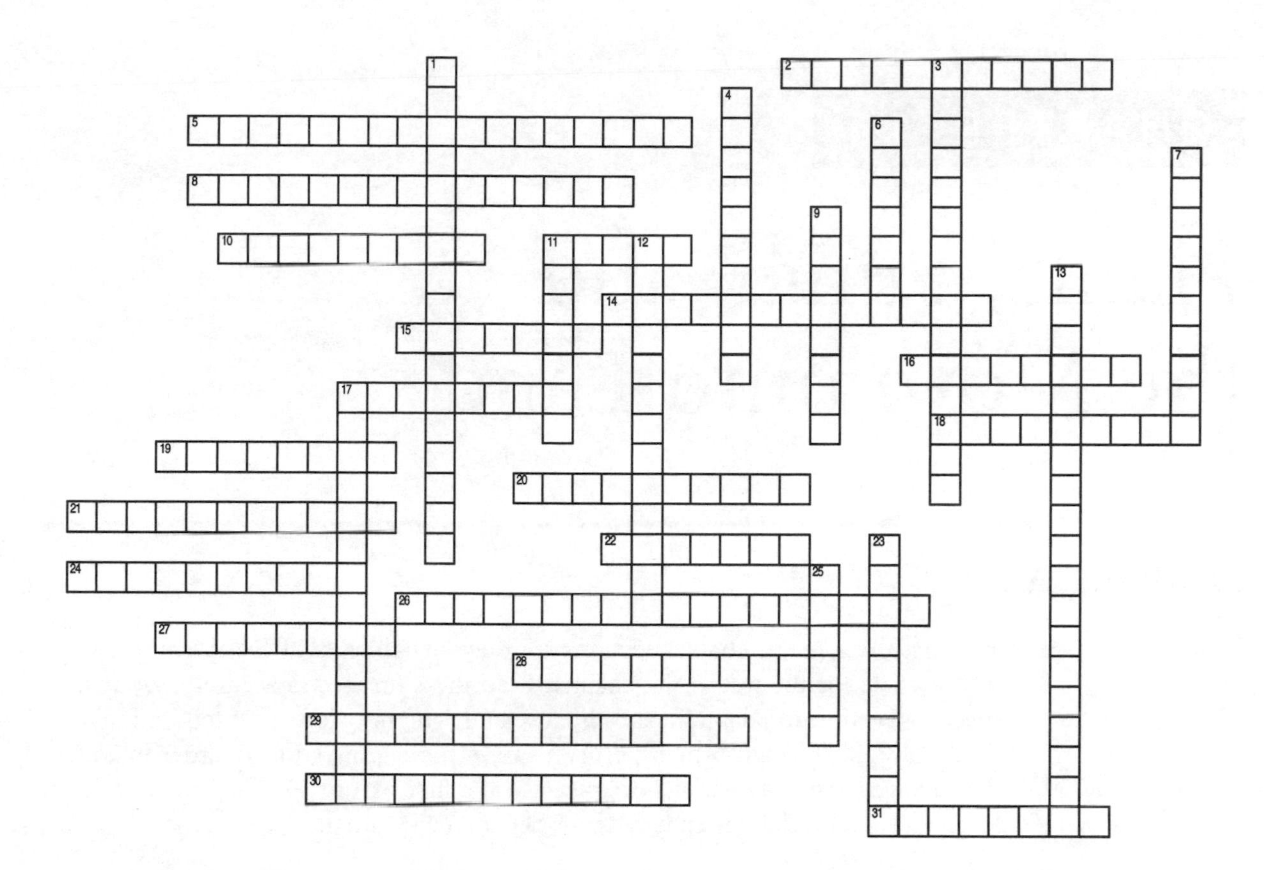
1
2
3
4
5
6
7
8
9
10
11
12
13
14
15
16
17
18
19
20
21
22
23
24
25
26
27
28
29
30
31

CHAPTER 16

Plants, Fungi, and the Move onto Land

Studying Advice

a. The first three major chapter sections (through "Highlights of Plant Evolution") set the stage for the rest of the chapter. Read these first sections carefully, paying special attention to adaptations to life on land. The next four chapter sections ("Bryophytes" through "Angiosperms") survey these groups in the order in which they evolved. The adaptations described in the first part of the chapter are referred to regularly. The chapter ends with a brief survey of fungi.

b. If you have ever wondered why ferns, trees, mosses, and mushrooms look so different, this chapter will be interesting. Maybe you have suffered from athlete's foot, ringworm, or a yeast infection. Plants and fungi are all around us. This interesting chapter is a starting point for appreciating their diverse structures and roles in our world.

Student Media

Biology and Society on the Web

Learn more about the impact of deforestation.

Activities

16A Terrestrial Adaptations of Plants
16B Highlights of Plant Evolution
16C Moss Life Cycle
16D Fern Life Cycle
16E Pine Life Cycle
16F Angiosperm Life Cycle
16G Madagascar and the Biodiversity Crisis
16H Fungal Reproduction and Nutrition

Case Studies in the Process of Science

What Are the Different Stages of a Fern Life Cycle?

How Are Trees Identified by Their Leaves?

How Does the Fungus *Pilobolus* Succeed as a Decomposer?

Evolution Connection on the Web

Discover more about lichens.

Organizing Tables

Describe the symbiotic relationships noted in the table below. Two cells are already filled in to help you.

TABLE 16.1

Type of Mutualistic Relationship	Organisms Involved?	Each Organism's Benefit in the Mutualistic Relationship
Mycorrhizae		
Plant Pollination		The pollinator gets nectar. Pollen is transferred between plants of the same species.
Lichens		
Endosymbiotic origin of mitochondria and chloroplasts (Chapter 15)	Mitochondria, chloroplasts, early eukaryotic cells	

Indicate the living group that best represents each stage in the evolution of plants. Check yes or no to indicate the specific traits found in each group.

TABLE 16.2

Four stages of plant evolution	Living Example	Gametangia	Vascular Tissue	Seeds Present and Enclosed in Special Chamber
First land plants		❑ Yes or ❑ No	❑ Yes or ❑ No	No seeds
Diversification of vascular plants		❑ Yes or ❑ No	❑ Yes or ❑ No	No seeds
Origin of seeds		❑ Yes or ❑ No	❑ Yes or ❑ No	❑ Yes or ❑ No
Flowering plants		❑ Yes or ❑ No	❑ Yes or ❑ No	❑ Yes or ❑ No

Content Quiz

Directions: Identify the *one* best answer for the multiple-choice questions. For true/false questions, determine if the statement is true or false. If false, change the underlined word(s) to make the statement true. Finally, add the correct word(s) to the fill-in-the-blank questions to make the statements true.

Biology and Society: The Balancing Act of Forest Conservation

1. Many scientists predict that global warming might hurt coniferous forests by
   A. causing sap to leak out through the bark.
   B. increasing the number of fungal diseases.
   C. causing the needles to fall off prematurely.
   D. affecting reproduction of some tree species.
   E. decreasing the amount of water available to the trees.

2. **True or False?** The original forests of North America were <u>less</u> biologically diverse than the forests that are now regrowing. ________________

3. Much of our lumber for building and wood pulp for paper production come from ________________ forests.

Colonizing Land

Terrestrial Adaptations of Plants

4. The authors note that water lilies are like porpoises because both
   A. need to come to the surface of the water to breathe.
   B. have very smooth surfaces.
   C. evolved from terrestrial ancestors.
   D. rely upon photosynthesis.

5. Plants are
   A. unicellular prokaryotes that make organic molecules by aerobic respiration.
   B. unicellular eukaryotes that make organic molecules by photosynthesis.
   C. multicellular prokaryotes that make organic molecules by photosynthesis.
   D. multicellular eukaryotes that make organic molecules by aerobic respiration.
   E. multicellular eukaryotes that make organic molecules by photosynthesis.

6. What single characteristic distinguishes plants from large algae?
   A. Plants use photosynthesis, and algae do not.
   B. Plants are eukaryotes, and algae are prokaryotes.
   C. Plants are multicellular, and algae are unicellular.
   D. Plants have terrestrial adaptations, and algae do not.
   E. Plants require oxygen, and algae do not.

7. The fungus in mycorrhizae give the plant
   A. water and essential minerals and in turn the fungus receives sugars from the plant.
   B. oxygen and in turn the fungus receives water and sugars from the plant.
   C. carbon dioxide and carbon and in turn the fungus receives sugars from the plant.
   D. sugars and in turn the fungus receives water and essential minerals from the plant.
   E. sugars and essential minerals and in turn the fungus receives water from the plant.

8. Vascular tissues in plants function most like
   A. highways for distributing resources within a country.
   B. the Internet for quick and reliable communication between parts of our country.
   C. the steel infrastructure of a high-rise office building.
   D. the extensive air duct system for moving air within a high-rise office building.
   E. an extensive sewage system draining the sewage systems of many apartments.

9. Lignin is to plant cells as
   A. muscle is to our bodies.
   B. calcium is to our bones.
   C. telephones are to society.
   D. blood is to our bodies.
   E. hemoglobin is to our blood.

10. The gametangia of plants function most like
    A. the plastic wrapper around a loaf of bread.
    B. the circulatory system of a mouse.
    C. windows that open up in an apartment.
    D. a can opener.
    E. paddles in a canoe.

11. In most plants,
    A. fertilization and development occur outside the female parent plant.
    B. fertilization occurs within, but development occurs outside the female parent plant.
    C. fertilization occurs outside, but development occurs within the female parent plant.
    D. fertilization and development occur within the female parent plant.

12. **True or False?** Mycorrhizae are found <u>on the roots</u> of some of the oldest plant fossils. ____________________

13. Most plants rely upon a symbiotic relationship between their roots and soil fungus, called ____________________.

14. Carbon dioxide and oxygen move between the atmosphere and the interior of leaves via ____________________, microscopic holes in the leaf surfaces.

15. Water loss is reduced from the leaf surface by the ____________________, a waxy coating.

16. Plants use ________________ to gather resources below ground and ________________ to gather resources from above ground.

17. The chemical ________________ hardens plant cell walls, making them rigid.

The Origin of Plants from Green Algae

18. The first terrestrial plants likely evolved
    A. from unicellular algae through a very rapid process.
    B. from unicellular algae through a slow and gradual process.
    C. from multicellular algae through a very rapid process.
    D. from multicellular algae through a slow and gradual process.
    E. from multicellular fungi through a very rapid process.

19. **True or False?** Gametangia would not have been adaptive in shallow water.

    ____

20. The living group most closely related to plants are the ________________.

Plant Diversity

Highlights of Plant Evolution

Matching: Match the group on the left to its best description on the right.

_____ 21. ferns	A. nonvascular early plants
_____ 22. angiosperms	B. vascular plants without seeds
_____ 23. gymnosperms	C. vascular plants with "naked" seeds
_____ 24. bryophytes	D. vascular plants with flowers

25. **True or False?** Most plants alive today are gymnosperms. ________________

26. An embryo packaged along with a store of food within a protective covering defines a(n) ________________.

Bryophytes

27. Mosses display two key adaptations to life on land. These adaptations are
    A. vascular tissue and lignin.
    B. vascular tissue and retention of embryos in the mother plant's gametangium.
    C. lignin and retention of embryos in the mother plant's gametangium.
    D. a waxy cuticle and retention of embryos in the mother plant's gametangium.
    E. a waxy cuticle and vascular tissue.

28. Examine Figure 16.10 of the life cycle of a moss. Meiosis
    A. occurs only in the gametophyte stage.
    B. occurs in the sporophyte stage.
    C. occurs in both the gametophyte and sporophyte stages.
    D. does not occur anywhere in the life cycle.

29. **True or False?** The cells of the sporophyte are haploid. ____________________

30. **True or False?** Mosses and other bryophytes are unique in that the sporophyte is the dominant generation. ____________________

31. Spores are produced by the ____________________, and gametes are produced by the ____________________.

32. A life cycle in which the gametophyte and sporophyte stages produce each other is called ____________________.

Ferns

33. Ferns are
    A. nonvascular, seedless plants with flagellated sperm.
    B. vascular, seedless plants with flagellated sperm.
    C. vascular, seedless plants that have pollen.
    D. vascular plants with seeds.
    E. vascular plants with seeds contained in special chambers.

34. **True or False?** The greatest amount of coal was deposited during the Carboniferous period. ____________________

35. Black sedimentary rock made up of fossilized plant material defines ____________________.

Gymnosperms

36. Which of the following types of climates favored the evolution of gymnosperms?
    A. wetter and warmer
    B. drier and warmer
    C. drier and colder
    D. wetter and colder

37. Which of the following adaptations to conserve water are found in gymnosperms?
    A. broad leaves, no stomata, and a thick cuticle
    B. broad leaves, stomata in pits, and a thin cuticle
    C. thin leaves, stomata in pits, and a thin cuticle
    D. thin leaves, stomata at the tips of bumps, and a thick cuticle
    E. thin leaves, stomata in pits, and a thick cuticle

38. Compared to ferns, which one of the following was *not* a new terrestrial adaptation of gymnosperms?
    A. vascular tissue
    B. pollen
    C. further reduction of the gametophyte
    D. seeds

39. Examine the three variations on alternation of generations in plants in Figure 16.14. Which one of the figures best represents gymnosperms?
    A. A
    B. B
    C. C

40. **True or False?** A conifer is really a <u>sporophyte</u> with tiny <u>gametophytes</u> living in cones. ____________________

41. **True or False?** Conifers and other gymnosperms <u>have</u> ovaries.

    ____

42. Some plant biologists believe that the shift toward the sporophyte as the dominant generation on land was an adaptation to best resist the damaging effects of

    ____________________.

43. The much reduced gametophyte stage that houses cells that will develop into sperm is called ____________________.

44. In gymnosperms, eggs develop within ____________________.

Angiosperms

Matching: Match the flower part on the left to its best description on the right.

_____ 45. sepals	A. a protective chamber containing one or more ovules
_____ 46. petals	B. a stalk bearing the anther
_____ 47. stamen	C. the style, with an ovary at the base and a stigma at its tip
_____ 48. anther	D. usually green, they enclose the flower before it opens
_____ 49. carpel	E. the male organ in which pollen grains develop
_____ 50. ovary	F. usually the most attractive part of a flower, attracting pollinators

51. Which one of the following traits is found in angiosperms but not gymnosperms?
    A. a seed enclosed within an ovary
    B. vascular tissue
    C. an adult gametophyte stage
    D. lignin supporting the cell walls
    E. stomata

52. Which one of the following statements about flowers is *false?*
    A. Flower shapes typically reflect the shape of the pollinator.
    B. Some flowers have markings visible only to animals that can see ultraviolet light.
    C. Flowers pollinated by birds are typically blue or green.
    D. The color and fragrance of flowers is usually keyed to the senses of the pollinator.
    E. Nectar is a high-energy fluid produced by plants to attract pollinators.

53. **True or False?** As fruits ripen, the sugar content increases and the fruit becomes softer. ____________________

54. **True or False?** Most of our food comes from gymnosperms.
    ____

55. **True or False?** It is the fruit that accounts for the unparalleled success of the angiosperms. ____________________

56. The ripened ovary of a flower is the ____________________.

57. Agriculture is a unique kind of evolutionary relationship between ____________________ and animals.

Plant Diversity as a Nonrenewable Resource

58. Which one of the following is the leading cause of forest destruction?
    - A. forest fires
    - B. hurricanes and tornadoes
    - C. human activities
    - D. erosion
    - E. pollution

59. **True or False?** Almost all food used by humans is based upon the cultivation of only about 25 species. ____________

60. **True or False?** Researchers have investigated the potential medical uses of fewer than 5,000 of the known plant species. ____________

Fungi

Characteristics of Fungi

61. Which one of the following statements about fungi is *false?*
    - A. Fungi are mostly multicellular, although yeast are unicellular fungi.
    - B. Fungi are eukaryotes.
    - C. Fungi are more closely related to plants than they are to animals.
    - D. Fungi decompose organic matter and recycle the nutrients into the soil.
    - E. Some parasitic fungi infect the lungs of humans.

62. Fungi reproduce by
    - A. producing spores by sexual or asexual reproduction.
    - B. using flowers.
    - C. producing seeds that are fertilized and develop within the female mushroom.
    - D. special types of pollen that fertilize each other and then develop underground.
    - E. budding or binary fission.

63. The relationship between a mushroom and its hyphae is most like the relationship between
    - A. a fish in a river.
    - B. a car on a highway.
    - C. a wristwatch and a person's arm.
    - D. a fire hydrant and the underground water pipes.
    - E. a book and a library.

64. **True or False?** The cell walls of fungi are made from cellulose.

    ____

65. The structures that account for wide dispersal of fungi are ________________.

66. Fungi feed by secreting ________________ into the environment and absorbing the digested compounds.

67. The bodies of most fungi are made from minute threads called ________________.

68. The feeding network of a fungus, composed of an interwoven mat of hyphae, is called a(n) ________________.

The Ecological Impact of Fungi

69. Which one of the following is *not* associated with fungi?
   A. mushrooms
   B. the production of the antibiotic penicillin
   C. athlete's foot
   D. ringworm
   E. vaginal yeast infections
   F. Dutch elm disease
   G. LSD (lysergic acid diethylamide)
   H. All of the above *are* associated with fungi.

70. **True or False?** Animals are <u>more</u> susceptible to parasitic fungi than are plants. ________________

71. Along with bacteria and some invertebrate animals, fungi play an important ecological role as ________________ in natural environments.

Evolution Connection: Mutual Symbiosis

72. Which one of the following relationships is a type of mutualism?
A. a thief stealing from a department store
B. a person picking mushrooms to eat
C. a hunter shooting ducks
D. a person buying corn from a farmer
E. a fungus growing on your shower curtain

73. **True or False?** A <u>parasitic</u> relationship is one in which both species in a relationship benefit. ________________

74. A(n) ________________ is composed of fungi and algae that live together in a mutualistic relationship.

Word Roots

angion = a container (*angiosperms:* flowering plants)

bryo = moss (*bryophytes:* a group of nonvascular plants that includes the mosses)

gamet = a wife or husband (*gametangia:* protective structures where plants produce their gametes)

phyte = plant (*gametophyte:* the moss stage with haploid cells)

gymno = naked; **sperma** = seed (*gymnosperms:* vascular plants such as conifers with seeds not enclosed in specialized chambers)

myce = fungus (*mycelium:* interwoven net of hyphae)

rhiza = root (*mycorrhizae:* symbiotic root-fungus combinations)

sporo = seed (*sporophyte:* the moss stage with diploid cells)

Key Terms

absorption
alternation of generations
angiosperms
anther
bryophytes
carpel
charophyceans
cuticle
double fertilization
endosperm
ferns
flower
fossil fuels
fruit
fungi
gametangia
gametophyte
germinate
gymnosperms
hyphae
lichens
lignin
mosses
mutualism
mycelium
mycorrhizae
ovary
ovules
petals
phloem
pollen
roots
seed
sepals
shoots
spores
sporophyte
stamen
stigma
stomata
style
symbiosis
vascular tissue
xylem

Crossword Puzzle

Use the Key Terms list from this chapter to fill in the crossword puzzle.

ACROSS

3. the sticky tip of a flower's carpel which traps pollen grains
7. aerial portions of a plant body, consisting of stems, leaves, and flowers
16. a life-cycle that switches between gametophyte and sporophyte stages
19. green algal groups that are considered to be the closest relative of land plants
23. a plant embryo packaged along with a food supply within a protective coat
24. a moss, liverwort, or hornwort
25. the male organ in a flower in which pollen grains develop
27. plant structures that anchor the plant in the soil, absorb and transport minerals and water, and store food
30. the portion of the vascular system in plants that transports sugar and other nutrients throughout the plant
31. flowering plants
33. a chemical that hardens the cell walls of plants
34. microscopic pores in leaf surfaces that facilitate gas exchange
35. the moss stage with diploid cells
36. typically the most striking parts of a flower, usually important in attracting insects
38. a pollen-producing part of a flower that consists of a stalk and an anther
39. in a seed plant, the male gametophytes that develop within the anthers of stamens
40. modified leaves, usually green, that enclose the flower before it opens

DOWN

1. a system of tube-shaped cells that branches throughout the plant
2. vascular plants that bear naked seeds
4. the group of vascular plants that bear naked seeds
5. a mature ovary of a flower that protects dormant seeds and aids in their dispersal
6. a symbiotic relationship in which both participants benefit
8. structures that develop in the plant ovary and contain the female gametophyte
9. a waxy coating of the leaves and other aerial parts of most plants
10. a short stem with four whorls of modified leaves
11. a mechanism of fertilization in angiosperms, in which two sperm cells unite with two cells in the embryo sac to form the zygote and endosperm
12. one of a group of seedless vascular plants
13. the female reproductive organ of a flower
14. energy deposits formed from the remains of extinct organisms
15. a mutualistic association of plant root and fungus
16. the process by which small organic molecules are brought in from the surrounding medium
17. a nutrient-rich tissue which provides nourishment to the developing embryo in angiosperm seeds
18. the densely branched network of hyphae in a fungus
20. a protective chamber containing one or more ovules
21. haploid cells that divide mitotically to produce the gametophyte without fusing with another cell
22. the tube-shaped, nonliving portion of the vascular system in plants that carries water and minerals from the roots to the rest of the plant
26. a moss stage with haploid cells
28. one of a group of seedless nonvascular plants
29. what a seed does when it first begins to grow
32. protective structures where plants produce their gametes
37. the stalk portion of the carpel

CHAPTER 17

The Evolution of Animals

Studying Advice

a. Take a big deep breath before starting this, the longest chapter in the textbook. Chapter 17 introduces animal diversity with a survey of the major invertebrate and vertebrate groups. There is much to consider. Two pieces of advice:

   1. Do not try to study this chapter a night or two before the exam. You will need time to read, understand, and process this information. Students who wait too long to get started often confuse details on exams.

   2. Try to break this chapter down into smaller, more manageable "bites." Consider reading the sections one at a time, and making note cards or other study tools as you read. Take short five-minute breaks between sections.

b. The two organizing tables will help you organize the information in many chapter sections into one clear table. Fill these tables in as you progress through the chapter.

Student Media

Biology and Society on the Web

Learn about how invasive species arrive in the United States.

Activities

17A Overview of Animal Phylogeny

17B Characteristics of Invertebrates

17C Characteristics of Chordates

17D Primate Diversity

17E Human Evolution

Case Studies in the Process of Science

How Are Insect Species Identified?

How Does Bone Structure Shed Light on the Origin of Birds?

Evolution Connection on the Web

Learn about what is being done to combat the effects of habitat destruction.

Organizing Tables

Examine Figure 17.6 and the corresponding text pages to complete the table below comparing invertebrate groups. Several cells are filled in to help you in this task.

TABLE 17.1

Major Animal Phyla	True Tissues	Type of Body Symmetry	Type of Body Cavity	Type of Digestive Tract	Special Features of the Group
Porifera	❑ Yes or ❑ No		None	None	
Cnidaria	❑ Yes or ❑ No		None		cnidocytes
Platyhelminthes	❑ Yes or ❑ No		None		
Nematoda	❑ Yes or ❑ No				
Mollusca	❑ Yes or ❑ No				
Annelida	❑ Yes or ❑ No				
Arthropoda	❑ Yes or ❑ No				
Echinodermata	❑ Yes or ❑ No				
Chordata	❑ Yes or ❑ No				

Examine Figure 17.31 and the corresponding text pages to complete the table below comparing the vertebrate classes. Some of the cells are already filled in to help you in this task.

TABLE 17.2

Vertebrate Classes	Are jaws present?	Is the skeleton made of bone or cartilage?	Do the adults have gills or lungs?	Do they have legs? If yes, how many do they walk on?	Do they have amniotic eggs?	Are they ectothermic or endothermic?
Agnatha		Cartilage	No	No		
Chondrichthyes						
Osteichthyes			Most with gills, a few with lungs			
Amphibia			Most with lungs, a few with gills			
Reptilia						
Aves						
Mammalia						

Content Quiz

Directions: Identify the *one* best answer for the multiple-choice questions. For true/false questions, determine if the statement is true or false. If false, change the underlined word(s) to make the statement true. Finally, add the correct word(s) to the fill-in-the-blank questions to make the statements true.

Biology and Society: Invasion of the Killer Toads

1. Which one of the following statements about cane toads is *false?* Cane toads
   A. are native to South America.
   B. have voracious appetites.
   C. are poisonous to almost all their predators, including quolls.
   D. were introduced by sugarcane growers to fight beetles that were damaging their sugar crops.
   E. were very effective at stopping beetles that damage sugar crops.

2. **True or False?** The introduction of damaging invasive non-native species has occurred all over the world. ________________

3. Native to Australia, ________________ are predators, hunting smaller animals, including many types of frogs.

The Origins of Animal Diversity

What Is an Animal?

4. Which one of the following sets of descriptions best describes animals? Animals are
   A. eukaryotic, unicellular, and heterotrophic organisms that use ingestion.
   B. eukaryotic, multicellular, and autotrophic organisms that use egestion.
   C. eukaryotic, multicellular, and autotrophic organisms that use ingestion.
   D. eukaryotic, multicellular, and heterotrophic organisms that use ingestion.
   E. prokaryotic, multicellular, and heterotrophic organisms that use ingestion.

5. Examine Figure 17.3 showing the development of a sea star. In the early stages of development, the overall size of the embryo grows very little. Therefore the average size of the embryonic cells
   A. decreases.
   B. stays about the same.
   C. increases.

6. **True or False?** The life histories of many animals include larval stages.

   ____

7. **True or False?** Animal development typically progresses from zygote, to gastrula, to blastula. ________________

8. A change in body form, called ________________, remodels a larval stage into an adult form.

9. Most animals have a muscular system and a ________________ system that controls it.

Early Animals and the Cambrian Explosion

10. The Cambrian explosion
    A. was a period of violent volcanic activity.
    B. resulted from the Earth's impact with a large comet or asteroid, which killed off the dinosaurs.
    C. refers to a time when animal diversity increased dramatically.
    D. was a huge internal blast deep within the Earth that triggered tremendous earthquakes.
    E. resulted in the loss of most animal species and the origin of most plant species.

11. **True or False?** The cause of the Cambrian explosion is unknown.

    ____

12. **True or False?** Most zoologists now agree that the Cambrian organisms can be regarded as the ancient representatives of modern phyla. ________________

Animal Phylogeny

13. Which one of the following phyla does *not* usually show bilateral symmetry?
    A. phylum Platyhelminthes
    B. phylum Arthropoda
    C. phylum Chordata
    D. phylum Annelida
    E. phylum Cnidaria

14. Which one of the following is *not* a function of a body cavity?
    A. cushions internal organs
    B. functions as a hydrostatic organ
    C. provides an extra space to store ingested food
    D. enables internal organs to move independently of the body surface

15. Which one of the following shows radial symmetry?
    A. an apple
    B. a Popsicle
    C. a submarine sandwich
    D. a pickle
    E. a cucumber

16. Most animals that move actively in their environment show
    ____________________ symmetry.

17. A body cavity that is only partially lined by tissues derived from mesoderm is a
    ____________________.

Major Invertebrate Phyla

Sponges

18. In sponges, food and other nutrients are distributed and wastes are removed by
    A. a primitive circulatory system with blood.
    B. a water vascular system.
    C. choanocytes.
    D. amoebocytes.
    E. tentacles.

19. Which one of the following is found in sponges?
    A. true tissues
    B. a coelom
    C. a digestive tract
    D. a skeleton
    E. gonads

20. **True or False?** Most sponges live in freshwater. ____________________

21. Sponge cells called ____________________ filter bacteria from water.

Cnidarians

22. Which one of the following statements about cnidarians is *false*? Cnidarians
    A. are carnivores.
    B. have a complete digestive tract with mouth and anus.
    C. occur in medusa and/or polyp form.
    D. have cnidocytes.
    E. are mostly marine.

23. Examine Figure 17.4. Which one of the following animals has a body plan that is basically the same as the final stage of evolution indicated on the right side of this figure?

    A. a jellyfish
    B. a shark
    C. an earthworm
    D. a grasshopper
    E. a sponge

24. **True or False?** The stinging cells of cnidarians are called <u>choanocytes</u>.

    ____

25. A sea anemone is an example of the ____________________ body plan.

26. A jellyfish is an example of the ____________________ body plan.

Flatworms

27. Flatworms are

    A. parasitic and free-living, bilaterally symmetrical, and unsegmented.
    B. free-living, bilaterally symmetrical, and segmented.
    C. free-living, radially symmetrical, and unsegmented.
    D. parasitic, bilaterally symmetrical, and segmented.
    E. parasitic and free-living, radially symmetrical, and segmented.

28. **True or False?** Tapeworms <u>do not</u> have a mouth or any digestive tract.

    ____

29. **True or False?** Parasitic flukes and tapeworms usually have <u>just one</u> host.

    ____

30. Tapeworms can infect humans who eat beef that is ____________________.

Roundworms

31. Roundworms are the first phylum studied to this point to have

    A. a gastrovascular cavity and a coelom.
    B. a complete digestive tract and a pseudocoelom.
    C. a gastrovascular cavity and bilateral symmetry.
    D. a mouth and a true coelom.
    E. an endoskeleton and a mouth.

32. Which one of the following items is shaped most like a roundworm?
    A. a train with hundreds of railroad cars
    B. a round toothpick, tapered at both ends
    C. a spoon
    D. an apple
    E. a pancake

33. **True or False?** A complete digestive tract allows food to move through the animal in only one direction. ________________

34. The body cavity of roundworms is a ________________.

Mollusks

35. The mollusk body consists of a
    A. segmented tail, a muscular shield, and a mantle.
    B. muscular mass, a segmented foot, and a shield.
    C. muscular mass, a segmented foot, and a mantle.
    D. muscular foot, a visceral mass, and a mantle.
    E. muscular foot, a segmented visceral mass, and a shield.

36. **True or False?** Most gastropods have a shell divided into two halves hinged together. ________________

37. The ________________ are a group of mollusks adapted for speed and agility.

38. Many mollusks use a(n) ________________, a straplike rasping organ, to scrape up food.

39. Most mollusks have a shell, secreted by the ________________.

Annelids

40. Which one of the following groups has segmental appendages that help the animals move and exchange gases?
    A. earthworms
    B. polychaetes
    C. leeches
    D. all of the groups

41. Which one of the following traits is found in annelids but *not* roundworms?
   A. an anus
   B. a complete digestive tract
   C. a mouth
   D. segmentation
   E. bilateral symmetry

42. **True or False?** Most leeches are parasites. ____________________

43. Earthworms eat ____________________ and eliminate ____________________, a mixture of undigested material and mucus.

44. Leech ____________________ is commercially produced because, in addition to other medical benefits, it can prevent ____________________.

Arthropods

45. Which one of the following is *not* a general characteristic of arthropods?
   A. segmentation
   B. jointed appendages
   C. exoskeleton
   D. six legs
   E. molting

46. Which one of the following habitats is *not* used by an arthropod?
   A. the surface of other animals
   B. the ocean
   C. underground
   D. the air
   E. All of these habitats *are* used by at least a few arthropod species.

47. Compare the antennae of a lobster and the insects in Figures 17.21 and 17.25. Which one of the following statements is *true*?
   A. Crustaceans and insects have two pairs of antennae.
   B. Crustaceans and insects have one pair of antennae.
   C. Crustaceans have two pairs of antennae, and insects have one pair of antennae.
   D. Crustaceans have one pair of antennae, and insects have two pairs of antennae.

Matching: Match the group on the left to its best description on the right.

____ 48. insects	A. adults with six legs and 1–2 pairs of wings
____ 49. crustaceans	B. usually four pairs of walking legs, the group includes spiders
____ 50. arachnids	C. multiple pairs of specialized appendages
____ 51. millipedes and centipedes	D. similar segments along length of long body

52. **True or False?** Most animal species on Earth are arthropods.

    ____

53. **True or False?** Insects with incomplete metamorphosis have a larval stage that looks entirely different from the adult stage. ____________________

54. Arthropods must molt their ____________________ in order to grow.

55. The most dominant arthropods in the oceans are the ____________________.

56. The branch of biology called ____________________ is the study of insects.

57. The many specialized body ____________________ of arthropods provide an efficient division of labor among body regions.

58. Most arthropods belong to the group called ____________________.

Echinoderms

59. Echinoderms are

    A. unsegmented marine animals with an endoskeleton and water vascular system.
    B. unsegmented freshwater animals with an endoskeleton and circulatory system.
    C. segmented freshwater and marine animals with an exoskeleton and circulatory system.
    D. segmented freshwater and marine animals with an endoskeleton and water vascular system.
    E. segmented marine animals with an exoskeleton and water vascular system.

60. **True or False?** Most echinoderms are sessile, or slow moving.

    ____

61. Most adult echinoderms have ____________________ symmetry, but larva typically have ____________________ symmetry.

Matching: Match the invertebrate group on the left to its best description on the right.

_____ 62. sponges	A. mollusk group with a single, spiraled shell
_____ 63. jellyfish	B. parasitic platyhelminth without a digestive tract
_____ 64. corals	C. unsegmented; a complete digestive tract; pseudocoelom
_____ 65. planarians	D. the group with the greatest number of species
_____ 66. blood flukes	E. segmented worm group that is mostly marine
_____ 67. tapeworms	F. a cnidarian group that uses the polyp body form
_____ 68. roundworms	G. segmented worm group including bloodsuckers
_____ 69. gastropods	H. mollusk group that includes squid and octopus
_____ 70. bivalves	I. segmented worm group that increases the fertility of soil
_____ 71. cephalopods	J. a cnidarian group that uses the medusa body form
_____ 72. leeches	K. mollusk group with two shells hinged together
_____ 73. earthworms	L. no true tissues; only freshwater or marine habitats
_____ 74. polychaetes	M. free-living platyhelminth group
_____ 75. arthropods	N. parasitic platyhelminths; live inside blood vessels

The Vertebrate Genealogy

Characteristics of Chordates

76. Which one of the following is *not* a chordate characteristic?
    A. ventral, solid nerve cord
    B. pharyngeal slits
    C. notochord
    D. post-anal tail

77. Examine the evolutionary relationships indicated in Figure 17.6. Which invertebrate phylum is most closely related to the Chordates?
    A. Mollusks
    B. Echinoderms
    C. Annelids
    D. Cnidarians
    E. Arthropods

78. **True or False?** The phylum Chordata <u>does not include</u> invertebrates.

    ____

79. Other than vertebrates, the phylum Chordata includes ____________________, which are long and thin, and ____________________, which as adults are sessile filter feeders.

Fishes

Matching: Match the group on the left to its most distinctive feature on the right.

_____ 80. ray-finned fishes	A. have no jaws
_____ 81. lobe-finned fishes	B. jaws; swim bladder; ancient animal; only one living species
_____ 82. lungfishes	C. jaws; swim bladder; greatest number of species
_____ 83. cartilaginous fishes	D. jaws; swim bladder; gulp air; live in Southern Hemisphere
_____ 84. agnathans	E. jaws; no swim bladder; skeleton made of cartilage

85. Which one of the following groups was the first to evolve?
    A. Chondrichthyes
    B. ray-finned fishes
    C. lungfishes
    D. agnathans
    E. lobe-finned fishes

86. **True or False?** If a shark stops swimming it tends to sink. ____________________

87. Sharks sense changes in water pressure by using their ____________________ system.

88. Bony fish have a protective covering over their gills called the ____________________.

89. When not moving, bony fish do not sink because they have a(n) ____________________.

Amphibians

90. Which one of the following characteristics is *not* lost during metamorphosis of a frog?
    A. tail
    B. gills
    C. legs
    D. lateral line system

91. During the life cycle of most frogs, the tadpoles
    A. and adults are herbivores.
    B. and adults are carnivores.
    C. are herbivores, and the adults are carnivores.
    D. are carnivores, and the adults are herbivores.

92. Which one of the following groups gave rise to the amphibians?
    A. cartilaginous fishes
    B. ray-finned fishes
    C. lungfishes
    D. agnathans
    E. lobe-finned fishes

93. In addition to lungs, the ________________ of most adult amphibians promotes gas exchange.

94. Animals with four limbs are called ________________.

Reptiles

95. Which of the following were new reptilian adaptations for living on land?
    A. amniotic egg and scales
    B. four legs and ears
    C. lungs and a tail
    D. lateral line system and ears

96. Examine the evolutionary relationships indicated in Figure 17.31. Within the vertebrates, which one of the following is a characteristic common and unique only to reptiles, birds, and mammals?
    A. vertebrae
    B. hair
    C. jaws
    D. legs
    E. amniotic eggs

97. **True or False?** Because they are ectotherms, reptiles survive on about 10 times the calories required by a mammal of similar size. ________________

98. The evolution of the ________________ egg allowed reptiles to reproduce without returning to water.

99. Reptiles are ________________, absorbing their body heat from the environment.

Birds

100. Which one of the following characteristics of birds is *not* found in reptiles?
    A. scales
    B. endothermic metabolism
    C. amniotic eggs
    D. lungs

101. Which one of the following is *not* an adaptation for flight in birds?
   A. honeycombed bones
   B. a single ovary
   C. loss of teeth
   D. ectothermic metabolism

102. **True or False?** One of the oldest known birds, *Archaeopteryx* is considered to be the direct ancestor of all birds. ________________

103. Feathers and the scales of reptiles are made of the protein ________________.

Mammals

104. Mammals are primarily
   A. terrestrial and endothermic.
   B. terrestrial and ectothermic.
   C. aquatic and endothermic.
   D. aquatic and ectothermic.

105. Which one of the following is *not* a characteristic of all mammals?
   A. hair
   B. mammary glands
   C. a placenta
   D. endothermy

106. **True or False?** A brief gestation followed by development in a pouch is characteristic of the eutherian mammals. ________________

107. The group of mammals that lays eggs are the ________________.

108. Humans, apes, and monkeys belong to the order ________________.

The Human Ancestry

The Evolution of Primates

109. Which one of the following is *not* a primate adaptation for living in trees?
   A. rigid shoulder joints
   B. eyes close together in the front of the face
   C. excellent hand-eye coordination
   D. extensive parental care
   E. dexterous hands

110. Which one of the following is a characteristic of Old World but *not* New World monkeys?

    A. ground dwellers
    B. prehensile tails
    C. single births with a long period of nurturing
    D. nails instead of claws

111. **True or False?** In general, apes <u>are larger than</u> monkeys.

    ____

112. **True or False?** Modern apes live only in tropical regions of the <u>New World</u>.

    ____

113. The oldest group of primates are the ____________________.

The Emergence of Humankind

Matching: Match the group on the left to its best description on the right.

_____ 114. *Homo habilis*	A. the species that includes Lucy, about 3.2 million years old
_____ 115. Neanderthals	B. the species of modern humans
_____ 116. *Homo erectus*	C. about 2.5 million years old, an early tool user
_____ 117. *Australopithecus afarensis*	D. the first species to extend humanity's range beyond Africa
_____ 118. *Homo sapiens*	E. Europeans, skilled toolmakers, 130–35,000 years old

119. Which one of the following hypotheses is correct about the multiregional and "Out of Africa" models of human evolution?

    A. Both models are still widely debated.
    B. Evidence now clearly shows that humans evolved according to the "Out of Africa" model.
    C. Evidence now clearly shows that humans evolved according to the multiregional model.
    D. Evidence now clearly shows that modern humans evolved by both models.

120. The pattern of the history of human evolution is most like

    A. a ladder.
    B. a series of steps.
    C. a bush.
    D. people lined up to go through a door.

121. Which one of the following is the correct sequence in the evolution of human culture?
   A. agriculture, hunting and gathering nomads, Industrial Revolution
   B. Industrial Revolution, agriculture, hunting and gathering nomads
   C. hunting and gathering nomads, Industrial Revolution, agriculture
   D. agriculture, Industrial Revolution, hunting and gathering nomads
   E. hunting and gathering nomads, agriculture, Industrial Revolution

122. **True or False?** Chimps <u>are</u> the parent species of humans.

   ____

123. **True or False?** Most of the distinct human traits evolved <u>at the same time</u>.

   ____

124. **True or False?** Upright posture evolved <u>before</u> our enlarged brain.

   ____

125. **True or False?** The brains of Neanderthals <u>were slightly larger</u> than modern humans. ____________________

126. Fossil evidence indicates that bipedalism evolved at least ____________________ million years ago.

127. Language, written and spoken, is the main way that ____________________ is transmitted.

Evolution Connection: Earth's New Crisis

128. Which one of the following human activities is having the greatest impact on other species?
   A. habitat destruction
   B. destruction of the ozone layer
   C. burning fossil fuels
   D. air pollution
   E. erosion

129. **True or False?** Human cultural evolution is occurring <u>much faster than</u> human biological evolution. ____________________

Word Roots

agnatha = without jaws (*agnathans:* vertebrates without jaws)

amphibious = living a double life (*amphibian:* the group of vertebrates that live life in and out of the water)

annel = ring (*Annelida:* the phylum of segmented worms)

arachn = spider (*arachnids:* the group of spiders, scorpions, mites, and ticks)

arthro = joint; **pod** = foot (*Arthropoda:* the phylum of animals with jointed legs)

bi = double; **valva** = leaf of a folding door (*Bivalvia:* the group of mollusks with two hinged shells)

centi = one hundred (*centipedes:* arthropods with many legs, one pair per body segment)

ceph(al) = head (*cephalopods:* the mollusk group that includes octopus and squid)

chondro = cartilage (*Chondrichthyes:* the group of cartilaginous fishes)

echino = spiny (*Echinodermata:* the phylum of sea stars that have spiny skin)

ecto = outside (*ectotherms:* animals that absorb heat instead of producing it internally)

endo = inside (*endothermic:* animals that produce heat internally)

entom = an insect (*entomology:* the branch of biology that studies insects)

exo = outside (*exoskeleton:* an external skeleton such as that found in all arthropods)

gaster = belly; **pod** = foot (*gastropods:* the group of mollusks with their guts near their muscular foot)

kephale = head; **pod** = foot (*cephalopods:* the mollusk group that includes octopus and squid)

milli = one thousand (*millipedes:* arthropods with many legs, two pairs per body segment)

mollusc = soft (*Mollusca:* the phylum of soft-bodied animals usually surrounded by hard shells)

mono = one (*monogenesis:* the type of theoretical model that modern humans evolved from a single archaic group)

noto = the back (*notochord:* a flexible, longitudinal rod characteristic of chordate animals)

osteo = bone (*Osteichthyes:* the group of bony fishes)

paleo = ancient; **anthrop** = a man (*paleoanthropology:* the study of human evolution)

platy = flat; **helminthes** = a worm (*Platyhelminthes:* the phylum of flatworms)

por = pore; **fer** = bearer (*Porifera:* the phylum of sponges)

post = behind (*post-anal tail:* the type of tail found in chordate animals)

pseudo = false (*pseudocoelem:* a body cavity that is not completely lined by tissue derived from mesoderm)

tetra = four; **pod** = foot (*tetrapods:* the group of terrestrial vertebrates with four legs)

Key Terms

agnathans
amniotic egg
amoebocytes
Amphibia
Annelida
anthropoids
arachnids
Arthropoda
Aves
bilateral symmetry
bivalves
Bivalvia
blastula
body cavity
bony fishes
cartilaginous fishes
centipedes
cephalopods
choanocytes
Chondrichthyes
Chordata
Cnidaria
cnidocytes
coelom
complete digestive tract
crustaceans
culture
dorsal
Echinodermata
ectotherms
endoskeleton
endotherms
entomology
eutherians
exoskeleton
gastropods
gastrovascular cavity
hollow nerve cord
hominids
ingestion
insects
invertebrates
lancelets
larva
lateral line system
lobe-finned fishes
lungfishes
Mammalia
mantle
marsupials
medusa
metamorphosis
millipedes
Mollusca
molting
monotremes
multiregional hypothesis
Nematoda
notochord
operculum
Osteichthyes
"Out of Africa" hypothesis
paleoanthropology
pharyngeal slits
placenta
Platyhelminthes
polyp
Porifera
post-anal tail
primates
prosimians
pseudocoelom
radial symmetry
radula
ray-finned fishes
replacement hypothesis
Reptilia
segmentation
swim bladder
tetrapods
tunicates
vertebrates
water vascular system

Crossword Puzzle

Use the Key Terms list from this chapter to fill in the crossword puzzle.

ACROSS

3. an embryonic stage in which the embryo is a hollow ball
7. group of bony fish with lungs
12. short, marine invertebrate chordates with a blade-like shape
13. types of fish with a skeleton made of cartilage
15. a fluid-filled space separating the digestive tract from the outer body wall
18. types of fish with a skeleton made of bone
20. pertaining to the back of a bilaterally symmetrical animal
22. members of a jawless class of vertebrates represented by lampreys and hagfishes
26. an external skeleton such as that found in all arthropods
27. the group of vertebrates that includes all birds
28. flagellated cells of sponges
31. a body cavity completely lined by tissue derived from mesoderm
32. the class of mollusks that includes clams, mussels, scallops, and oysters
34. type of body symmetry in which only a single plane cuts the animal into mirror images
35. the phylum of segmented worms
41. arthropods with many legs, one pair per body segment
42. the group of scorpions, spiders, ticks, and mites
44. the group of cartilaginous fishes
45. an internal skeleton
46. the group of mollusks that includes squids and octopuses
47. the phylum of jellyfish, sea anemones, corals, and hydras
48. eating food

DOWN

1. members of the human family
2. the floating body plan of a cnidarian
4. the primate group that includes monkeys, apes, and humans
5. the phylum of snails, oysters, squids, octopuses, and clams
6. the group of terrestrial vertebrates with four legs
8. the phylum of roundworms
9. the study of insects
10. type of animals that produce heat internally
11. mollusks with a single, spiraled shell
14. arthropods with many legs, two pair per body segment
16. the phylum of animals that includes sea urchins, starfish, and sea cucumbers
17. the class of vertebrates that includes humans
19. animals that absorb heat instead of producing it internally
21. animals without backbones
22. amoebalike cells that move by pseudopodia
23. the type of egg enclosed in a shell and used by reptiles and birds
24. a member of the vertebrate class that includes frogs and salamanders
25. the group of placental mammals
29. the sheet of tissue that secretes the shell of a mollusk
30. the stinging cells of cnidarians
33. the group of crabs, lobsters, crayfish, shrimp, and barnacles
36. a flexible, longitudinal rod characteristic of chordate animals
37. the largest phylum which includes animals with jointed legs and exoskeletons
38. an internal organ in a female mammal that is used to nurture the embryo
39. a sexually immature animal
40. the phylum of animals that includes humans
43. the sessile body plan of a cnidarian

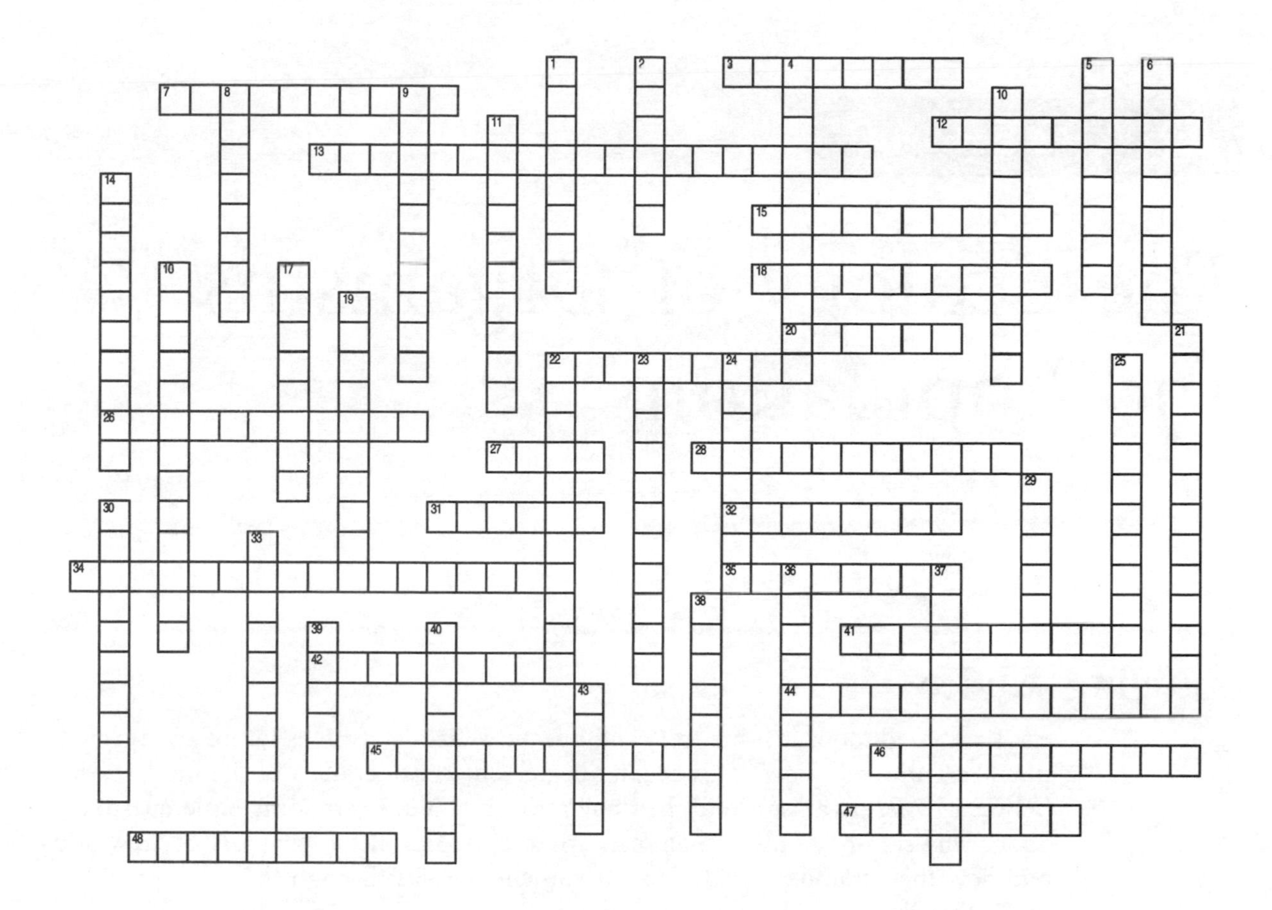

CHAPTER 18

The Ecology of Organisms and Populations

Studying Advice

Look out a window at some trees and bushes. What determines where any type of plant can and will grow? This chapter begins our consideration of the interaction between living organisms and their environments. It answers some basic questions about why certain plants and animals are restricted to just a few parts of the world and how their numbers are limited. If you enjoy walks through the park or wild areas, this chapter will surely be of interest.

Student Media

Biology and Society on the Web

This U.S. census website will provide you with some impressive numbers.

Activities

18A DDT and the Environment
18B Evolutionary Adaptations
18C Techniques for Estimating Population Density and Size
18D Human Population Growth
18E Analyzing Age-Structure Diagrams
18F Investigating Life Histories

Case Studies in the Process of Science

Do Pillbugs Prefer Wet or Dry Environments?
How Do Abiotic Factors Affect the Distribution of Organisms?

Evolution Connection on the Web

Learn why the dominant male may not be the best choice when it comes to mating success.

Organizing Tables

Using the textbook, define each of the following levels of ecology. These definitions will be a helpful reference as you read and review the chapter.

TABLE 18.1

Level of Ecology	Textbook Definition
Organismal	
Population	
Community	
Ecosystem	

Compare the three patterns of dispersion in the table below.

TABLE 18.2		
Pattern	**Factors That Contribute to This Pattern**	**Examples**
Clumped		
Uniform		
Random		

Content Quiz

Directions: Identify the *one* best answer for the multiple-choice questions. For true/false questions, determine if the statement is true or false. If false, change the underlined word(s) to make the statement true. Finally, add the correct word(s) to the fill-in-the-blank questions to make the statements true.

Biology and Society: The Human Population Explosion

1. The continued increase of the human population in the face of limited resources has led to
   A. global warming.
   B. civil strife aggravated by depressed economics.
   C. conflicts over oil in the Middle East.
   D. declining health of the oceans.
   E. accumulation of toxic waste.
   F. All of the above have occurred.

2. **True or False?** Earth's most significant biological phenomenon is now the human population explosion. ____________________

3. In 1999, the human population grew past ____________________ people on Earth.

An Overview of Ecology

Ecology as Scientific Study

4. Ecology is the scientific study of interactions between
   - A. organisms and their environment.
   - B. different types of animals.
   - C. animals and plants.
   - D. abiotic factors and the environment.
   - E. plants and the organisms that pollinate them.

5. **True or False?** Many ecologists <u>are able to</u> conduct experiments in the field of ecology. ____________________

6. The nonliving chemical and physical factors of the environment are the ____________________ component.

A Hierarchy of Interactions

Matching: Match the levels of ecology on the left to its best description on the right.

____	7. organismal ecology	A. the study of all the interactions between the abiotic factors and the community of species that exists in a certain area
____	8. community ecology	B. the study of the evolutionary adaptations that enable individual organisms to meet the challenges posed by their abiotic environments
____	9. ecosystem ecology	C. the study of the factors that affect population size, growth, and composition
____	10. population ecology	D. the study of all the organisms that inhabit a particular area

11. Which one of the following correctly lists the levels of ecology in order of increasingly comprehensive levels?
    - A. organismal ecology, community ecology, population ecology, ecosystem ecology
    - B. population ecology, organismal ecology, community ecology, ecosystem ecology
    - C. ecosystem ecology, community ecology, organismal ecology, population ecology
    - D. organismal ecology, community ecology, ecosystem ecology, population ecology
    - E. organismal ecology, population ecology, community ecology, ecosystem ecology

12. **True or False?** The distribution of organisms is limited by the abiotic conditions they can tolerate. ____________________

13. The ____________________ is the sum of all the planet's ecosystems, or all of life and where it lives.

Ecology and Environmentalism

14. The publication of Rachel Carson's book, *Silent Spring* was an important event in the field of ecology because it
    A. detailed the dramatic increases in farm productivity that enabled the United States to grow surplus food and market it overseas.
    B. highlighted the worldwide decreases in malaria and other insect-borne diseases due to DDT spraying.
    C. focused attention on the genetic resistance to pesticides that evolved in an increasing number of pest populations.
    D. detailed the environmental damage caused by the use of DDT.

15. **True or False?** The immediate results of spraying DDT were decreases in farm productivity. ____________________

16. A basic understanding of the field of ____________________ is required to analyze environmental issues and plan for better practices.

Abiotic Factors of the Biosphere

17. The patchy nature of the biosphere exists mainly because of
    A. the types of predatory animals that live in a particular region.
    B. differences in climate and other abiotic factors.
    C. random events due to the chance evolution of a new species in a particular region.
    D. the types of plants that grow in a particular region.

18. Which one of the following is *not* an adaptation to conserve water?
    A. a waxy coating on the leaves and other aerial parts of most plants
    B. a dead layer of outer skin containing a waterproofing protein.
    C. the ability of human kidneys to excrete a very concentrated urine
    D. All of the above are adaptations to conserve water.

19. Which one of the following is *not* an abiotic factor?
    A. wind
    B. fires, hurricanes, tornadoes, and volcanic eruptions
    C. temperature
    D. predation
    E. water
    F. sunlight

20. **True or False?** Most organisms can not maintain a sufficiently active metabolism at temperatures close to 0°C. ________________

21. **True or False?** Birds and mammals can remain considerably warmer than their surroundings. ________________

22. **True or False?** Most photosynthesis occurs near the bottom of a body of water. ________________

23. **True or False?** In some communities, catastrophic disturbances such as fires are necessary in order to maintain a community. ________________

24. Pollen dispersal, openings in forests, and increased loss of water by evaporation are all consequences of high ________________.

The Evolutionary Adaptations of Organisms

Matching: Match the term on the left to its example on the right.

____ 25.	short-term physiological response	A.	the gradual growth of a heavy coat of fur on a rabbit as winter approaches
____ 26.	anatomical acclimation	B.	producing extra blood cells after a person moves to a home at a higher elevation
____ 27.	physiological acclimation	C.	the sudden appearance of goose bumps as a cold winter wind blasts your skin
____ 28.	behavioral response	D.	migrating great distances to reach the best nesting grounds

29. **True or False?** Ectotherms are better able to tolerate the greatest temperature extremes. ________________

30. **True or False?** In general, animals are more anatomically plastic than plants. ________________

31. Individual adaptations that are reversible but take days or weeks to complete are examples of ________________.

32. Adaptations that enable plants and animals to adjust to changes in their environments occur during a period called ________________ time.

33. The flagging of the tree in Figure 18.10 is an example of a(n) ________________ response.

What is Population Ecology?

Population Density

Matching: Match the terms on the left to their best description on the right.

_____ 34. population ecology	A. a group of individuals of the same species living in a given area at a given time
_____ 35. mark-recapture method	B. the field that examines the factors that influence a population's size, density, and characteristics
_____ 36. population density	C. a sampling technique used to estimate characteristics of a wildlife populations
_____ 37. population	D. the number of individuals of a species per unit area or volume

38. A biologist is trying to determine the number of carp in a pond. She uses a net to capture 100 carp, marks each fish with a tag, and returns them to the pond. A week later she nets out 100 more carp and finds that 50 of them had a tag and were caught the previous week. How many carp does she estimate are in the pond?
    A. 100
    B. 150
    C. 200
    D. 250
    E. 300

39. A biology graduate student uses the mark-recapture method to estimate the population of squirrels living in the middle of the university campus. He sets out 100 traps and finds that each trap has one squirrel inside. He tags the squirrels and releases them. A week later he sets out 100 traps and again catches one squirrel in each trap. In this second set, 75 of the squirrels had tags and were caught the previous week. When analyzing his data, his research advisor tells him that many students have done this study before and that many of these squirrels seem to be attracted to the traps to get a free meal. The bright graduate student thus concludes that when he makes his calculations using the formula for the mark-recapture method, his estimate will be:
    A. very close to the actual population size.
    B. too high compared to the actual population size.
    C. too low compared to the actual population size.

40. **True or False?** When estimating populations, the <u>larger</u> the number and size of sample plots, the more accurate the estimates of population size.

    ____

Patterns of Dispersion

41. During the springtime, tiger salamanders migrate to small ponds to reproduce. During these mating seasons, these animals show which type of dispersion pattern?
    A. clumped
    B. scattered
    C. uniform
    D. random
    E. irregular

42. **True or False?** Interactions among individuals of a population often produce a random pattern of dispersion. ________________

43. In the absence of strong attractions or repulsions among individuals in a population we expect to find a(n) ________________ pattern of dispersion.

Population Growth Models

44. Exponential growth typically produces a curve shaped like the letter ________________, while logistic growth typically produces growth shaped most like the letter ________________.
    A. S, J
    B. J, S
    C. L, S
    D. U, J
    E. S, L

45. Examine Figure 18.19. From 1940–1950, the male fur seal population
    A. declined.
    B. reached the carrying capacity.
    C. increased.
    D. experienced exponential growth.
    E. crashed.

46. **True or False?** Growth rate is greatest in the logistic growth model at the highest population level. ________________

47. **True or False?** Growth rate during exponential growth is greatest at the highest population level. ________________

48. The rate of expansion of a population under ideal conditions is called ________________.

49. A description of idealized population growth that is slowed by limiting factors is the ________________ model.

50. Environmental factors that restrict population growth are called ________________ factors.

Regulation of Population Growth

51. Which one of the following is most likely a density-dependent factor?
    A. hurricane damage to the nesting sites of birds
    B. a long, very cold winter that kills many bison
    C. a shortage of good browsing vegetation because of overeating
    D. a limited number of nesting sites for birds because of forest fires

52. When deer populations are low,
    A. density-dependent factors have their greatest impact.
    B. many twins are produced.
    C. food quality is poor.
    D. many females fail to reproduce.
    E. density-independent factors have their greatest intensity.

53. Aphids and many other insects often show exponential growth in the spring and then rapid die-offs. These populations are
    A. crashing when they reach their carrying capacities.
    B. limited by factors closely related to population density.
    C. crashing when they are overcome by disease.
    D. most likely limited by density-independent factors.

54. A population that grows exponentially, but is eventually limited by density-dependent factors, will have a growth chart most similar to the shape of the letter
    A. J.
    B. U.
    C. L.
    D. S.
    E. V.

55. The boom-and-bust cycles of snowshoe hares is most likely caused by
    A. competition with other grazing mammals.
    B. fluctuations in the lynx population.
    C. the effects of predation.
    D. fluctuations in the hare's food sources.
    E. the combined effects of predation and fluctuations in the hare's food sources.

56. **True or False?** A density-dependent factor intensifies as the population decreases in size. __________

57. **True or False?** Density-independent factors are unrelated to population density.

____

58. **True or False?** In many populations, density-independent factors limit population size before density-dependent factors have their greatest impact.

____

59. A population that remains near its carrying capacity for a long period of time is most likely limited by __________ factors.

60. The reliance of individuals of the same species on the same limited resources is called __________.

Human Population Growth

61. Human population growth
    - A. is closest to the model for exponential growth.
    - B. is closest to the model for logistic growth.
    - C. has remained steady over the course of human history.
    - D. is now on the decline, due to density-dependent factors.
    - E. has experienced a pattern of boom and bust.

62. In Italy, populations are stable because
    - A. birth rates continue to decline while death rates remain steady.
    - B. birth rates continue to increase while death rates are decreasing.
    - C. birth rates remain steady while death rates continue to decline.
    - D. birth rates and death rates remain in balance.

63. A unique feature of human population growth is
    - A. the independence of factors affecting the birth rates.
    - B. the independence of factors affecting the death rates.
    - C. our ability to voluntarily control it.
    - D. that it can not be limited by density-dependent factors.
    - E. that it can not be limited by density-independent factors.

64. **True or False?** Since the Industrial Revolution, exponential growth of the human population has resulted mainly from an increase in death rates.

____

65. **True or False?** The population of the United States continues to increase.

____

66. **True or False?** Delayed reproduction dramatically <u>decreases</u> population growth rates. ____________________

67. **True or False?** Carrying capacity <u>has changed</u> with human cultural evolution. ____________________

68. Human population growth is based on the same two general parameters that affect other animal and plant populations: ____________________ rates and ____________________ rates.

69. The proportion of individuals in different age groups defines the ____________________ of that population.

Life Histories and Their Evolution

Life Tables and Survivorship Curves

70. The survivorship curve for humans has
    A. the highest mortality later in life.
    B. the highest mortality in the middle of the potential life span.
    C. the highest mortality earliest in life.
    D. an even mortality rate throughout the life span.
    E. high rates of mortality early and late in life and low mortality in the middle of the lifespan.

71. **True or False?** Oysters have a Type III survivorship curve with the <u>highest</u> mortality early in life. ____________________

72. Survivorship and mortality in a population can be tracked by a(n) ____________________.

73. A plot of the number of people still alive at each age in a life table is called a(n) ____________________.

74. Traits that affect an organism's schedule of reproduction and death make up its ____________________.

Life History Traits as Evolutionary Adaptations

75. Populations that exhibit an equilibrial life history
    A. have a Type III survivorship curve.
    B. are mainly smaller-bodied species such as insects.
    C. mature later.
    D. produce many offspring.
    E. do not care for their young.

76. **True or False?** Life history traits, like anatomical features, are shaped by adaptive evolution. ____________________

77. Organisms that exhibit a(n) ____________________ life history tend to grow exponentially when conditions are favorable.

Evolution Connection: Testing a Darwinian Hypothesis

78. In the locations where guppies were preyed upon by pike-cichlids, which eat mostly large and mature guppies, guppies tend to
    A. be larger.
    B. mature later.
    C. produce more offspring each time they give birth.
    D. develop more slowly.

79. **True or False?** Reznick and Endler were able to conduct experiments that showed that the main variable affecting the guppy life histories was the presence of different predators. ____________________

Word Roots

equi = equal (*equilibrial life history:* the type of life history in which individuals usually mature later and produce few offspring but care for their young)

intra = within (*intraspecific competition:* the reliance of individuals of the same species on the same limited resources)

Key Terms

abiotic component
acclimation
age structure
biosphere
biotic component
carrying capacity
clumped
community
community ecology
density-dependent factor
density-independent factor
dispersion pattern
ecology
ecosystem
ecosystem ecology
equilibrial life history
exponential growth model
habitats
intraspecific competition
life history
life table
logistic growth model
mark-recapture method
opportunistic life histories
organismal ecology
population
population density
population ecology
population-limiting factors
random
survivorship curve
uniform

Crossword Puzzle

Use the Key Terms list from this chapter to fill in the crossword puzzle.

ACROSS

1. reversible, long term physiological responses to the environment
3. the global ecosystem
5. all the organisms in a given area, along with the nonliving factors with which they interact
6. the type of component in the environment that is alive
8. the number of individuals in a population that an environment can sustain
9. describing a dispersion pattern in which individuals are aggregated in patches
10. the type of ecology concerned with energy flow and the cycling of chemicals among the various biotic and abiotic factors
12. all the organisms living together and potentially interacting in a particular area
13. the type of ecology concerned with evolutionary adaptations of individual organisms
14. a characteristic of a population referring to the proportion of individuals in different age groups
15. a type of sampling method used to estimate wildlife populations
16. the type of ecology concerned with interactions between species
19. the type of mathematical model of idealized population growth that is restricted by limiting factors
20. a group of individuals of the same species living in a particular geographic area
21. environmental situations in which organisms live
22. the rate of expansion of a population under ideal conditions
23. the type of life history in which individuals usually mature later and produce few offspring, but care for their young
24. the type of pattern that often results from interactions among individuals of a population
25. the type of life history in which natural selection has reinforced quantity of reproduction more than individual survivorship

DOWN

1. a type of component in the environment that is not alive
2. a type of population-limiting factor that intensifies as the population increases in size
4. a listing of survival and death in a population in a particular time period
7. the way individuals are spaced within their area
11. the type of pattern in which individuals in a population are spaced in a patternless, unpredictable way
17. the study of interactions between organisms and their environments
18. the traits that affect an organism's schedule of reproduction and death

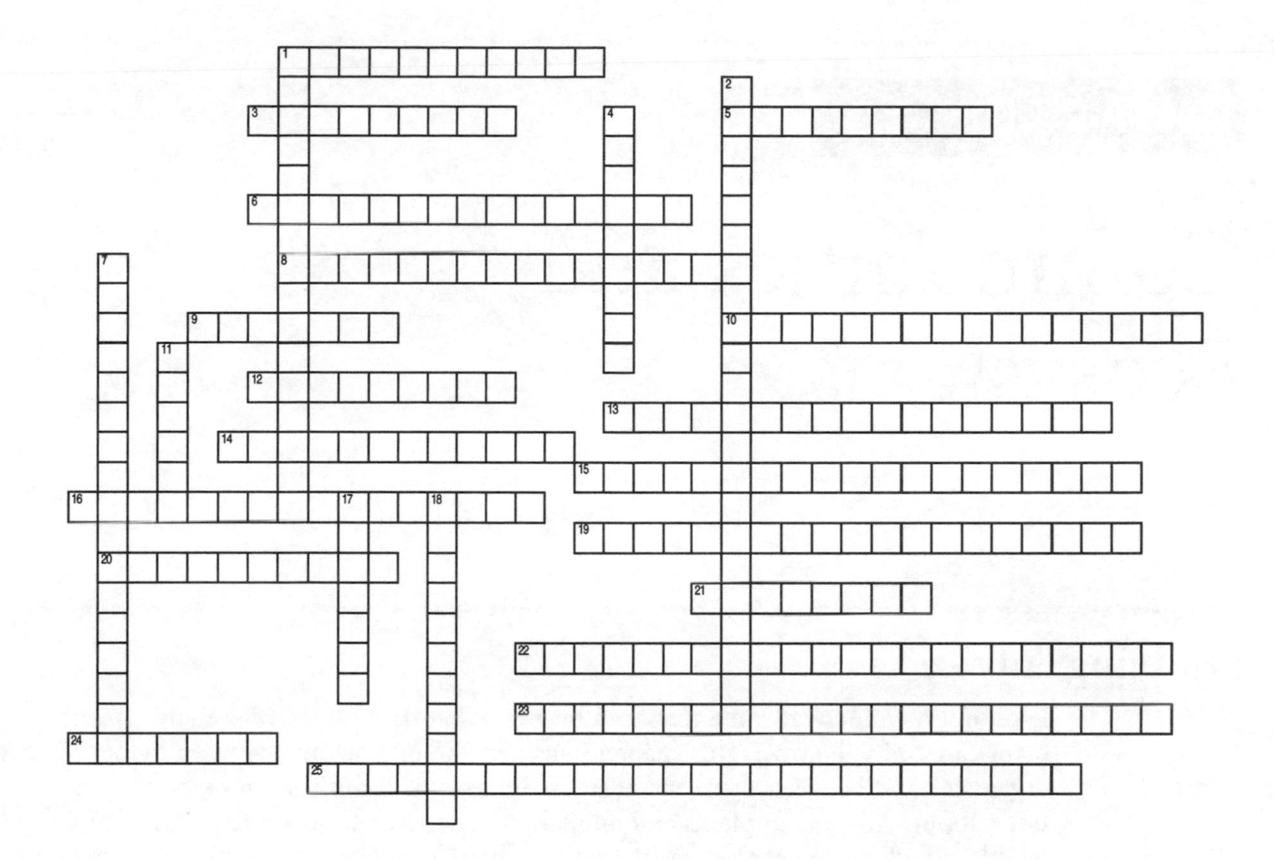

1
2
3
4
5
6
7
8
9
10
11
12
13
14
15
16
17
18
19
20
21
22
23
24
25

CHAPTER 19

Communities and Ecosystems

Studying Advice

Look out a window at some trees and bushes. What determines where any type of plant can and will grow? This chapter considers the interactions between living organisms and between them and their environments. It answers some basic questions about why certain plants and animals are restricted to just a few parts of the world and why some ecosystems that are far apart still look very similar. (Could you tell the difference between a rain forest in South America and a rain forest in Africa?) If you enjoy walks through the park or wild areas, this chapter will surely be of interest.

The many organizing tables provide a place to arrange the ideas and definitions discussed in this chapter.

Student Media

Biology and Society on the Web

Find out how old subway cars can be made into reefs.

Activities

19A Interspecific Interactions
19B Primary Succession
19C Energy Flow and Chemical Cycling
19D Food Webs
19E Energy Pyramids
19F The Carbon Cycle
19G The Nitrogen Cycle
19H Terrestrial Biomes
19I Aquatic Biomes

Case Studies in the Process of Science

How Are Impacts on Community Diversity Measured?

How Does Light Affect Primary Productivity?

Evolution Connection on the Web

Learn more about coevolution.

Organizing Tables

Describe each of the four properties of a community in Table 19.1 below.

TABLE 19.1

	Description
Diversity	
Prevalent form of vegetation	
Stability	
Trophic structure	

In Table 19.2, distinguish between the following pairs of terms.

TABLE 19.2	
Batesian mimicry vs. Müllerian mimicry	
Parasitism vs. mutualism	
Primary succession vs. secondary succession	

Define and give an example of each of the four trophic levels noted in Table 19.3 below. Try to use examples that could all be found in one food chain.

TABLE 19.3		
	Definition	**Example**
Producer		
Primary consumer		
Secondary consumer		
Detritivore		

In Table 19.4, describe examples of human behavior that disrupt each of the following nutrient cycles.

TABLE 19.4

	Human Activities Disruptive to the Cycle
Carbon cycle	
Nitrogen cycle	
Phosphorus cycle	
Water cycle	

Define and compare the parts of freshwater biomes in the table below, using information from the textbook. Also indicate examples of organisms that live in each zone. One cell in the table is already filled in to help you.

TABLE 19.5

Freshwater zone	Definition	Examples of Organisms That Live in This Zone
Photic zone		
Aphotic zone		No plants. Fish and other swimming animals may pass through.
Benthic zone		

In Table 19.6 below, compare the properties of a river or stream near its source versus its properties near the lake or ocean where it empties.

TABLE 19.6

Location in the Stream or River	Water Clarity	Water Temperature	Location of Algae	Types of Benthic Animals	Types of Predators
Near its source					
Where it dumps into a lake or the ocean					

Compare the marine biomes in Table 19.7 below.

TABLE 19.7

Marine Biomes	Location	Special Characteristics
Estuaries		
Intertidal zones		
Pelagic zone		
Benthic zone		

TABLE 19.7, *continued*		
Marine Biomes	**Location**	**Special Characteristics**
Hydrothermal vent communities		
Coral reefs		

Content Quiz

Directions: Identify the *one* best answer for the multiple-choice questions. For true/false questions, determine if the statement is true or false. If false, change the underlined word(s) to make the statement true. Finally, add the correct word(s) to the fill-in-the-blank questions to make the statements true.

Biology and Society: Reefs: Coral and Artificial

1. Which one of the following statements about corals is *false*?
   A. Corals dominate coral reef systems.
   B. Corals secrete soft external skeletons made of sodium chloride.
   C. Corals feed on microscopic organisms and particles of organic debris.
   D. Corals obtain organic molecules from the photosynthesis of symbiotic algae that live in their tissues.

2. **True or False?** Some coral reefs cover enormous expanses of <u>deep</u> ocean.

   ____

3. Corals are also subject to damage from both native and introduced

   ____________________.

4. Coral reefs are distinctive and complex ____________________.

Key Properties of Communities

Diversity; Prevalent Form of Vegetation; Stability; Trophic Structure

Matching: Match the property on the left to its best description on the right.

____ 5. trophic structure	A. the variety of organisms that make up a community
____ 6. stability	B. the types and structural features of plants
____ 7. diversity	C. the community's ability to resist change and recover
____ 8. prevalent form of vegetation	D. the feeding relationships among the members of the community

9. What property of a community determines the passage of energy and nutrients from plants and other photosynthetic organisms to herbivores and then to carnivores?
   A. stability
   B. trophic structure
   C. diversity
   D. prevalent form of vegetation

10. **True or False?** A forest dominated by cedar and hemlock trees is a <u>highly stable</u> community because these trees can withstand most lightning-caused fires.

    ____

11. The term ____________________, as used by ecologists, considers *both* diversity factors: richness and relative abundance.

12. An organism's ____________________ includes other individuals in its population and populations of other species living in the same area.

13. An assemblage of species living close enough together for potential interaction is called a(n) ____________________.

Interspecific Interactions in Communities

Competition Between Species

14. What will happen if two species in an ecosystem have identical niches?
    A. Both species will be driven to local extinction.
    B. One species will be driven to local extinction.
    C. One species will be driven to local extinction or one of the species may evolve enough to use a different set of resources.
    D. Both species will evolve enough to use a different set of resources.
    E. Both species will be driven to local extinction or both species will evolve enough to use a different set of resources.

15. Which one of the following typically results when population density increases and nears carrying capacity?
    A. Population growth increases.
    B. Birth rates increase.
    C. Mortality increases.
    D. Individuals have access to a larger share of some limiting resource.

16. Which of the following statements is true according to the competitive exclusion principle?
    A. As the number of species in a community increases, interspecific competition decreases.
    B. Species will share the community resources to decrease interspecific competition.
    C. When interspecific competition increases, population densities decrease.
    D. If two species compete for the same limiting resources, they cannot coexist in the same place.

17. **True or False?** The population growth of a species may be limited by the density of competing species. ________________

18. Interactions between species are called ________________ interactions.

19. When populations of two or more species in a community rely on similar limiting resources, they may be subject to ________________.

20. An organism's ecological role is the same as its ecological ________________.

21. The differentiation of niches that enables similar species to coexist in a locale is called ________________.

Predation

22. Which one of the following statements about predation, according to your chapter, is *false*?
    A. In a predator-prey relationship, the consumer is the predator.
    B. In a predator-prey relationship, the food species is the prey.
    C. In herbivory, the prey is an animal.
    D. Predation is a form of interspecific competition.
    E. Natural selection refines the adaptations of predators and prey.

23. Which one of the following is characteristic of predators?
    A. acute senses
    B. thorns
    C. poisons such as strychnine, mescaline, and tannins
    D. alarm calls
    E. mobbing behavior

24. Which one of the following defense mechanisms would be used by a mother bird to draw attention to herself to keep a predator from attacking her chicks?
    A. alarm calls
    B. active self defense
    C. fleeing
    D. distraction displays
    E. mobbing

25. If a harmless species is protected by appearing to be a harmful model, it is an example of
    A. warning coloration.
    B. cryptic coloration.
    C. Müllerian mimicry.
    D. Batesian mimicry.

26. A moth that has a color pattern that makes it appear to blend into its environment is showing
    A. warning coloration.
    B. cryptic coloration.
    C. Müllerian mimicry.
    D. Batesian mimicry.

27. **True or False?** Predators that <u>ambush</u> their prey are generally fast and agile.

    ____

28. **True or False?** Many animals use <u>warning</u> coloration to make it difficult to spot them in their environment. ____________________

29. **True or False?** Predator-prey relationships <u>can preserve</u> species diversity.

    ____

30. **True or False?** In <u>Batesian</u> mimicry, two or more unpalatable species resemble each other. ____________________

31. A toxic animal will often have a ____________________ coloration to caution predators not to attack.

32. A species that reduces the density of the strongest competitors in a community is called a(n) ____________________ predator.

Symbiotic Relationships; The Complexity of Community Networks

33. Which one of the following is *not* an example of a parasitic relationship?
    A. mosquitoes and people
    B. aphids and plants
    C. root-fungus associations called mycorrhizae
    D. leech and fish
    E. tapeworms and cattle

34. Which one of the following is the best example of a parasitic relationship?
    A. a person shopping for groceries
    B. a student writing a paper for a course
    C. a sailor using wind to propel a sailboat
    D. a student stealing a book from another student
    E. a physician treating the wound of a patient

35. Which one of the following is the best example of a mutualistic relationship?
    A. a person collecting and eating nuts that have fallen from the trees
    B. giving someone $20 to help you change a flat tire
    C. a person moving from one apartment to another
    D. a cow eating grass
    E. bird migrations across over great distances

36. **True or False?** In the symbiotic relationship termed mutualism, one organism benefits at the expense of the other. ____________

37. **True or False?** Many mutualistic relationships may have evolved from predator-prey or host-parasite interactions. ____________

38. **True or False?** Biologists have sorted out the complex networks of many biological communities. ____________

39. In a symbiotic relationship, the ____________ lives in or on another species, the ____________.

Disturbance of Communities

Ecological Succession

40. Which one of the following is the most common sequence of appearance of organisms during primary succession?
    A. lichens and mosses, grasses, autotrophic microorganisms, shrubs, trees
    B. lichens and mosses, grasses, shrubs, trees, autotrophic microorganisms
    C. autotrophic microorganisms, lichens and mosses, grasses, shrubs, trees
    D. autotrophic microorganisms, grasses, shrubs, trees, lichens, mosses

41. What has happened to a community to cause secondary succession? A disturbance has
    A. destroyed most of the grasses, shrubs, and trees but left most microorganisms, lichens, and mosses intact.
    B. destroyed an existing community but left the soil intact.
    C. completely eliminated one level of a community and the soil.
    D. completed destroyed a community and all the components of its environment.

42. **True or False?** Ecological disturbances are <u>more common</u> than stability in most communities. ________________

43. The process of community change is called ________________.

A Dynamic View of Community Structure

44. Species diversity appears to be greatest in communities with ________________ amounts of disturbance.
    A. virtually no
    B. very small
    C. intermediate
    D. very large
    E. continual

45. **True or False?** Frequent fires prevent grassland communities from becoming <u>forests</u>. ________________

46. According to the ________________ hypothesis, species diversity appears to be greatest in communities with intermediate degrees of disturbance.

An Overview of Ecosystem Dynamics

47. The two key processes of ecosystem dynamics are
    A. photosynthesis and metabolism.
    B. energy flow and phase changes.
    C. energy flow and chemical recycling.
    D. phase changes and chemical recycling.

48. **True or False?** Energy reaches most ecosystems in the form of <u>heat</u>.

    ____

49. **True or False?** Unlike matter, <u>energy</u> cannot be recycled. ________________

50. The highest level of biological organization is a(n) ________________.

51. Chemical elements can be recycled between an ecosystem's living community and the __________________ environment.

52. Plants and other producers acquire their carbon, nitrogen, and other chemical elements in inorganic form from the __________________ and __________________.

53. Energy flow and chemical recycling in an ecosystem depend on the transfer of substances in the __________________, or feeding relationships.

Trophic Levels and Food Chains; Food Webs

Matching: Match the terms on the left to their best description on the right.

_____ 54. detritivore	A. organisms that use photosynthesis
_____ 55. primary consumer	B. animals that eat herbivores
_____ 56. producer	C. decomposers
_____ 57. secondary consumer	D. herbivores

58. Which one of the following *never* functions as a producer in an ecosystem?
   A. terrestrial plants
   B. fungi
   C. phytoplankton
   D. multicellular algae and aquatic plants

59. All organisms in trophic levels above the producers are
   A. autotrophic producers.
   B. autotrophic consumers.
   C. heterotrophic producers.
   D. heterotrophic consumers.

60. Examine the food web in Figure 19.23. Which one of the following changes would likely decrease the number of hawks?
   A. an increase in the mouse population
   B. an increase in the snake population
   C. an increase in the owl population
   D. an increase in the lizard population

61. **True or False?** The trophic level that supports all others are the <u>consumers</u>.

   ____

62. Ecologists divide the species of an ecosystem into different __________________ based on their main sources of nutrition.

63. The sequence of food transfer from trophic level to trophic level is called a(n)

_______________.

64. An ecosystem's main detritovores are _______________ and

_______________.

65. The feeding relationships in an ecosystem are usually woven into elaborate food

_______________.

Energy Flow in Ecosystems

Productivity and the Energy Budgets of Ecosystems; Energy Pyramids

66. Which one of the following statements about energy pyramids is *true*?
    A. Primary consumers form the lowest level of an energy pyramid.
    B. The highest level of an energy pyramid represents the producers.
    C. Most energy pyramids have 10–15 levels.
    D. Herbivores usually appear in the second level of an energy pyramid.

67. Examine the killer whale food chain in Figure 19.21. Within this ecosystem, how would the weight of all the killer whales compare to the collective weight of the zooplankton?
    A. The weight of the whale would be about 100 times greater than the weight of the zooplankton.
    B. The weight of the whale would be about the same as the weight of the zooplankton.
    C. The weight of the zooplankton would be about 10 times greater than the weight of the whale.
    D. The weight of the zooplankton would be about 100 times greater than the weight of the whale.
    E. The weight of the zooplankton would be about 1,000 times greater than the weight of the whale.

68. **True or False?** Only <u>about 1%</u> of the visible light that reaches producers is converted to chemical energy by photosynthesis. _______________

69. **True or False?** On average, only <u>about 10%</u> of the energy in the form of organic matter at each trophic level is stored as biomass in the next level of the food chain. _______________

70. The amount of living organic material in an ecosystem is the

_______________.

71. The rate at which producers build organic matter is the ecosystem's

_______________.

Ecosystem Energetics and Human Nutrition

72. When we eat beef, we are
    A. producers.
    B. herbivores.
    C. primary consumers.
    D. secondary consumers.

73. **True or False?** It takes about the same amount of photosynthetic productivity to produce 10 pounds of corn or one pound of hamburger. ________________

74. Most humans are ________________, eating both meat and plant material.

Chemical Cycles in Ecosystems

The General Scheme of Chemical Cycling; Examples of Biogeochemical Cycles

75. A chemical's specific route through an ecosystem depends upon the
    A. particular element and the trophic structure of the ecosystem.
    B. the amount of rainfall and the variation of the seasons.
    C. the number of producers and consumers.
    D. the amount of carbon and nitrogen in the atmosphere.

76. Which one of the following nutrients is not very mobile and is mostly cycled locally?
    A. carbon
    B. nitrogen
    C. phosphorus
    D. water

77. **True or False?** Plants get most of their nitrogen from the air.

    ____

78. **True or False?** Human activities have intruded into the cycling of local and global chemicals. ________________

79. **True or False?** Each chemical that moves through an ecosystem cycles through an abiotic reservoir. ________________

80. Chemical cycles in an ecosystem are also called ________________ cycles.

Biomes

Terrestrial Biomes

81. Most biomes are named
    A. according to the type of wind and soil type that is predominant in the region.
    B. for major physical or climatic features and for their predominant vegetation.
    C. after the largest geological feature that is near them, such as a lake, mountain, or ocean.
    D. based upon their physical position on Earth.

82. **True or False?** If the climate in two geographically separate areas is similar, the same type of biome may occur in them. ______________

83. **True or False?** The same type of biome throughout the world has the same species living in it. ______________

84. Species living in the same type of biome in different parts of the world may look similar to each other because of ______________.

Freshwater Biomes

____ 85.	photic zone	A.	the substrate at the bottom of all aquatic biomes
____ 86.	aphotic zone	B.	communities of organisms that live in the benthic zone
____ 87.	benthic zone	C.	the area where light levels are too low for photosynthesis
____ 88.	benthos	D.	shallow water near shore and the upper stratum of water away from shore
____ 89.	detritus	E.	a major source of food for the benthos

90. Which one of the following types of biomes occupies the largest part of the biosphere?
    A. savannas
    B. estuaries
    C. aquatic biomes
    D. grasslands
    E. temperate deciduous forest

91. Which one of the following traits is *not* characteristic of a river where it runs into a lake or the ocean?
    A. slow moving water
    B. warmer water (compared to the water at the source of the river)
    C. clearer water (compared to the water at the source of the river)
    D. predators such as catfish that find food more by scent and taste than sight
    E. benthic worms and insects that burrow into the muddy bottom

92. **True or False?** Lakes and ponds that receive large inputs of nitrogen and phosphorus often have <u>low</u> populations of algae. ____________________

93. **True or False?** A stream or river near its source is usually <u>colder</u> and <u>faster</u> than where it reaches a lake or the ocean. ____________________

Marine Biomes

94. Which one of the following statements about marine biomes is *false*?
    A. Evaporation from the oceans provides most of the Earth's rainfall.
    B. Life originated in the sea.
    C. Photosynthesis by zooplankton supplies a substantial portion of the biosphere's oxygen.
    D. Ocean temperatures have a major effect on climate and wind patterns.
    E. Aquatic biomes occupy the largest part of the biosphere.

95. Which one of the following is *not* characteristic of intertidal zones?
    A. twice-daily alternations of submergence in seawater and exposure to air
    B. strong wave action
    C. relatively even temperatures
    D. wide fluctuations in the availability of nutrients
    E. organisms that can burrow or that can cling to rocks or vegetation

96. **True or False?** Estuaries are crucial feeding areas for <u>many birds</u>.

    ____

97. **True or False?** Many fish and invertebrates use estuaries <u>as breeding grounds</u>.

    ____

98. **True or False?** The major producers in estuaries are <u>saltmarsh grasses and algae</u>. ____________________

99. **True or False?** Marine life is distributed according to <u>different factors than</u> those that affect the distribution of freshwater life. ____________________

100. **True or False?** The ocean's main photosynthetic producers are <u>zooplankton</u>.

    ____

101. The environment where a freshwater stream or river merges with the ocean is called a(n) ____________________.

102. Fishes, squids, and marine mammals spend most of their time in the ____________________ zone.

103. The most extensive part of the biosphere is the dark ________________ zone, occupying the deepest parts of the ocean.

104. Unusual ________________ communities are powered by chemical energy from the Earth's interior instead of sunlight.

Evolution Connection: Coevolution in Biological Communities

105. Which one of the following relationships is *not* an example of coevolution?
   A. defenses of plants against herbivores
   B. animal defenses against predators
   C. root-soil relationships
   D. parasite-host relationships
   E. mutual symbiosis

106. The yellow spots on the passionflower vine appear to serve two purposes. Which one of the following pairs *are* functions of these yellow spots? The yellow spots
   A. attract ants and wasps that prey on *Heliconius* eggs or larvae, and the spots discourage *Heliconius* butterflies from laying eggs on the leaves.
   B. attract ants and wasps that prey on *Heliconius* eggs or larvae, and the sacs are bags of toxin that kill *Heliconius* butterflies feeding on the leaves.
   C. are sacs of toxin that kill *Heliconius* butterflies feeding on the leaves, and the spots discourage *Heliconius* butterflies from laying eggs on the leaves.
   D. are sacs of toxin that kill *Heliconius* butterflies feeding on the leaves, and the spots promote a fungus that kills any *Heliconius* eggs laid on the leaves.
   E. attract ants and wasps that prey on *Heliconius* eggs or larvae, and the spots promote a fungus that kills any *Heliconius* eggs laid on the leaves.

107. **True or False?** Community interactions plus the interface of the community with soil, water, and other abiotic factors in the environment defines <u>ecosystems</u>. ________________

108. The term ________________ is used for those cases in which the adaptations of two species are closely connected.

Word Roots

detrit = wear off (*detritivores:* decomposers in an ecosystem)

eco = house (*ecosystem:* an ecological system)

inter = between (*interspecific competition:* competition between species)

crypt = hidden (*cryptic coloration:* another name for camouflage)

mutu = reciprocal (*mutualism:* a symbiosis that benefits both partners)

para = near; **site** = food (*parasitism:* when one organism benefits at the expense of another)

sym = together; **bio** = life (*symbiont:* the name given to a species that lives on another species)

troph = food (*trophic structure:* the feeding relationships among the various species making up a community)

Key Terms

abiotic reservoir
aphotic zone
Batesian mimicry
benthic zone
biogeochemical cycling
biomass
biomes
carnivores
chemical cycling
coevolution
community
competitive exclusion principle
cryptic coloration
decomposers
detritivores
detritus
disturbance
ecological niche
ecological succession
ecosystems
energy flow
energy pyramid
estuary
food chain
food webs
herbivores
host
hydrothermal vent communities
intermediate disturbance hypothesis
interspecific competition
interspecific interactions
intertidal zone
keystone predator
Müllerian mimicry
mutualism
omnivores
parasite
parasitism
pelagic zone
photic zone
phytoplankton
predation
predator
prey
primary consumers
primary productivity
primary succession
producers
quaternary consumers
resource partitioning
secondary consumers
secondary succession
species diversity
stability
symbiont
symbiotic relationship
temperate zones
tertiary consumers
trophic levels
trophic structure
tropics
warning coloration
zooplankton

Crossword Puzzle

Use the Key Terms list from this chapter to fill in the crossword puzzle.

ACROSS

1. a force that changes a biological community and usually removes organisms from it
3. the number and relative abundance of species in a biological community
5. nonliving organic matter
6. the smaller participant in a symbiotic relationship, living in or on the host
13. networks of interconnecting food chains
15. animals that eat both plants and animals
19. the process of biological community change resulting from disturbance
22. the type of mimicry in which a species that a predator can eat looks like a species that is harmful
24. the passage of energy through the components of an ecosystem
27. the amount of organic material in an ecosystem
32. a type of chemical cycling occurring in an ecosystem, involving both biotic and abiotic components
33. animals that drift in the pelagic zone of an aquatic environment
34. major types of ecosystem that covers a large geographic region
37. the type of zone that includes the substrate of a lake, pond, or seafloor
38. the type of zone where land meets the sea
39. a diagram depicting the cumulative loss of energy from a food chain

DOWN

2. a type of cycling that reuses chemical elements
4. organisms that make organic food molecules from carbon dioxide and water and other inorganic molecules
7. a type of symbiosis that benefits both species
8. the reciprocal evolutionary influence between two species
9. an interaction between species in which one species, the predator, eats the other, the prey
10. organisms that promote the breakdown of organic materials into inorganic ones
11. animals that eat other animals
12. the type of zone which includes water near shore and at the surface exposed to light
14. all the organisms living together and potentially interacting in a particular area
16. the type of zone that includes the open ocean
17. the larger participant in a symbiotic relationship, serving as home and feeding ground to the symbiont
18. the sequence of food transfer from producers through several levels of consumers in an ecosystem
20. a type of coloration that is a form of camouflage
21. a type of mimicry that is mutual by two species, both of which are harmful to a predator
23. the type of zone which includes water that does not get exposed to light
24. a population's role in its community
25. the diverse algae and cyanobacteria that drift passively in the pelagic zone
26. organisms that derive their energy from organic wastes and dead organisms; also called decomposer
28. animals that eat plants, algae, or autotrophic bacteria
29. a symbiotic relationship in which the symbiont benefits at the expense of the host
30. all the abiotic factors in addition to the community of species that exists in certain areas
31. the tendency of a biological community to resist change and return to its original species composition after being disturbed
35. an area where fresh water merges with seawater
36. an organism eaten by a predator

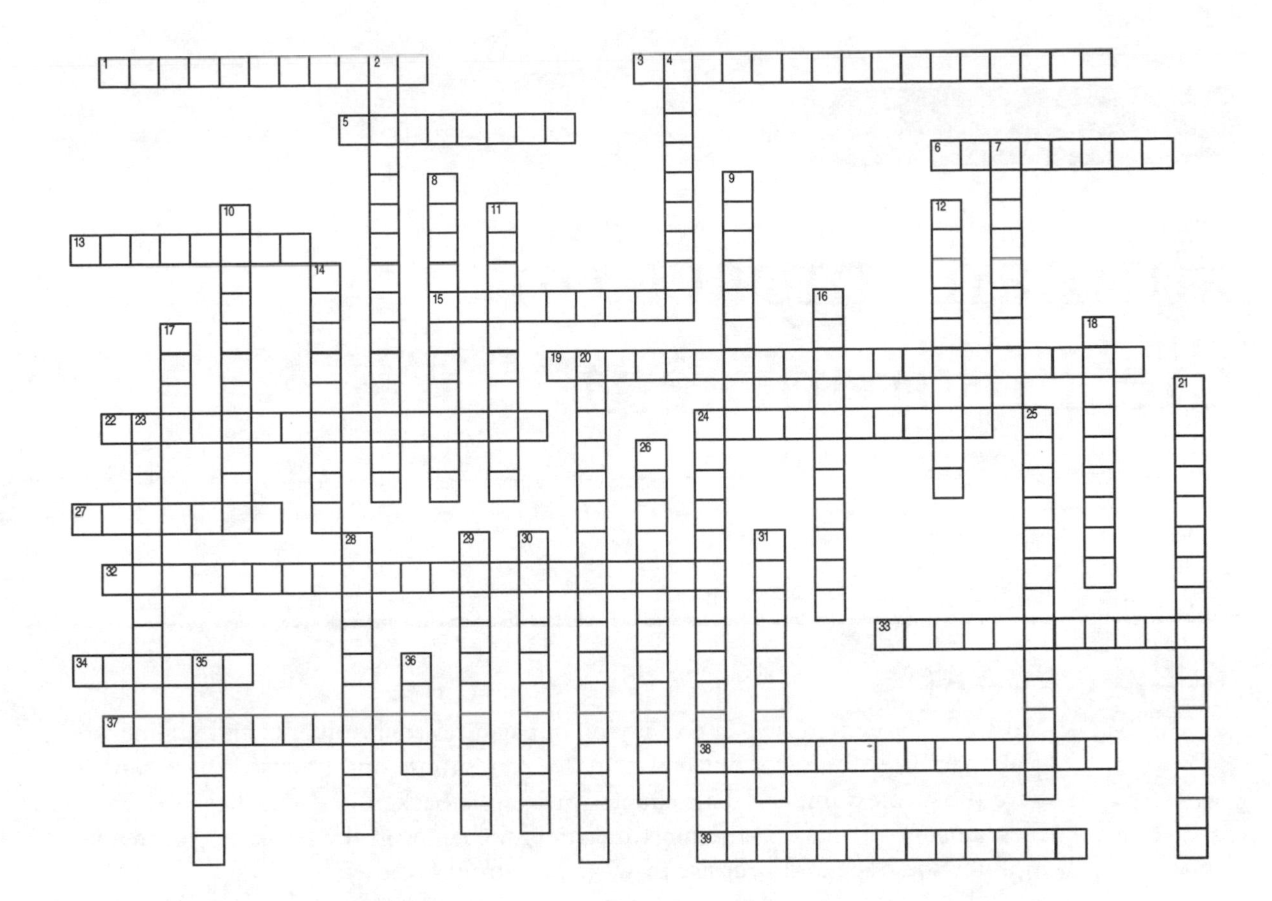

CHAPTER 20

Human Impact on the Environment

Studying Advice

Most of us are interested in conserving our natural environments and willing to make sacrifices to do so. But how shall we concentrate our efforts? Where can we have the greatest impact? This chapter provides a background that helps us find these answers. It may be the most meaningful chapter in the book. As you read, look for the hope and promise of what we can do together.

Student Media

Biology and Society on the Web

Learn what the U.S. government has to say about introduced species.

Activities

20A Fire Ants: An Introduced Species
20B Water Pollution from Nitrates
20C DDT and the Environment
20D The Greenhouse Effect
20E Madagascar and the Biodiversity Crisis
20F Conservation Biology Review

Case Study in the Process of Science

How Are Potential Prairie Restoration Sites Analyzed?

Evolution Connection on the Web

Read more on biophilia.

Organizing Tables

Describe the three main types of biological diversity in the table below, and provide an example of each.

TABLE 20.1		
	Description	**Examples**
Diversity of ecosystems		
Variety of species		
Genetic variation within each species		

Describe the three main causes of the biodiversity crisis in Table 20.2 below, and describe an example of each.

TABLE 20.2		
	Description	**Examples**
Habitat destruction		
Introduced species		
Overexploitation		

Content Quiz

Directions: Identify the *one* best answer for the multiple-choice questions. For true/false questions, determine if the statement is true or false. If false, change the underlined word(s) to make the statement true. Finally, add the correct word(s) to the fill-in-the-blank questions to make the statements true.

Biology and Society: Aquarium Menaces

1. Which one of the following was introduced into the Mediterranean Sea and has already caused severe disruptions of that ecosystem?
   A. the alga *Caulerpa*
   B. morel mushrooms
   C. the bacterium *E. coli*
   D. the northern snakehead fish, *Channa argus*
   E. loggerhead sea turtles

2. **True or False?** Humans have a disproportionately high impact on the environment. ____________________

3. Human activities are putting communities, ecosystems, and biological ____________________ at risk.

Human Impact on Biological Communities

Human Disturbance of Communities

4. Which one of the following statements is *false*?
   A. Much of the United States is now a hodgepodge of early successional growth.
   B. Weedy and shrubby vegetation often colonizes an area after forests are clear-cut and abandoned.
   C. Humans currently use about 30% of Earth's land, mostly for homes and cities.
   D. Most crops are grown in monocultures.
   E. Tropical rain forests are quickly disappearing as a result of clear-cutting for lumber and pastureland.

5. **True or False?** Human disturbance of communities is limited to the United States and Europe. ____________________

6. The greatest impact on natural communities is caused by ____________________.

Introduced Species

7. Which one of the following organisms was accidentally introduced into a new part of the world?
   A. kudzu
   B. starlings
   C. Australia's rabbits
   D. zebra mussels
   E. All of the above were intentionally introduced into a new part of the world.

8. The Columbian Exchange was
   A. the exchange of plants and animals between Europe and the North American colonies in the fifteenth and sixteenth centuries.
   B. a system of trading pelts and agricultural products between British Columbia and the United States.
   C. the transport of plants and animals between Columbia and the North American colonies in the fifteenth and sixteenth centuries.
   D. the exchange of Columbian coffee for North American species of plants and animals during the 18th century.
   E. the introduction into North America of many new species by Christopher Columbus.

9. **True or False?** Most introduced species fail to thrive in their new homes.

   ____

10. **True or False?** In many cases, introduced species encounter more pathogens, parasites, and predators than native species. ________________

11. The most common cause of extinction is ________________.

12. The second most common cause of extinction is ________________.

13. The U.S. Department of Agriculture encouraged the import of ________________ to the American South in the 1930s to help control erosion.

Human Impact on Ecosystems

Impact on Chemical Cycles; Deforestation and Chemical Cycles: A Case Study

Matching: Match the cycles to examples of human activities that disrupt these cycles. Each cycle matches just one choice on the right.

_____ 14. carbon cycle

_____ 15. nitrogen and phosphorus cycles

_____ 16. water cycle

A. fertilizer and sewage contamination of streams and lakes

B. pumping groundwater for irrigation

C. burning of fossil fuels

17. After logging and spraying herbicides on a section of the Hubbard Brook Experimental Forest,

    A. net loss of the minerals from the forest remained about the same.

    B. nitrate levels in the drainage creek decreased.

    C. water runoff from the watershed increased by 30%–40%.

    D. the root systems of the logged trees still prevented water loss from the soil.

18. **True or False?** Humans have managed to intrude in one way or another into the dynamics of most ecosystems. __________________

19. Sewage outflow into lakes and rivers can cause "over-fertilization," or __________________, which can lead to heavy algal growth that suffocates the aerobic life of these ecosystems.

The Release of Toxic Chemicals to Ecosystems

20. Because of biological magnification, which one of the following members of a food chain will be most affected by the introduction of a toxin into an ecosystem?

    A. producer

    B. herbivore

    C. detritivore

    D. primary consumer

    E. secondary consumer

21. **True or False?** Bacteria in the bottom mud of rivers and the ocean convert mercury into an extremely toxic and soluble compound. __________________

22. The toxin __________________ accumulates in fatty tissues and has caused serious declines in populations of pelicans, ospreys, and eagles.

Human Impact on the Atmosphere and Climate

23. Which one of the following will likely *not* occur because of global warming?
    A. Plant and animal populations will spread to new locations.
    B. Global air temperatures will rise.
    C. Sea levels will decrease as the oceans start to evaporate.
    D. Climate patterns will shift.

24. Which one of the following will likely *not* result from the destruction of the ozone?
    A. an increase in the number of skin cancers
    B. an increase in the number of cataracts
    C. damage to crops and phytoplankton
    D. an increase in global atmospheric temperatures

25. **True or False?** Most of the increase in CO_2 levels comes from burning <u>trees</u>.

    ____

26. **True or False?** If all chlorofluorocarbons were banned today, the chlorine molecules already in the atmosphere <u>would continue</u> to decrease atmospheric ozone levels. ____________________

27. Because of increased atmospheric levels of carbon dioxide and water vapor in the atmosphere, the process of global warming is called the ____________________.

28. Scientists now routinely find a hole in the ozone layer over the continent of

    ____________________.

The Biodiversity Crisis

The Three Levels of Biodiversity; The Loss of Species

29. Which one of the following is *not* one of the three main components of biodiversity?
    A. genetic variation within each species
    B. the diversity of ecosystems
    C. the variety of species
    D. the number of individuals of a species

30. Compared to the Cretaceous mass extinction, the current mass extinction is
    A. broader and faster.
    B. narrower and faster.
    C. narrower and slower.
    D. broader but slower.

31. Which one of the following statements is *false*?
    A. Fewer than 10% of all species have been described by scientists.
    B. About 50% of all known bird species in the world are endangered.
    C. Some researchers estimate that at the current rate of destruction, over half of all plant and animal species will be gone by the end of this new century.
    D. About 20% of the known freshwater fishes in the world have either become extinct during historical times or are seriously threatened.

32. **True or False?** The current mass extinction of species is due to the evolution of humans. ____________________

The Three Main Causes of the Biodiversity Crisis; Why Biodiversity Matters

33. Which one of the following is the single greatest threat to biological diversity?
    A. global warming
    B. depletion of the ozone layer
    C. habitat destruction
    D. introduction of nonnative species
    E. overexploitation

34. Pigeons, house sparrows, and starlings are species that have
    A. gone extinct due to overexploitation.
    B. been impacted by the introduction of the Nile perch.
    C. been introduced by humans.
    D. been severely harmed by habitat destruction.

35. Whales, the American bison, and the Galápagos tortoises are species that have been
    A. overexploited.
    B. impacted by the introduction of non-native species.
    C. greatly reduced because of the loss of habitat.
    D. poisoned by the introduction of environmental toxins.

36. **True or False?** Scientists estimated the average annual value of ecosystem dynamics in the biosphere at $33 million. ____________________

37. About 25% of all prescription drugs contain substances derived from ____________________.

38. The introduction of nonnative, or ____________________, species often results in the elimination of native species.

Conservation Biology

Biodiversity "Hot Spots"

39. The two main goals of conservation biology are to
    A. preserve individual species and sustain ecosystems.
    B. preserve individual species and create new ecosystems.
    C. identify new species and restore ecosystems.
    D. identify new species and sustain ecosystems.

40. Examine the map of hot spots in Figure 20.17. In general, most hot spots are located
    A. in the northern hemisphere.
    B. in the southern hemisphere.
    C. near the poles.
    D. near the equator.

41. **True or False?** Biodiversity hot spots are also hot spots of <u>extinction</u>.

    ____

42. A species found nowhere else is a(n) ____________________ species.

Conservation at the Population and Species Levels

43. Which one of the following does *not* usually result from habitat fragmentation?
    A. increased gene flow
    B. a decrease in the overall size of populations
    C. subpopulations are separated into habitat patches that vary in quality
    D. the number of sink habitats increases

44. Conservation biology often highlights the relationships between
    A. an organism and its environment.
    B. biology and society.
    C. the biotic and abiotic factors of the environment.
    D. biology and the Earth.

45. **True or False?** The populations of many species were subdivided into groups <u>after</u> humans began altering habitats significantly. ____________________

46. **True or False?** An area of habitat where a subpopulation's reproductive success exceeds its death rate is called a <u>sink</u> habitat. ____________________

47. **True or False?** The spotted owl situation illustrates the importance of protecting source habitats. ________________

48. **True or False?** Controlled fires were needed to help maintain mature pine and help the red-cockaded woodpecker recover from near-extinction.

____

49. **True or False?** Habitat use is almost always at issue in conflicts involving conservation biology. ________________

50. A(n) ________________ species is in danger of extinction throughout all or a significant portion of its range.

51. A(n) ________________ species is likely to become endangered in the foreseeable future throughout all or a significant portion of its geographic range.

Conservation at the Ecosystem Level; The Goal of Sustainable Development

52. Which one of the following statements about the edges between ecosystems is *false*?
    A. Edges are prominent features of landscapes.
    B. Edges have their own sets of physical conditions.
    C. Edges often have their own type and amount of disturbance.
    D. Edges can have both positive and negative effects on biodiversity.
    E. Landscapes edges produced by human activities often have more species.

53. Which one of the following statements about movement corridors is false? Movement corridors
    A. promote dispersal of members of a population.
    B. help populations when they connect source and sink habitats.
    C. are important to species that seasonally migrate between different habitats.
    D. can promote the spread of disease.

54. **True or False?** Cowbird populations have decreased due to human activities.

____

55. **True or False?** Streamside habitats often serve as natural corridors.

____

56. **True or False?** The areas surrounding a zoned reserve cannot to be used to support the local human population. ________________

57. The application of ecological principles to the study of land-use patterns is called ________________.

58. A narrow strip or series of small clumps of quality habitat connecting otherwise isolated patches is a(n) ________________.

59. An extensive region of land that includes one or more areas undisturbed by humans is a(n) ________________.

60. The goals of ________________ are balancing human needs with the health of the biosphere.

Evolution Connection: Biophilia and an Environmental Ethic

61. The concept of ________________ reflects a human desire to affiliate with other life in its many forms.

Word Roots

bio = life; **geo** = the Earth; **chemo** = chemical (*biogeochemical cycles:* cycles in an ecosystem that involve both biotic and abiotic components)

eu = good; **troph** = food (*eutrophication:* overfertilization of a lake)

Key Terms

biodiversity
biodiversity crisis
biodiversity hot spot
biological magnification
conservation biology
endangered species
endemic species
eutrophication
greenhouse effect
introduced species
landscape ecology
movement corridor
ozone layer
population fragmentation
sink habitat
source habitat
sustainable development
threatened species
zoned reserve

Crossword Puzzle

Use the Key Terms list from this chapter to fill in the crossword puzzle.

ACROSS

1. a small geographic area with an exceptional concentration of species
2. an increase in productivity of an aquatic ecosystem
4. the accumulation of persistent chemicals in the living tissues of consumers in food chains
5. the type of species that is in danger of extinction throughout all or a significant portion of its range
6. a series of small clumps or a narrow strip of quality habitat that connects otherwise isolated patches of quality habitat
7. a species that has a distribution limited to a specific geographic area
8. the type of ecology that examines the application of ecological principles to the study of land-use patterns
12. the type of biology that studies ways to counter the loss of biodiversity
13. the band of O_3 in the upper atmosphere that protects life on Earth from the harmful ultraviolet rays in sunlight
14. process of atmospheric carbon dioxide trapping heat
15. the type of species that is moved from its native location to a new geographic region
16. the type of development which produces long-term prosperity of human societies and the ecosystems that support them

DOWN

1. the current rapid decline in the variety of life on Earth, largely due to the effects of human culture
3. the type of fragmentation in which populations are split and isolated
9. a type of habitat where a species' death rate exceeds its reproductive success
10. a type of habitat where a species' reproductive success exceeds its death rate and from which new individuals often disperse to other areas
11. an extensive region of land that includes one or more areas undisturbed by humans

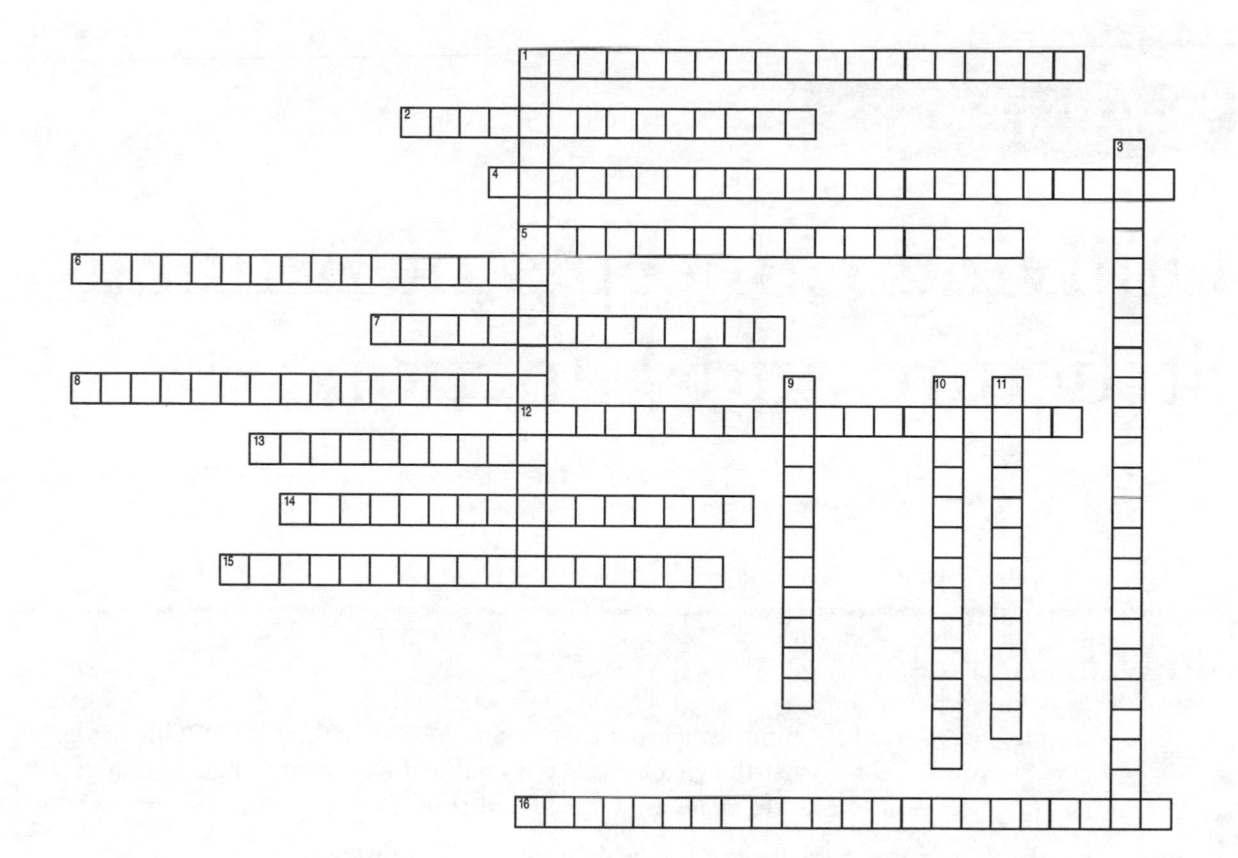
1
2
3
4
5
6
7
8
9
10
11
12
13
14
15
16

CHAPTER 21

Unifying Concepts of Animal Structure and Function

Studying Advice

a. Before reading, flip through the chapter and look over the figures. This will give you a feel for what the chapter is about and help you identify figures that you will want to refer to again as you read and review.

b. This chapter relates to many activities that occur inside your body. Pause occasionally while reading to reflect on those events that you have noticed going on inside your body.

c. Complete the organizing tables as you read to help define your studies.

Student Media

Biology and Society on the Web

Learn why someone would bother to study sweat.

Activities

21A Correlating Structure and Function of Cells
21B The Levels of Life Card Game
21C Overview of Animal Tissues
21D Epithelial Tissue
21E Connective Tissue
21F Muscle Tissue
21G Nervous Tissue
21H Regulation: Negative and Positive Feedback
21I Structure of the Human Excretory System
21J Nephron Function
21K Control of Water Reabsorption

Case Studies in the Process of Science

How Does Temperature Affect Metabolic Rate in Daphnia?

What Affects Urine Production?

Evolution Connection

Examine what is known about the evolution of whales.

Organizing Tables

Compare the definitions of the following sets of terms.

TABLE 21.1

Negative Feedback	Positive Feedback
Endotherm:	Ectotherm:
Anatomy:	Physiology:
Open system:	Closed system:
Osmoconformers:	Osmoregulators:
Reabsorption:	Secretion:

In Table 21.2, define each of the following types of tissues and provide at least two examples of each.

TABLE 21.2

Tissue Type	Definition/Key Features	Examples
Epithelia		
Connective		
Muscular		
Nervous		

Describe the main events of the excretory functions performed by the kidneys.

TABLE 21.3

Step	Events
Filtration	
Reabsorption	
Secretion	
Excretion	

Content Quiz

Directions: Identify the *one* best answer for the multiple-choice questions. For true/false questions, determine if the statement is true or false. If false, change the underlined word(s) to make the statement true. Finally, add the correct word(s) to the fill-in-the-blank questions to make the statements true.

Biology and Society: Keeping Cool

1. Which one of the following will *not* help to cool down the body of an exercising athlete?
   A. lowered blood pressure
   B. widened surface blood vessels
   C. panting
   D. evaporation of sweat

2. **True or False?** Exercising in a hot environment can lead to intense sweating, dehydration, an eventual drop in blood pressure, and then fainting in a condition called heat stroke. ____________________

3. Under extreme conditions of intense heat, a person may suffer from ____________________, in which the body temperature rises dangerously high, the skin becomes hot and dry, and organs start to fail.

The Structural Organization of Animals

Form Fits Function

4. **True or False?** The form/function relationship of biological equipment is a result of natural selection. ____________________

5. The study of the *structure* of an organism and its parts defines ____________________ and the study of the *function* of an organism's structural equipment defines ____________________.

Tissues

Matching: Match the tissue on the left to its description on the right.

____ 6.	muscular tissue	A. covers the surface of the body and lines organs and cavities within the body; also forms glands
____ 7.	epithelial tissue	B. found in the brain and spinal cord; cells in this tissue carry signals rapidly over long distances
____ 8.	nervous tissue	C. a sparse population of cells scattered through an extensive extracellular matrix
____ 9.	connective tissue	D. cells have specialized proteins arranged into a structure that contracts when the cell is stimulated

10. Which one of the following is a connective tissue with a matrix that is liquid instead of solid?
    A. adipose tissue
    B. blood
    C. cartilage
    D. bone
    E. fibrous connective tissue

11. Which one of the following is an involuntary muscle tissue found in the walls of the digestive tract and blood vessels?
    A. smooth muscle
    B. cardiac muscle
    C. skeletal muscle
    D. fibrous muscle

12. **True or False?** The basic unit of nervous tissue is the spinal cord.

    ____

13. The most widespread connective tissue in the vertebrate body is

    ____________________.

Organs and Organ Systems

14. Which one of the following is *not* a body organ?
    A. brain
    B. blood
    C. stomach
    D. liver
    E. lungs

15. **True or False?** All of the body's organ systems play a role in homeostasis.

    ____

16. Teams of organs that work together to perform a vital bodily function define

    ____________________.

Exchanges with the External Environment

Body Size and Shape

17. Which one of the following statements is *false*?
    A. In an animal body, every living cell must be bathed in water.
    B. Exchange with the environment is easier for multicellular animals than single-celled organisms.
    C. Simple body forms, such as two-layered sacs and flat shapes, do not allow much complexity in internal organization.
    D. The epithelium of the human lungs has a very large total surface area.
    E. The digestive, respiratory, and excretory systems exchange materials with the external environment.

18. **True or False?** Every organism is a closed system because all life continuously exchanges chemicals and energy with its surroundings. ____________________

19. The cells of more complex animals indirectly exchange materials with the environment by connecting exchange surfaces with body cells through a(n) ____________________ system.

Regulating the Internal Environment

Homeostasis

20. Which one of the following is the best example of homeostasis?
    A. the elevation of hormones in the blood during puberty
    B. the elevation of hormones in the blood during pregnancy
    C. an increase in body temperature in response to a bacterial infection
    D. the water content of your cells stays about the same no matter what you drink

21. **True or False?** Homeostasis in the human body permits small changes to occur.
    ____

22. The body's tendency to maintain relatively constant conditions in the internal environment despite changes in the external environment defines ____________________.

Negative and Positive Feedback

23. The part of a negative feedback mechanism in a household furnace that monitors temperature and switches the heater on and off is the:
    A. control center.
    B. end zone.
    C. set point.
    D. detector.
    E. heater.

24. Which of the following represent(s) a positive feedback mechanism?
    A. a healing wound itches, then you scratch it, only to injure the skin again
    B. shivering to warm up the body and then stopping after you warm up
    C. eating when you are hungry and stopping when you are full
    D. A and B are examples of a positive feedback mechanism.
    E. B and C are examples of a positive feedback mechanism.
    F. A, B, and C are examples of a positive feedback mechanism.

25. **True or False?** Negative feedback occurs when the results of some process <u>promote</u> that very process. ____________________

26. In ____________________ feedback, the results of some process inhibit that very process.

Thermoregulation

27. Which one of the following represent behavioral adaptations that assist thermoregulation during cold winter periods?
    A. growing thick fur in the winter
    B. accumulating a thick layer of fat under the skin
    C. increasing the metabolic rate
    D. migrating to suitable climates
    E. None of the above are behavioral adaptations that assist thermoregulation during cold winter periods.

28. Which one of the following does *not* occur when we develop a fever due to a bacterial infection?
    A. Bacteria release chemicals called pyrogens.
    B. Pyrogens travel through the bloodstream to the brain's control center.
    C. Pyrogens stimulate the control center to raise the body's internal temperature.
    D. A mild fever develops, which can discourage bacterial growth, promote phagocytosis, and speed repair of damaged tissues.
    E. All of the above occur when we develop a fever due to a bacterial infection.

29. **True or False?** Animals that derive the majority of their body heat from their metabolism are called <u>ectotherms</u>. ____________________

30. The maintenance of internal body temperature within defined limits defines ____________________.

Osmoregulation

31. Living cells depend on a precise balance of ________________ to maintain the integrity of cell membranes and to provide an appropriate environment for metabolic reactions.
    A. salts and minerals
    B. chemical and solar energy
    C. heat and cold
    D. sugars and carbohydrates
    E. water and solutes

32. **True or False?** Most marine invertebrates are osmoconformers.

    ____

33. **True or False?** All land animals are osmoconformers. ________________

34. The control of the gain or loss of water and dissolved solutes in animals defines

    ________________.

Homeostasis in Action: The Kidneys

Matching: Match the structure in the left column to its description on the right.

____	35. nephron	A. urine leaves the kidneys in this
____	36. ureter	B. urine is stored in this
____	37. bladder	C. the functional unit of the kidneys
____	38. urethra	D. urine leaves the bladder in this

39. The excretory system plays a central role in homeostasis, forming and excreting ________________ while regulating the amount of ________________ in body fluids.
    A. urine, water and salts
    B. blood, sugars
    C. feces, blood
    D. urine, sugars
    E. sweat, water and sugars

40. Which one of the following statements is *false*?
    A. The spleen is the main processing center in the human excretory system.
    B. The excretory system plays a central role in homeostasis.
    C. Every day, the total volume of blood in the body passes into the kidneys hundreds of times.
    D. The body uses hormones to control the internal concentration of water and dissolved molecules.
    E. Humans with one functioning kidney can lead a normal life.

41. Which of the following is the correct sequence of excretory functions performed by the kidneys?
   A. reabsorption, filtration, secretion, excretion
   B. secretion, excretion, reabsorption, filtration
   C. secretion, filtration, excretion, reabsorption
   D. filtration, reabsorption, secretion, excretion
   E. filtration, reabsorption, excretion, secretion

42. **True or False?** The number of kidneys available for transplant is <u>not enough</u> to meet the current demand for transplants. ____________________

43. A person who suffers from kidney failure can be treated by using a blood filtration machine in a process called ____________________.

Evolution Connection: How Physical Laws Constrain Animal Form

44. The size of single cells is limited by:
   A. physical law.
   B. the size of the other organisms living in the region.
   C. the amount of available sunlight.
   D. the amount of salt in the surrounding water.
   E. the amount of sugar available in the environment.

45. **True or False?** The independent development of similar forms defines <u>emergent</u> evolution. ____________________

46. The ____________________ body shape, tapered at both ends, is common in fast-swimming aquatic animals.

Word Roots

endo = inside (*endotherms:* organisms whose bodies are warmed by heat generated by metabolism; this heat is used to maintain a body temperature higher than that of the external environment)

fibro = a fiber (*fibrous connective tissue:* a dense tissue with large numbers of collagenous fibers organized into parallel bundles)

homeo = same; **stasis** = standing (*homeostasis:* the steady-state physiological condition of the body)

inter = between (*interstitial fluid:* the fluid that fills the space between cells)

Key Terms

adipose tissue
anatomy
blood
bone
bladder
cardiac muscle
cartilage
connective tissues
control center
convergent evolution
dialysis
endotherms
epithelium
epithelial tissue
excretion
exotherms
fever
fibrous connective tissue
filtrate
filtration
homeostasis
interstitial fluid
loose connective tissue
muscle tissue
negative feedback
nephron
nervous tissue
neuron
open system
organ
organ systems
osmoconformers
osmoregulation
osmoregulators
physiology
positive feedback
reabsorption
secretion
set point
skeletal muscle
smooth muscle
thermoregulation
tissue
tubules
ureter
urethra
urine

Crossword Puzzle

Use the Key Terms list from this chapter to fill in the crossword puzzle.

ACROSS

2. in the vertebrate kidney, the extraction of water and small solutes, including metabolic wastes, from the blood by the nephrons
3. a control mechanism in which the products of a process stimulate the process that produced them
7. the fundamental structural and functional unit of the nervous system
9. the tubular excretory unit and associated blood vessels of the vertebrate kidney
11. a duct that conveys urine from the kidney to the urinary bladder
13. the study of the functions of an organism
16. in the vertebrate kidney, the discharge of wastes from the blood into the filtrate from the nephron tubules
17. The study of the structure of an organism
19. an aqueous solution that surrounds body cells and through which materials pass back and forth between the blood and the body tissues
20. striated muscle that forms the contractile tissue of the heart
21. the steady-state physiological condition of the body
22. a type of connective tissue with a fluid matrix called plasma in which blood cells are suspended
23. It binds epithelia to underlying tissues and nfunctions as packing material, holding organs in place
24. a type of connective tissue, consisting of living cells held in a rigid matrix of collagen fibers embedded in calcium salts
26. in the vertebrate kidney, the reclaiming of water and valuable solutes from the filtrate
27. a type of connective tissue, consisting of living cells embedded in a rubbery matrix with collagenous fibers
30. separation and disposal of metabolic wastes from the blood by mechanical means
31. the control of the gain and loss of water and dissolved solutes in a organism
32. tissue consisting of cells held in an abundant extracellular matrix, which they produce
33. a control mechanism in which a chemical reaction, metabolic pathway, or hormone-secreting gland is inhibited by the products of the reaction, pathway, or gland

DOWN

1. species from different evolutionary lineages come to resemble each other as a result of living in very similar environments
4. a group of organs that work together in performing vital body functions
5. a dense tissue with large numbers of collagenous fibers organized into parallel bundles
6. the disposal of nitrogen-containing metabolic wastes
8. an organism whose body fluids have a solute concentration different from that of its environment and that must use energy in controlling water loss or gain
10. an organism whose body fluids have a solute concentration equal to that of its surroundings
12. animals that use metabolic energy to maintain a constant body temperature
14. a structure consisting of several tissues adapted as a group to perform specific functions
15. a sheet of tightly packed cells lining organs and cavities
17. a type of connective tissue whose cells contain fat
18. tissue consisting of long muscle cells that are capable of contracting when stimulated by nerve impulses
25. fluid extracted by the excretory system from the blood or body cavity
28. a duct that conveys urine from the urinary bladder to the outside
29. a cooperative unit of many similar cells that perform a specific function within a multicellular organism

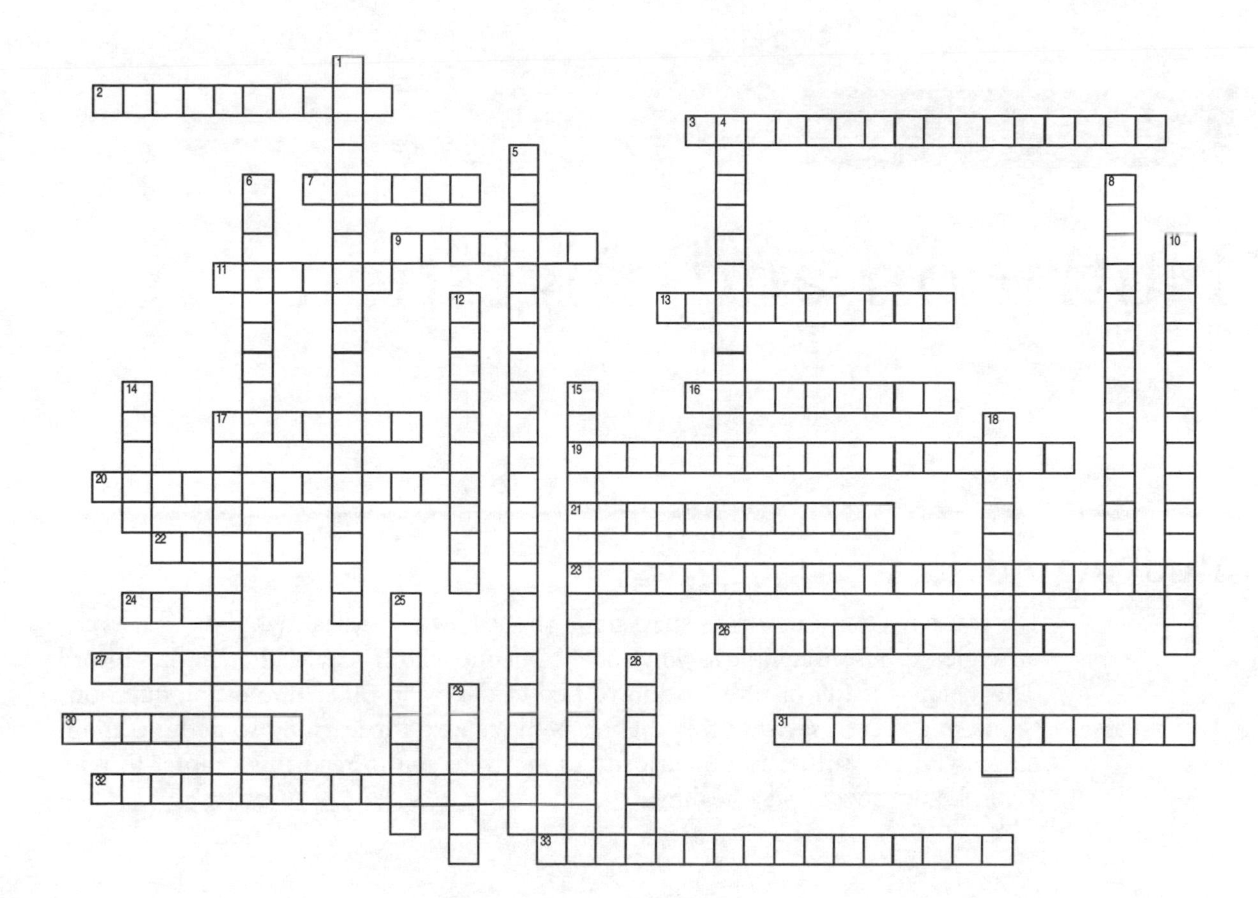
1
2
3
4
5
6
7
8
9
10
11
12
13
14
15
16
17
18
19
20
21
22
23
24
25
26
27
28
29
30
31
32
33

CHAPTER 22

Nutrition and Digestion

Studying Advice

Do you pay attention to the latest news about how to be healthy? Have you ever struggled to lose weight? Do you know someone who has battled an eating disorder? This chapter is full of information related to these and other interesting questions. Some might even say that this chapter is the reward for learning something about chemistry! Use your natural curiosity to motivate you to read this chapter to understand better how to stay healthy.

Student Media

Biology and Society on the Web

Learn how eating disorders are diagnosed and treated.

Activities

22A How Animals Eat Food

22B Human Digestive System

22C Analyzing Food Labels

22D Case Studies of Nutritional Disorders

Case Study in the Process of Science

What Role Does Salivary Amylase Play in Digestion?

Evolution Connection on the Web

See how study of the Pima is contributing to our knowledge of the relationship between diet, health, and genetics.

Organizing Tables

Compare the definitions and list examples for each of the following types of diets.

TABLE 22.1

Type of Diet	Definition	Examples of Organisms Using This Type of Diet
Herbivore		
Carnivore		
Omnivore		

Compare the main events and locations of each of the four stages of food processing in Table 22.2 below.

TABLE 22.2

Stage	Description of This Process	Digestive Compartment(s) Where This Occurs
Ingestion		
Digestion		
Absorption		
Elimination		

Compare the different types of essential human nutrients in the table below.

TABLE 22.3

Type of Nutrient	Description/Definition	Number that Are Essential to Humans
Essential amino acids		
Vitamins		
Minerals		

Content Quiz

Directions: Identify the *one* best answer for the multiple-choice questions. For true/false questions, determine if the statement is true or false. If false, change the underlined word(s) to make the statement true. Finally, add the correct word(s) to the fill-in-the-blank questions to make the statements true.

Biology and Society: Eating Disorders

1. Which one of the following statements is *false*?
   A. Bulimia involves binge eating followed by purging.
   B. A person with bulimia is usually emaciated like an anorexic person.
   C. The causes of bulimia are unknown.
   D. Bulimia can result in serious health problems.
   E. Bulimia is associated with an extreme concern with body weight and shape.

2. **True or False?** We do not know what causes anorexia and bulimia.

____

3. The condition associated with self-starvation due to an intense fear of gaining weight is called _______________.

Overview of Animal Nutrition

Matching: Match the term on the left to its best definition on the right.

____	4. elimination	A.	eating
____	5. absorption	B.	the breakdown of food to relatively small nutrient molecules
____	6. ingestion	C.	the uptake of the small nutrient molecules by the body's cells
____	7. digestion	D.	the disposal of undigested materials from meals

8. Which one of the following ingests plants and animals?
   A. herbivore
   B. omnivore
   C. carnivore
   D. endovore
   E. maximore

9. Digestive sacs, with a single opening that functions as both the entrance for food and the exit for undigested wastes, are found in
   A. jellyfish and their relatives.
   B. birds.
   C. many single-celled organisms and some multicellular organisms like sponges.
   D. herbivores, but not carnivores.

10. **True or False?** Mechanical digestion is the chemical breakdown of food by digestive enzymes. _______________

11. In the hydrolysis reaction, polymers are broken down into _______________.

12. Enzymes that catalyze digestive hydrolysis reactions are called _______________.

13. The simplest digestive compartments are _______________, intracellular membrane-bound organelles filled with digestive enzymes.

A Tour of the Human Digestive System

Matching: Match each item on the left to its best description on the right.

_____ 14. pepsin	A. the cause of most stomach ulcers
_____ 15. esophagus	B. a large sac that can store enough food to sustain us for hours
_____ 16. *Helicobacter pylori*	C. the first section of the small intestine
_____ 17. gastric juice	D. an intersection of the food and breathing pathways
_____ 18. small intestine	E. where bile is produced
_____ 19. alimentary canal	F. a muscular tube that connects the pharynx to the stomach
_____ 20. duodenum	G. site of water resorption and the formation of feces
_____ 21. pancreas	H. a digestive fluid secreted by the cells lining the stomach
_____ 22. stomach	I. an enzyme that breaks pieces of protein into polypeptides
_____ 23. liver	J. another name for the entire digestive tube
_____ 24. large intestine	K. the longest part of the alimentary canal
_____ 25. pharynx	L. a large gland that secretes pancreatic juice into the duodenum
_____ 26. microvilli	M. where bile is stored
_____ 27. gallbladder	N. microscopic projections on intestinal cells

28. What keeps the stomach from just eating itself from the inside out?
    A. A protective coating of mucus limits the contact between the cells lining the stomach and the gastric juices.
    B. Nerves and hormones limit the secretion of gastric juice to periods when food is in the stomach.
    C. About once every three days new cells must replace the stomach lining that still gets damaged.
    D. All of the above are correct.
    E. None of the above are correct.

29. Today, most ulcers are effectively cured by
    A. antacids.
    B. Pepto Bismol.
    C. antibiotics.
    D. milk.
    E. vitamins.

30. **True or False?** Occasional backflow of acid chyme into the esophagus causes ulcers. ________________

31. The enzyme salivary ________________ hydrolyzes starch.

32. Food is moved through the esophagus and intestines by ________________, rhythmic waves of muscular contractions that squeeze the food ball along the esophagus.

33. Rings of muscle that control the movement of materials into and out of the stomach are called ________________.

34. The last 6 inches of the large intestine is called the ________________, where feces are stored until they can be eliminated.

Human Nutritional Requirements

35. Cellular respiration uses ________________ to break down sugar and other food molecules, generating many molecules of ________________ for cells to use as a direct source of energy.
    A. oxygen, ATP
    B. carbon dioxide, oxygen
    C. ATP, carbon dioxide
    D. water, oxygen
    E. water, ATP

36. **True or False?** Dietary calories listed on food labels are actually kilocalories.
    ____

37. **True or False?** Most animal proteins are deficient in one or more of the essential amino acids. ________________

38. **True or False?** Overdoses of certain vitamins (such as A, D, and K) can be harmful. ________________

39. **True or False?** Minerals are organic substances required in a healthy diet.
    ____

40. The amount of energy it takes just to maintain your basic body support functions is called the ________________.

41. Essential amino acids must be obtained in the ________________.

42. Most vitamins function as assistants to ____________________ in catalyzing metabolic reactions.

43. The minimal standards established by nutritionists for preventing nutrient deficiencies are called the ____________________.

44. Cells use cellular respiration to extract ____________________ stored in the organic molecules of food and use it to do work.

45. About 60% of our food energy is lost as ____________________ that dissipates to the environment.

Nutritional Disorders

46. Which one of the following statements is *false*?
    A. In humans, obesity increases the risk of heart attack, diabetes, and several other diseases.
    B. The best way to maintain a healthy weight is to exercise and eat a balanced diet.
    C. The most reliable sources of essential amino acids are animal products.
    D. Most victims of protein deficiency are adults.
    E. Eating disorders can also cause undernutrition.

47. **True or False?** <u>Malnutrition</u> is a deficiency in calories. ____________________

48. In Africa, the syndrome named kwashiorkor refers to a disease caused by a deficiency of ____________________.

Evolution Connection: Fat Cravings

49. Fat cravings likely evolved in our human ancestors because they were always
    A. in danger of not finding enough water.
    B. in danger of starvation.
    C. running short of protein.
    D. suffering from obesity.
    E. reproducing.

50. **True or False?** The majority of Americans consume too many <u>fatty</u> foods, which contributes to obesity. ____________________

51. For our ancestors, foods that were ____________________ or sweet were probably hard to come by.

Word Roots

herb = grass; **vora** = eat (*herbivore:* a heterotrophic animal that eats plants)

omni = all (*omnivore:* a heterotrophic animal that consumes both meat and plant material)

peri = around; **stalsis** = a constriction (*peristalsis:* rhythmic waves of contraction of smooth muscle that push food along the digestive tract)

Key Terms

absorption
acid chyme
alimentary canal
anus
appendix
basal metabolic rate (BMR)
bile
calorie
carnivores
chemical digestion
digestion
digestive sacs
digestive tubes
duodenum
elimination
esophagus
essential amino acids
essential nutrients
feces
food vacuoles
gallbladder
gastric juice
herbivores
hydrolases
ileum
ingestion
jejunum
kilocalorie
large intestine
liver
malnutrition
mechanical digestion
metabolic rate
minerals
mouth (oral cavity)
obesity
omnivores
pancreas
pepsin
peristalsis
pharynx
Recommended Daily Allowances (RDAs)
rectum
salivary amylase
small intestine
sphincters
stomach
tongue
undernutrition
vitamins

Crossword Puzzle

Use the Key Terms list from this chapter to fill in the crossword puzzle.

ACROSS

1. a chemical element that is required for a plant to grow from a seed and complete the life cycle, producing another generation of seeds
6. the collection of fluids secreted by the epithelium lining the stomach
8. a mixture of recently swallowed food and gastric juice
11. the number of kilocalories a resting animal requires to fuel its essential body processes for a given time
14. a salivary gland enzyme that hydrolyzes starch
15. the amount of heat energy required to raise the temperature of 1 gram of water by 1 Celsius
16. the middle section of the small intestine involved in the absorption of nutrients and water
17. animals that eat both plants and animals
20. a digestive tract consisting of a tube running between a mouth and an anus
22. the opening through which undigested materials are expelled
23. heterotrophic animals that eats plants
24. chemical elements other than carbon, hydrogen, oxygen, or nitrogen that an organism requires for proper body functioning
25. the act of eating; the first main stage of food processing
26. the first portion of the vertebrate small intestine after the stomach, where acid chyme from the stomach is mixed with bile and digestive enzymes
28. an enzyme present in gastric juice that begins the hydrolysis of proteins
29. the channel through which food passes in a digestive tract; usually receives food from the pharynx
31. rhythmic waves of contraction of smooth muscles
32. a thousand calories
33. the tubular portion of the vertebrate alimentary tract between the small intestine and the anus

DOWN

1. the passing of undigested material out of the digestive compartment
2. the terminal portion of the large intestine where the feces are stored until they are eliminated
3. a pouchlike organ in a digestive tract, which grinds and churns food and may store it temporarily
4. a solution of bile salts secreted by the liver, which emulsifies fats and aids in their digestion
5. membranous sacs formed by phagocytosis
7. the mechanical and chemical breakdown of food into molecules small enough for the body to absorb
8. the uptake of small nutrient molecules by an organism's own body
9. the waste of the digestive tract
10. the last of three sections of the small intestine primarily involved in the absorption of nutrients and water
12. the amino acids that an animal cannot synthesize itself and must obtain from food
13. an organ that stores bile and releases it as needed into the small intestine
18. organic molecules required in the diet in very small amounts
19. the longest section of the alimentary canal
21. animals that eat other animals
27. an opening through which food is taken into an animal's body
30. the organ in a digestive tract that receives food from the oral cavity

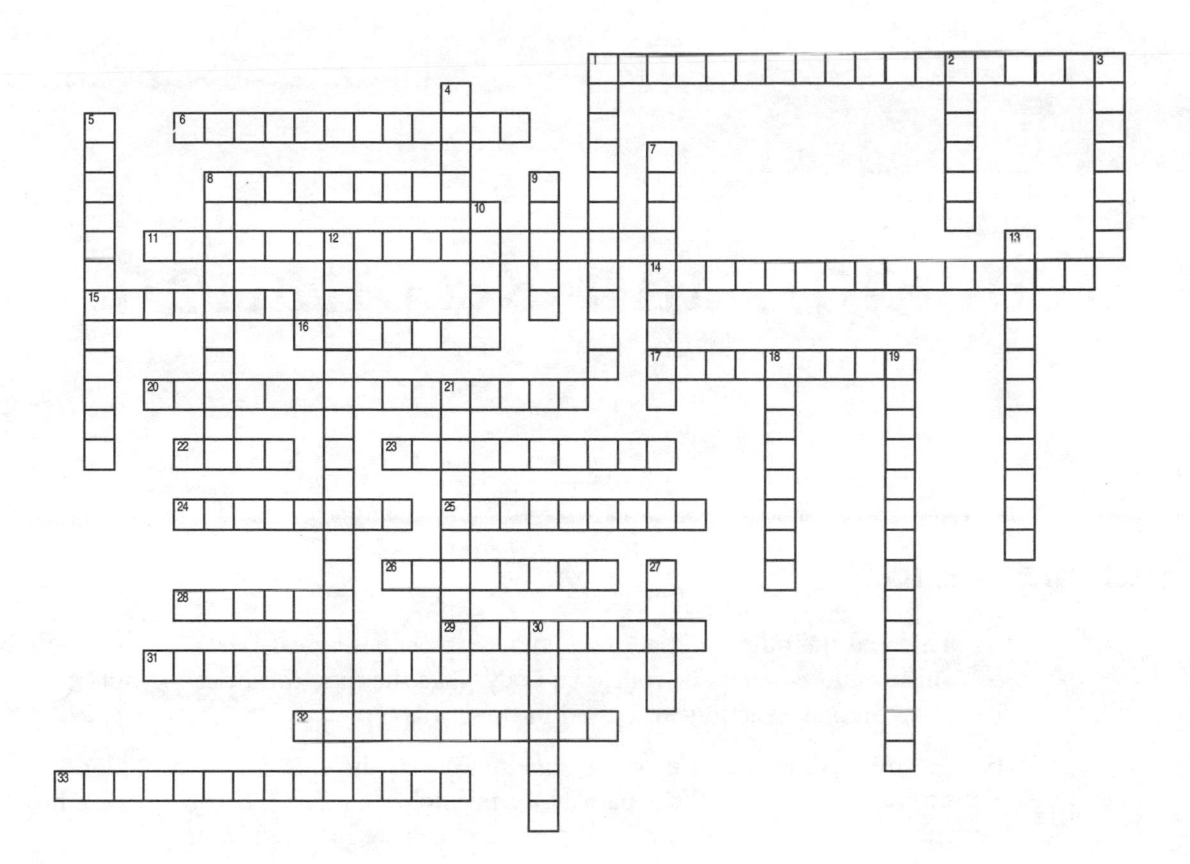

1
2
3
4
5
6
7
8
9
10
11
12
13
14
15
16
17
18
19
20
21
22
23
24
25
26
27
28
29
30
31
32
33

CHAPTER 23

Circulation and Respiration

Studying Advice

a. This and the other animal physiology chapters have great relevance to the mechanisms and wonders of your own body. Take the time to notice in yourself the many human reactions discussed in these chapters.

b. As you read, create your own simple flowcharts showing the path of blood through the body and the path of air through the lungs. Then keep these flowcharts handy to clarify events discussed elsewhere in the chapter.

Student Media

Biology and Society on the Web

Learn what to do if your cat or dog stops breathing.

Activities

23A Path of Blood Flow

23B Cardiovascular System Structure

23C Cardiovascular System Function

23D The Human Respiratory System

23E Transport of Respiratory Gases

Case Study in the Process of Science

How Is Cardiovascular Fitness Measured?

Evolution Connection

Explore the relationship between lungfishes and four-limbed animals.

Organizing Tables

Compare open and closed circulatory systems in the table below.

TABLE 23.1

	Open Circulatory System	**Closed Circulatory System**
The path of blood		
Phyla with this system		

Compare the functions of the four chambers in a human heart in Table 23.2 below.

TABLE 23.2

Heart Chamber	Receives Blood from	Sends Blood to
Right atrium		
Right ventricle		
Left atrium		
Left ventricle		

In Table 23.3 below, compare the structures and functions of blood vessels.

TABLE 23.3

Type of Blood Vessel	Structure of the Walls	Which Ones Have One-Way Valves?
Capillaries		
Arteries		
Veins		

Compare the structures and functions of the components of human blood using the Table 23.4 below.

TABLE 23.4

Component	Structure	Function(s)
Red blood cells		
White blood cells		
Platelets		

Compare the four main types of respiratory surfaces found in animals using the table below.

TABLE 23.5		
Type of Surface	**Structure**	**Examples of Animals with This System**
Skin		
Gills		
Tracheae		
Lungs		

Content Quiz

Directions: Identify the *one* best answer for the multiple-choice questions. For true/false questions, determine if the statement is true or false. If false, change the underlined word(s) to make the statement true. Finally, add the correct word(s) to the fill-in-the-blank questions to make the statements true.

Biology and Society: The ABCs of Saving Lives

1. Which one of the following statements about the ABCs of life is *false*?
   A. "A" stands for airway.
   B. "B" stands for blood.
   C. "C" stands for circulation.
   D. CPR is used to temporarily force blood through the body.
   E. The first priority is to establish an open airway.

2. **True or False?** Electric shocks and drugs can be used to restart breathing.

   ____

3. The ________________ system transports gases and other substances throughout the body.

Unifying Concepts of Animal Circulation

Open and Closed Circulatory Systems

4. Which one of the following is *not* a main component of a circulatory system?
   A. a vascular system
   B. a circulating fluid
   C. lungs or gills
   D. a central pump

5. Which one of the following is the functional center of the circulatory system?
   A. arteries
   B. capillary beds
   C. arterioles
   D. veins
   E. heart

6. **True or False?** Blood is confined to vessels in <u>a closed</u> circulatory system.

   ____

7. **True or False?** Metabolic wastes, such as <u>oxygen</u>, diffuse from cells into the circulatory system. ________________

8. The closed circulatory system in humans is called a(n) ________________ system.

The Human Cardiovascular System

The Path of Blood

9. Which one of the following represents the correct sequence of flow of blood through the human cardiovascular system?
   A. left ventricle → pulmonary veins → capillaries in the lungs → pulmonary arteries → right atrium
   B. left ventricle → pulmonary arteries → pulmonary veins → capillaries in the lungs → right atrium
   C. right ventricle → pulmonary veins → capillaries in the lungs → pulmonary arteries → right atrium
   D. right ventricle → pulmonary arteries → capillaries in the lungs → pulmonary veins → left atrium
   E. right atrium → right ventricle → left atrium → capillaries in the lungs → left ventricle

10. **True or False?** The systemic circuit carries blood from the heart to organs in the rest of the body. ____________________

11. Humans and other terrestrial vertebrates have a(n) ____________________ circulatory system with two distinct circuits of blood flow.

How the Heart Works

12. Which one of the following statements about the control of heart rate is *false*?
    A. The SA node generates electrical impulses in the right atrium.
    B. The signals from the SA node can be detected by an EKG.
    C. During a heart attack, the pacemaker is often unable to maintain normal rhythm.
    D. The SA node receives an impulse from the AV node.
    E. The pacemaker of the heart is composed of specialized muscle cells.

13. **True or False?** Blood in the heart moves only in one direction due to one-way valves. ____________________

14. **True or False?** Caffeine and epinephrine make the heart rate slow down.
    ____

15. The relaxation phase of the heart cycle is called ____________________.

16. A defect in one or more heart valves can lead to a hissing heart sound called a(n) ____________________.

Blood Vessels

17. Which one of the following statements is *false*?
    A. Veins, but not arteries, have one-way valves.
    B. Veins carry blood that is under very little blood pressure.
    C. Veins have the thinnest walls of any blood vessel.
    D. Arteries have layers of elastic connective tissue and smooth muscle.
    E. Blood pressure is the main force driving the blood from the heart through blood vessels.

18. Blood is primarily moved through veins back to the heart by
    A. the contraction of surrounding skeletal muscles and one-way valves in veins.
    B. smooth muscle contractions in the walls of capillaries and veins.
    C. the blood pressure created by heart contractions.
    D. the pull of gravity on the blood.
    E. the production of blood throughout the body.

19. **True or False?** Arteries carry blood away from the heart. ________________

20. **True or False?** During strenuous exercise, blood is diverted from the digestive tract to skeletal muscles and the skin. ________________

21. Persistent systolic blood pressure above 140 and diastolic pressure above 90 defines ________________.

Blood

22. Which one of the following is *not* true about red blood cells? Red blood cells
    A. are the most numerous type of blood cell.
    B. do not have nuclei and other organelles.
    C. are adapted to transport oxygen.
    D. have carbohydrates on their surface that determine the blood type.
    E. contain fibrinogen, an iron-containing molecule that transports oxygen.

23. Which one of the following is *not* true about platelets? Platelets
    A. are the primary carriers of carbon dioxide in the blood.
    B. adhere to damaged tissue to form a sticky cluster that seals minor breaks.
    C. release clotting factors that convert fibrinogen into fibrin.
    D. help to plug leaks by forming clots.
    E. are bits of cytoplasm pinched off from larger cells in the bone marrow.

24. Compared to red blood cells, white blood cells are
    A. smaller.
    B. have hemoglobin.
    C. lack nuclei.
    D. are about 1,000 times more abundant.
    E. are found outside the circulatory system.

25. **True or False?** Leukemia is cancer of leukocytes. ________________

26. **True or False?** An abnormally low amount of hemoglobin or a low number of red blood cells is called anemia. ________________

27. A(n) ________________ is a blood clot that forms in the absence of injury.

28. About half of the blood volume is a watery fluid called ________________.

29. Blood doping and abusing erythropoietin are artificial ways to increase athletic stamina by increasing the number of ________________ cells.

The Role of the Cardiovascular System in Homeostasis; Cardiovascular Disease

30. Which one of the following is *not* a risk factor for developing cardiovascular disease?
    A. a diet high in cholesterol, trans fat, and saturated fat
    B. a diet high in fruits and vegetables
    C. smoking
    D. lack of exercise

31. **True or False?** The leading cause of death in the United States is stroke.

    ____

32. **True or False?** The complete blockage of a coronary artery will likely lead to a(n) stroke. ____________________

33. The circulatory system contributes to homeostasis by exchanging nutrients and wastes with ____________________.

34. The chronic cardiovascular disease that results in a narrowing of blood vessels is called ____________________.

Unifying Concepts of Animal Respiration

The Structure and Function of Respiratory Surfaces

35. Which one of the following respiratory systems uses a system of branching tubes to transport oxygen to virtually every cell in the body?
    A. lungs
    B. tracheae
    C. gills
    D. body surface

36. **True or False?** In general, air holds less oxygen than water.

    ____

37. Insects breathe using ____________________.

38. Extensions, or outfoldings, of the body surface typically used by aquatic organisms are called ____________________.

39. Cellular respiration uses ____________________ and glucose to produce energy-carrying ____________________ molecules.

The Human Respiratory System

Structure and Function of the Human Respiratory System

40. As air passes into our respiratory system, it moves from:
    A. pharynx → larynx → trachea → bronchi → bronchioles → alveoli.
    B. pharynx → trachea → larynx → bronchioles → bronchi → alveoli.
    C. larynx → pharynx → trachea → bronchioles → bronchi → alveoli.
    D. pharynx → larynx → trachea → bronchioles → alveoli → bronchi.
    E. trachea → pharynx → larynx → bronchioles → alveoli → bronchi.

41. Which one of the following is *not* considered one of the three phases of gas exchange in humans?
    A. transport of O_2 from the extensively branched lungs to the rest of the body via the circulatory system
    B. production of ATP by aerobic metabolism consuming oxygen and producing carbon dioxide
    C. breathing
    D. the transport of O_2 from the extensively branched lungs to the rest of the body via the circulatory system

42. **True or False?** The inner surface of each bronchus is the respiratory surface.

    ____

43. Humans vocal sounds are produced by flexing muscles in the voice box as air rushes by, which stretch the ____________________ and make them vibrate.

Taking a Breath; The Role of Hemoglobin in Gas Transport; How Smoking Affects the Lungs

44. Which one of the following is *not* true about blood?
    A. Each molecule of hemoglobin consists of four polypeptide chains, each with a heme at the center of which is an atom of iron.
    B. Most of the oxygen in blood is carried in hemoglobin molecules.
    C. Oxygen readily dissolves in blood.
    D. A shortage of iron causes a decrease in the rate of synthesis of hemoglobin.
    E. Iron deficiency is the most common cause of anemia.

45. Which one of the following statements about smoking is *false*?
    A. One of the worst sources of airborne pollutants is tobacco smoke.
    B. The carbon particles in cigarette smoke contain over 4,000 different chemicals.
    C. Tobacco smoke kills about 430,000 Americans every year.
    D. Smoking causes emphysema.
    E. Secondhand cigarette smoke is not considered a health hazard.

46. **True or False?** Hiccups are caused by sudden contractions of the <u>lungs</u>.

____

47. Expanding the chest cavity to inhale air is known as ____________________ pressure breathing.

48. Automatic control centers in the brain regulate breathing rate in response to levels of ____________________ in the blood.

Evolution Connection: The Move onto Land

49. Which one of the following was an adaptation that helped animals meet the high-oxygen demands of living on land?
   A. gill breathing evolved into lung breathing
   B. gills became more elaborate and highly branched
   C. fish fins increasingly became leglike, able to support the bodies on land
   D. the skin of the animals became scaly to resist drying
   E. the animals used heat from metabolic reactions to maintain stable, warm internal temperatures

50. **True or False?** Some modern lungfish inhabit stagnant ponds and swamps that have <u>low</u> levels of oxygen. ____________________

51. Survival and propagation on land calls for a ____________________ metabolic rate and faster delivery of ____________________ to support cellular respiration.

Word Roots

alveol = a cavity (*alveoli:* one of the dead-end, multilobed air sacs that constitute the gas exchange surface of the lungs)

atrio = a vestibule; **ventriculo** = ventricle (*atrioventricular node:* a region of specialized muscle tissue between the right atrium and right ventricle; it generates electrical impulses that primarily cause the ventricles to contract)

cardi = heart; **vascula** = a little vessel (*cardiovascular system:* the closed circulatory system characteristic of vertebrates)

fibrino = a fiber; **gen** = produce (*fibrinogen:* the inactive form of the plasma protein that is converted to the active form *fibrin,* which aggregates into threads that form the framework of a blood clot)

Key Terms

- alveoli
- anemia
- arterioles
- arteries
- atherosclerosis
- atrium
- AV (atrioventricular) node
- blood pressure
- breathing
- bronchi
- bronchioles
- capillary beds
- cardiac cycle
- cardiovascular disease
- cardiovascular system
- circulatory system
- closed circulatory system
- coronary arteries
- diaphragm
- diastole
- double circulation
- electrocardiogram (ECG or EKG)
- embolus
- emphysema
- erythrocytes
- fibrin
- fibrinogen
- gills
- heart attack
- heart murmur
- heart rate
- hemoglobin
- hypertension
- larynx
- leukemia
- leukocytes
- lungs
- negative pressure breathing
- open circulatory system
- pacemaker
- pharynx
- plasma
- platelets
- pulmonary circuit
- pulse
- red blood cells
- respiratory surface
- respiratory system
- SA (sinoatrial) node
- systemic circuit
- systole
- thrombocytes
- thrombus
- trachea
- veins
- ventricle
- venules
- vocal cords
- white blood cells

Crossword Puzzle

Use the Key Terms list from this chapter to fill in the crossword puzzle.

ACROSS

9. the sheet of muscle that expands the chest cavity and reduces it; it separates the chest cavity from the abdominal cavity
10. organ system that transports materials like nutrients, oxygen, and hormones to body cells and transports carbon dioxide and other wastes from body cells
11. an iron containing protein in red blood cells that reversibly binds oxygen and transports it to body tissues
13. the stage of the heart cycle in which the heart muscle is relaxed, allowing the chambers to fill with blood
16. the SA node
18. death of cardiac muscle cells and the resulting failure of the heart to deliver enough blood to the body
19. vessels that carry blood away from the heart to other parts of the body
20. a circulatory system in which blood is confined to vessels and is kept separate from the interstitial fluid
21. white blood cells; typically functions in immunity, such as phagocytosis or antibody production
22. the force that blood exerts against the walls of blood vessels
24. the organ system that functions in exchanging gases with the environment
26. a heart chamber that receives blood from the veins
27. the organ in a digestive tract that receives food from the oral cavity
28. a heart chamber that pumps blood out of a heart
29. a record of the electrical impulse that travel through cardiac muscle during the heart cycle
30. a circulatory system in which blood is pumped through open ended vessels and out among the body cells

DOWN

1. internal sacs, lined with moist epithelium, where gases are exchanged between inhaled air and the blood
2. pieces of membrane-enclosed cytoplasm from a large cell in the bone marrow of a mammal; blood clotting element
3. the plasma protein that is activated to form a clot when a blood vessel is injured
4. the liquid matrix of the blood in which the blood cells are suspended
5. the windpipe
6. a breathing system in which air is pulled into the lungs
7. networks of capillaries that infiltrate every organ and tissue in the body
8. the rate of heart contraction
10. a closed circulatory system with a heart and branching network of arteries, capillaries, and veins
12. a pair of breathing tubes that branch from the trachea into the lungs
14. the contraction stage of the heart cycle, when the hear chambers actively pump blood
15. in animals, vessels that returns blood to the heart
17. one of two main blood circuits in terrestrial vertebrates; conveys blood between the heart and the lungs
19. millions of tiny sacs within the vertebrate lungs where gas exchange occurs
23. a respiratory disease caused by smoking, in which the alveoli become brittle and rupture, reducing the lungs' capacity for gas exchange
25. the voicebox, containing the vocal cords

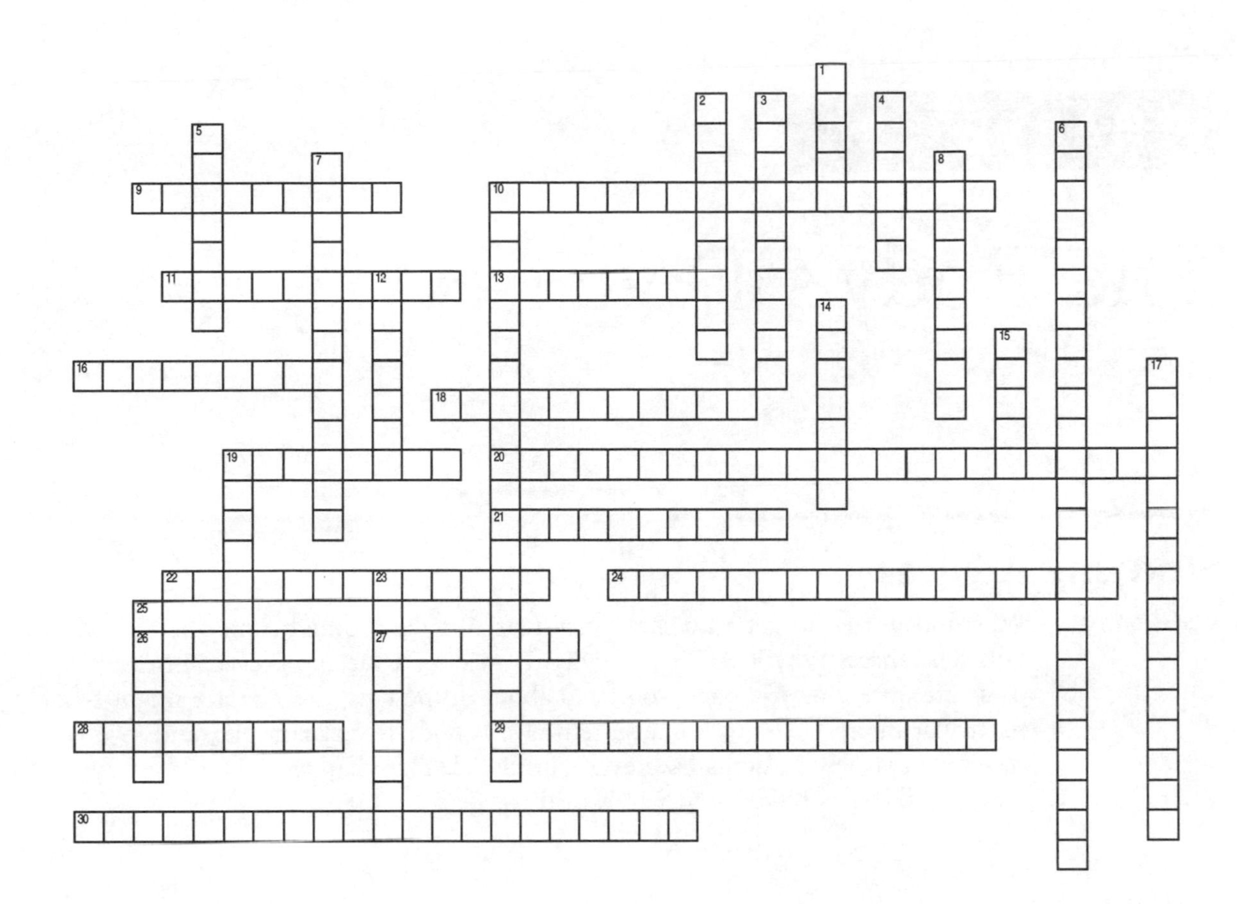
1
2
3
4
5
6
7
8
9
10
11
12
13
14
15
16
17
18
19
20
21
22
23
24
25
26
27
28
29
30

CHAPTER 24

The Body's Defenses

Studying Advice

Why do we need to get a flu shot every year? Why are some people allergic to certain substances? Why is AIDS so deadly? How does a vaccine prevent disease? If these are questions you have wondered about in the past, this chapter might be of particular interest. The immune system of our body is very complex and involves reactions that few students have ever considered. This chapter is your chance to get a peak inside this hidden but vital world!

Student Media

Biology and Society on the Web

Learn how HIV affects immune system function.

Activities

24A Immune System Responses

24B HIV Reproductive Cycle

Case Studies in the Process of Science

Why Do AIDS Rates Differ Across the U.S.?

What Causes Infections in AIDS Patients?

Evolution Connection on the Web

Read the latest news for the CDC about SARS, another viral threat to health.

Organizing Table

Describe the development and functions of the following cells of the immune system in the table below.

TABLE 24.1

Cell Type	How It Develops	How It Functions
B cell		
T cell		
Effector cells		
Memory cells		
Helper T cells		
Cytotoxic T cells		

Content Quiz

Directions: Identify the *one* best answer for the multiple-choice questions. For true/false questions, determine if the statement is true or false. If false, change the underlined word(s) to make the statement true. Finally, add the correct word(s) to the fill-in-the-blank questions to make the statements true.

Biology and Society: The Discovery of AIDS

1. AIDS is caused by
   A. a bacterium.
   B. a retrovirus.
   C. a nanovirus.
   D. yeast.
   E. mycorrhizae.

2. **True or False?** HIV selectively destroys cytotoxic T cells. ____________________

3. **True or False?** A vaccine to prevent AIDS has not been produced.
   ____

4. Before the discovery of AIDS, Kaposi's sarcoma and pneumocystis pneumonia almost exclusively affected people with severely depleted ____________________ systems.

Nonspecific Defenses

External Barriers

5. What happens to particles that are trapped in mucus lining the respiratory tract?
   A. Mucous cells digest the particles.
   B. Acids released by other cells lining the respiratory tract destroy the particles.
   C. The particles are surrounded by a tough protective capsule that prevents infection.
   D. The particles are moved upward until they are swallowed or expelled by coughing, sneezing, etc.
   E. Helper T cells lining the respiratory tract remove the particles.

6. **True or False?** Concentrated stomach acids kill most of the bacteria swallowed with food or saliva. ____________________

7. Lining the mucous membranes of the respiratory and digestive tracts, ____________________ traps bacteria, dust, and other particles.

8. Sweat, saliva, and tears contain ____________________, an enzyme that disrupts the cell walls of bacteria.

Internal Nonspecific Defenses

9. Which one of the following are white blood cells that penetrate the plasma membranes of virus-infected cells, causing them to burst?
   A. neutrophils
   B. macrophages
   C. natural killer cells
   D. helper T cells
   E. none of the above

10. **True or False?** Interferon produced by a virus-infected cell <u>protects other cells</u> from all kinds of viruses. ________________

11. Proteins produced by virus-infected body cells that help other cells resist viruses are called ________________.

12. Some ________________ proteins coat the surfaces of microbes, making them easier for macrophages to engulf.

The Inflammatory Response

13. Which one of the following causes blood vessels to dilate and leak fluid into wounded tissue, causing it to swell?
    A. histamine
    B. prostaglandins
    C. interferons
    D. pyrogens
    E. ibuprofen

14. Which one of the following stimulates nerves to send pain signals to the brain?
    A. histamine
    B. prostaglandins
    C. interferons
    D. pyrogens
    E. ibuprofen

15. **True or False?** Aspirin and ibuprofen treat the symptoms of an illness and <u>address</u> the underlying cause. ________________

16. Chemicals that travel through the bloodstream to the hypothalamus to stimulate a fever are called ________________.

The Lymphatic System

17. Which one of the following does *not* occur when your body is fighting an infection?
    A. Lymph nodes fill with a huge number of lymphocytes.
    B. Lymph can pick up microbes from infection sites just about anywhere in the body.
    C. Blood delivers microbes to the lymphatic nodes and organs.
    D. Macrophages in lymphatic tissue engulf invaders.
    E. Lymphocytes may be activated to mount a specific immune response.

18. The lymphatic vessels carry fluid called ____________________, which is similar to interstitial fluid.

19. The lymphatic system includes saclike organs called ____________________ that are packed with macrophages and lymphocytes.

Specific Defenses: The Immune System

20. In the United States, widespread vaccination of children has virtually eliminated all of the following except
    A. smallpox.
    B. polio.
    C. mumps.
    D. measles.
    E. AIDS.

21. In the process of passive immunity, a person becomes resistant to disease by receiving
    A. antigens.
    B. antibodies.
    C. a live virus.
    D. a harmless variant of a disease-causing microbe or one of its components.
    E. cells that are producing the correct type of antibody for that particular infection.

22. **True or False?** Antibodies produced against one particular antigen are usually effective against any other foreign substance. ____________________

23. **True or False?** Active immunity is temporary, usually lasting a few weeks or months. ____________________

24. A foreign substance that elicits an immune response is a(n) ____________________.

25. When the immune system detects an antigen, it produces defensive proteins called ________________.

26. The term ________________ means resistance to specific invaders.

27. In the proccss of ________________, the immune system is confronted with a vaccine composed of a harmless variant of a disease-causing microbe or one of its components.

Recognizing the Invaders

28. Which one of the following cell types secretes antibodies?
    A. T cells
    B. B cells
    C. helper T cells
    D. cytotoxic T cells
    E. none of the above

29. Antibodies are carried to sites of infection in the body by
    A. humoral secretions.
    B. T cells.
    C. blood.
    D. lymph.
    E. lymph and blood.

30. Which one of the following are involved in cell-mediated immunity *and* humoral immunity?
    A. T cells
    B. B cells
    C. memory cells
    D. effector cells
    E. none of the above

31. T cells circulate in the blood and lymph, attacking
    A. body cells that have been infected with bacteria or viruses.
    B. fungi and protozoa.
    C. cancerous cells of our own body.
    D. none of the above
    E. all of the above

32. Which one of the following statements about antibodies is *false*?
    A. At the tip of each arm of an antibody, a pair of variable regions forms an antigen-binding site.
    B. Each antibody molecule recognizes and binds to just one specific type of antigen.
    C. An antigen-binding site and the antigen it binds have complementary shapes.
    D. The structural variety of antibodies limits the humoral immune system's ability to react to only a few kinds of antigen.
    E. Antibody binding might cause viruses, bacteria, or foreign eukaryotic cells to form large clumps.

33. **True or False?** Cell-mediated immunity is produced by effector cells.

    ____

34. **True or False?** Humoral immunity is produced by T cells. ____________________

35. Lymphocytes that develop in the bone marrow become ____________________ cells. Those that develop in the thymus become ____________________ cells.

36. The overall shape of an antibody is most like the letter ____________________.

Responding to the Invaders

37. Clonal selection produces a clone of
    A. helper T cells.
    B. cytotoxic T cells.
    C. memory cells.
    D. effector cells.
    E. none of the above

38. Which one of the following statements about helper T cells is *false*?
    A. Helper T cells bind to other white blood cells that have previously encountered an antigen.
    B. Helper T cells are activated by the binding of a T cell receptor to a self/nonself complex.
    C. Receptors embedded in a helper T cell's plasma membrane recognize and bind to a complex of self-protein and a foreign antigen displayed on a macrophage.
    D. An activated helper T cell grows and divides, producing more active helper T cells and memory cells.
    E. An activated helper T cell inhibits the activity of cytotoxic T cells.

39. Which one of the following statements about cytotoxic T cells is *false*?
   A. Cytotoxic T cells are the only cells in the body that kill other cells.
   B. Cytotoxic T cells identify infected body cells through binding of a membrane receptor to a self/nonself complex.
   C. An activated cytotoxic T cell discharges the protein perforin that makes holes in the infected cell's plasma membrane.
   D. Another activated cytotoxic T cell protein enters the infected cell and triggers programmed cell death.

40. **True or False?** The first exposure of lymphocytes to an antigen is called the primary immune response. ________________

41. **True or False?** Clonal selection also produces cytotoxic T cells, which can last decades in the lymph nodes. ________________

42. Effector cells produce ________________, specialized for defending against the very antigen that triggered the response.

43. A faster secondary immune response is produced when ________________ bind an antigen and multiply quickly, producing a large new clone of lymphocytes.

Immune Disorders

Allergies

44. Which one of the following does *not* occur during sensitization?
   A. An allergen enters the bloodstream.
   B. The allergen binds to B cells with complementary receptors.
   C. The B cells then proliferate through clonal selection.
   D. The B cells secrete large amounts of antibodies to that allergen.
   E. The B cells secrete large amounts of histamine that triggers the inflammatory response.

45. **True or False?** Mast cells release allergens, which causes local inflammation that produces sneezing, coughing, and itching. ________________

46. **True or False?** Allergic symptoms are most common in the nose and throat because allergens usually enter the body through these passageways.

   ____

47. Antigens that cause allergies are called ________________.

48. Some people are so extremely sensitive to certain allergens that any contact with them might cause a life-threatening response called ________________.

Autoimmune Diseases

Matching: Match the autoimmune disease on the left to its description on the right.

_____ 49. rheumatoid arthritis	A. antibodies are produced against histones and DNA
_____ 50. insulin-dependent diabetes	B. damage and painful inflammation of joints
_____ 51. multiple sclerosis	C. of myelin sheaths
_____ 52. lupus	D. of cells of the pancreas

53. To prevent organ transplant rejection, doctors
    A. employ cytotoxic T cells to destroy the B cells that will in turn destroy the transplant.
    B. promote anaphylactic shock in the organ recipient.
    C. look for a donor with self-proteins matching the recipient's as closely as possible.
    D. first expose the person about to receive the transplant to antigens found on the organ to be transplanted.
    E. do none of the above.

54. **True or False?** Every individual has a unique set of self-proteins.

    ____

55. When the immune system turns against the body's own molecules,

    ____________________ diseases occur.

Immunodeficiency Diseases; AIDS

56. Which one of the following statements about AIDS is *false*?
    A. HIV most often attacks helper memory cells.
    B. AIDS is an immunodeficiency disease.
    C. Most people with AIDS die from another infectious agent or from cancer.
    D. Education is the best weapon against the spread of AIDS.
    E. Since 1981, AIDS has killed more than 20 million people worldwide.

57. **True or False?** People with SCID are born with a marked deficit in both T cells and B cells. ____________________

58. Hodgkin's disease, radiation therapy, and drug treatments used against many cancers can all cause harm by affecting ____________________, key cells of the immune system.

59. ____________________ diseases occur in people who lack one or more of the components of the immune system.

60. HIV is deadly because it destroys the ____________________ system, leaving the body defenseless against most invaders.

Evolution Connection: The Yearly Battle with the Flu

61. The best way to protect yourself from the influenza virus is by:
    A. frequently consuming a clear hot beverage with lemon.
    B. frequently washing your hands.
    C. getting a yearly flu vaccine.
    D. eating many foods containing plenty of vitamin C.
    E. bundling up when you go outside in the cold.

62. **True or False?** The flu virus has an RNA genome that mutates rapidly.

    ____

63. The ____________________ virus is responsible for some of the twentieth century's most deadly worldwide epidemics.

Word Roots

an = without; **aphy** = suck (*anaphylactic shock:* an acute, life-threatening allergic response)

anti = against; **gen** = produce (*antigen:* a foreign macromolecule that does not belong to the host organism and that elicits an immune response)

macro = large; **phage** = eat (*macrophage:* an amoeboid cell that moves through tissue fibers, engulfing bacteria and dead cells by phagocytosis)

neutro = neutral; **phil** = loving (*neutrophils:* the most abundant type of leukocyte; neutrophils tend to self-destruct as they destroy foreign invaders, limiting their life span to a few days)

Key Terms

active immunity
allergens
allergies
anaphylactic shock
antibody
antigen
autoimmune diseases
B cells
cell-mediated immunity
clonal selection
complement proteins
cytotoxic T cells
effector cells
helper T cells
histamine
humoral immunity
immune system
immunity
immunodeficiency diseases
inflammatory response
interferons
lymph
lymph nodes
lymphatic system
lymphocytes
macrophages
memory cells
monoclonal antibodies
natural killer cells
neutrophils
passive immunity
primary immune response
prostaglandins
pyrogens
secondary immune response
T cells
vaccination

Crossword Puzzle

Use the Key Terms list from this chapter to fill in the crossword puzzle.

ACROSS

2. type of lymphocytes that matures in the thymus and is responsible for cell-mediated immunity
4. a procedure that presents the immune system with a harmless variant or derivative of a pathogen; stimulating the immune system to mount a long-term defense against the pathogen
13. immunity conferred by recovering from an infectious disease
15. a chemical alarm signal released by injured cells that causes blood vessels to dilate during an inflammatory response
16. the organ system that protects the body by recognizing and attacking specific kinds of pathogens and cancer cells
17. the production of a lineage of genetically identical cells that recognize and attack the specific antigen that stimulated their proliferation
18. type of lymphocytes that mature in the bone marrow and later produce antibodies
19. type of lymphocytes that attack body cells infected with pathogens
21. nonspecific defensive proteins produced by virus-infected cells and capable of helping other cells resist viruses
22. type of lymphocytes that help activate other types of T cells and may help stimulate B cells to produce antibodies
23. temporary immunity obtained by acquiring ready-made antibodies or immune cells
24. a nonspecific body defense caused by a release of histamine and other chemical alarm signals, which trigger increased blood flow, a local increase in white blood cells, and fluid leakage from the blood
25. the immune response elicited when an animal encounters the same antigen at some later time

DOWN

1. a large family of local regulators secreted by virtually all tissues and performing a wide variety of regulatory functions
3. the type of specific immunity brought about by T cells
5. immunological disorders in which the immune system lacks one or more components, making the body susceptible to infectious agents that would ordinarily not be pathogenic
6. disorders of the immune system caused by an abnormal sensitivity to an antigen
7. a protein dissolved in blood plasma that attaches to a specific kind of antigen and helps counter its effects
8. a foreign molecule that elicits an immune response
9. type of white blood cells that is chiefly responsible for the immune response
10. large, amoeboid, phagocytic white blood cells that develop from a monocyte
11. the initial immune response to an antigen, which appears after a lag of several days
12. a potentially fatal allergic reaction caused by extreme sensitivity to an allergen
13. antigens that causes allergies
14. one of a clone of long-lived lymphocytes formed during the primary immune response
20. a fluid similar to interstitial fluid that circulates in the lymphatic system

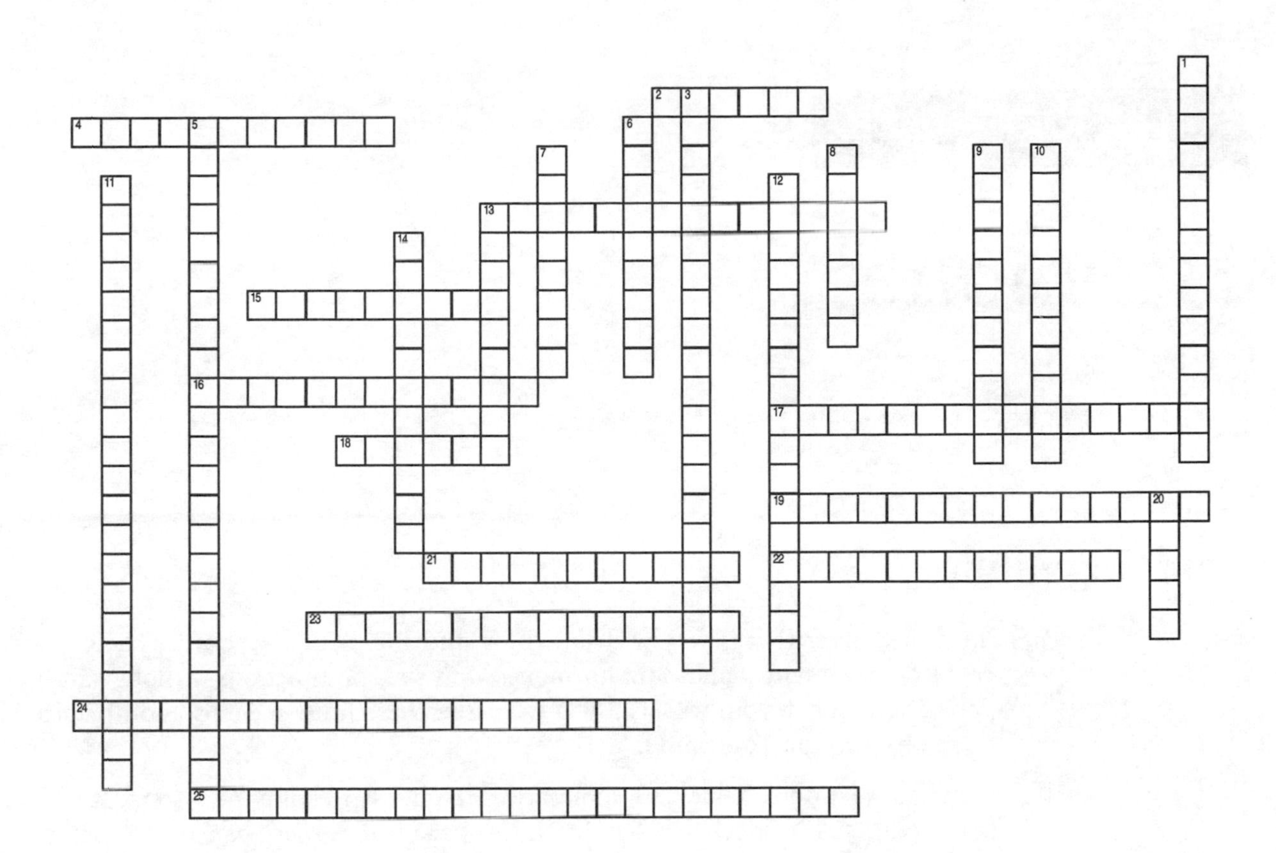

1
2
3
4
5
6
7
8
9
10
11
12
13
14
15
16
17
18
19
20
21
22
23
24
25

CHAPTER 25

Hormones

Studying Advice

a. This is a chapter that is very well suited for studying using note cards. The chapter introduces many glands and hormones with specific functions. A note card type of studying system will help you to master these fundamentals and organize the information in your mind.

b. As the authors note, Table 25.1 summarizes the human endocrine system. It is a very useful table to refer to often. Mark this page with a piece of paper to make finding it easier as you read.

c. Complete the organizing table as you read to help you remember the key functions of each gland. Check the accuracy of what you have written against the information in textbook Table 25.1.

Student Media

Biology and Society on the Web

Learn what the National Cancer Institute has to say about hormone replacement therapy.

Activities

25A Signal Transduction Pathway

25B Action of Amino-Acid-Based Hormones

25C Action of Steroid Hormones

25D Human Endocrine Glands and Hormones

Case Study in the Process of Science

How Does a Thyroid Hormone Affect Metabolism?

Evolution Connection on the Web

Explore the history of estrogen.

Organizing Table

As you read, complete the table below by describing the functions of each gland.

TABLE 25.1

Gland	Functions
Hypothalamus	
Posterior pituitary	
Anterior pituitary	
Thyroid	
Parathyroid	
Islet cells of the pancreas	
Adrenal medulla	
Adrenal cortex	
Gonads	

Content Quiz

Directions: Identify the *one* best answer for the multiple-choice questions. For true/false questions, determine if the statement is true or false. If false, change the underlined word(s) to make the statement true. Finally, add the correct word(s) to the fill-in-the-blank questions to make the statements true.

Biology and Society: Hormone Replacement Therapy

1. Which one of the following is *not* associated with menopause?
   A. increased risk of stroke
   B. increased risk of lung cancer
   C. increased risk of heart disease
   D. vaginal dryness
   E. hot flashes

2. Menopause is caused by decreasing levels of
   A. estrogen and progesterone.
   B. testosterone and progesterone.
   C. androgen and estrogen.
   D. testosterone and estrogen.
   E. progesterone and androgen.

3. **True or False?** Hormones help maintain homeostasis within the human body.

   ____

4. Recent studies suggest that hormone replacement therapy results in an increased risk of ____________________ cancer.

Hormones: An Overview

5. Which one of the following statements about how a steroid hormone works is *false*?
   A. A steroid hormone enters a cell by diffusing through the plasma membrane.
   B. The hormone binds to a receptor protein in the cytoplasm or nucleus.
   C. The hormone triggers a signal-transduction pathway.
   D. The hormone-receptor complex attaches to specific sites on the cell's DNA in the nucleus.
   E. All steroid hormones act by turning genes on or off.

6. **True or False?** Hormones are made and secreted mainly by organs called exocrine glands. ____________________

7. Derived from cholesterol, __________________ hormones bind to receptors inside a cell.

8. A regulatory chemical that travels from its production site through the blood to affect other sites in the body is called a(n) __________________.

9. Cells that respond to a hormone are called __________________ cells.

10. The main body system for internal chemical regulation is the __________________ system.

The Human Endocrine System

11. Which one of the following has endocrine and nonendocrine functions?
    A. pancreas
    B. pituitary gland
    C. thyroid
    D. cartilage
    E. skeletal muscle

12. **True or False?** Sex hormones affect most of the tissues of the body.
    ____

13. The pituitary gland is controlled by the __________________, another endocrine gland that is part of the brain.

The Hypothalamus and Pituitary Gland

14. Which one of the following is *not* released by the pituitary gland?
    A. endorphins
    B. growth hormone
    C. luteinizing hormone
    D. follicle-stimulating hormone
    E. calcitonin

15. Which one of the following statements is *false*?
    A. The anterior pituitary is composed of non-nervous, glandular tissue.
    B. The posterior pituitary is composed of nervous tissue.
    C. The pituitary is the master control center of the entire endocrine system.
    D. Releasing hormones make the anterior pituitary secrete hormones.
    E. Inhibiting hormones make the anterior pituitary stop secreting hormones.

16. **True or False?** The mammary glands are stimulated to produce milk by endorphins. __________________

17. The hypothalamus makes ________________ hormone, which helps cells of the kidney reabsorb water.

18. Serving as the body's painkillers, ________________ are made by the anterior pituitary gland.

19. Too little growth hormone during development can lead to ________________.

The Thyroid and Parathyroid Glands

20. Calcium in the body is needed to allow
    A. muscles to function properly.
    B. nerve signals to be transmitted from cell to cell.
    C. cells to transport molecules across their membranes.
    D. blood to clot.
    E. all of the above

21. Which of the following requires iodine to produce its hormones?
    A. adrenal medulla
    B. thyroid
    C. parathyroid
    D. pituitary
    E. all of the above

22. Which of the following might result from excessive secretion of parathyroid hormone?
    A. fatal convulsions known as tetany
    B. gigantism
    C. hypothyroidism
    D. uncontrollable muscle contractions
    E. loss of calcium in bones

23. **True or False?** Hypothyroidism can result from dietary deficiencies or from a defective thyroid gland. ________________

24. Many industrialized nations have reduced the frequency of ________________ by the incorporation of iodine into table salt.

25. Calcitonin and PTH are said to be ________________ hormones because they have opposite effects.

26. The hormone ________________ causes calcium to be deposited in bones and makes the kidneys reabsorb less calcium as urine is formed.

The Pancreas

27. Which of the following best describes the relationship of insulin to glucagon?
    A. They work together to prepare the body to deal with stress.
    B. Insulin stimulates the pancreas to secrete glucagon.
    C. High levels of insulin inhibit pancreatic secretion of glucagon.
    D. They are antagonistic hormones.
    E. Insulin is a steroid hormone; glucagon is a protein hormone.

28. **True or False?** When the concentration of glucose in the blood rises following the digestion of a meal, insulin is released. ____________________

29. A rise in the concentration of sugar in the blood is caused by the hormone ____________________.

30. In the disease ____________________, body cells are unable to absorb glucose from the blood.

31. Type I diabetes requires regular supplements of ____________________.

32. Treatment for type II diabetes includes exercise and controlling the intake of ____________________ in the diet.

The Adrenal Glands

33. Which one of the following is *not* a function of epinephrine?
    A. release of glucose from the liver
    B. increased heart rate
    C. increased blood pressure
    D. increased absorption of glucose by the digestive tract
    E. increased breathing rate

34. Which one of the following glands is located nearest to the kidneys?
    A. pancreas
    B. pituitary gland
    C. parathyroid glands
    D. testes
    E. adrenal glands

35. **True or False?** Epinephrine and norepinephrine are secreted by the cells in the adrenal cortex. ____________________

36. Receiving ____________________ for a long period of time makes a person highly susceptible to infection.

The Gonads

37. Which, if any, of the following is *not* a category of sex hormone?
    A. prolactin
    B. estrogens
    C. androgens
    D. progestins
    E. All of the choices *are* sex hormones.

38. Which, if any, of the following would *not* be secreted in a sexually mature, healthy woman before she reaches menopause?
    A. testosterone
    B. estrogens
    C. progestins
    D. luteinizing hormone
    E. All of the choices *would* be secreted.

39. **True or False?** Females and males have all three types of sex hormones, but in different proportions. ________________

40. The uterus is prepared to support a developing embryo by the hormones called ________________.

41. The female reproductive system is maintained by the hormones called ________________.

42. The development and maintenance of the male reproductive system is stimulated by the hormones called ________________.

Matching: Match the gland on the left to its best description on the right.

____ 43. parathyroid	A.	stores and releases antidiuretic hormone
____ 44. anterior pituitary	B.	releases PTH to help regulate calcium levels in blood
____ 45. gonad	C.	produces insulin and glucagon
____ 46. posterior pituitary	D.	produces glucocorticoids
____ 47. adrenal medulla	E.	produces hormones that regulate metabolic rate
____ 48. thyroid	F.	synthesizes and releases FSH, LH, GH, and PRL
____ 49. islet cell	G.	secretes androgens, estrogens, and progestins
____ 50. adrenal cortex	H.	secretes epinephrine and norepinephrine

Evolution Connection: The Changing Roles of Hormones

51. The hormone prolactin is present in many kinds of vertebrates, but it controls different processes in different vertebrate groups. What does this situation say about the probable evolutionary history of this hormone?
    A. Prolactin has probably evolved several times among the vertebrates.
    B. Prolactin is probably evolutionarily ancient and has been turned to different uses in different groups.
    C. Prolactin was probably the "master hormone" of primitive vertebrates.
    D. The prolactin gene is probably highly susceptible to mutation.
    E. The prolactin gene probably resides within mitochondria.

52. Which one of the following is *not* a function of prolactin?
    A. It stimulates mammary glands to grow and produce milk in mammals.
    B. In amphibians, it stimulates movement toward water and affects metamorphosis.
    C. It helps to regulate salt and water balance in fish that migrate between salt water and fresh water.
    D. In birds, it regulates calcium and glucose levels in the blood.

53. **True or False?** Hormones play important roles in <u>virtually all</u> organisms.

    ____

Word Roots

andro = male; **gen** = produce (*androgens:* the principal male steroid hormones, such as testosterone, which stimulate the development and maintenance of the male reproductive system and secondary sex characteristics)

endo = inside (*endorphin:* a hormone produced in the brain and anterior pituitary that inhibits pain perception)

epi = above, over (*epinephrine:* a hormone produced as a response to stress; also called adrenaline)

lut = yellow (*luteinizing hormone:* a gonadotropin secreted by the anterior pituitary)

para = beside, near (*parathyroid:* four endocrine glands, embedded in the surface of the thyroid gland, that secrete parathyroid hormone and raise blood calcium levels)

pro = before; **lact** = milk (*prolactin:* a hormone produced by the anterior pituitary gland; it stimulates milk synthesis in mammals)

tri = three; **iodo** = violet (*triiodothyrodine:* one of two very similar hormones produced by the thyroid gland and derived from the amino acid tyrosine)

Key Terms

adrenal cortex	endocrine glands	growth hormone (GH)	parathyroid glands
adrenal glands	endocrine system	hormone	parathyroid hormone (PTH)
adrenal medulla	endorphins	hypothalamus	pituitary gland
androgens	epinephrine	insulin	posterior pituitary
antagonistic hormones	estrogens	islet cells	progestins
anterior pituitary	follicle-stimulating hormone (FSH)	luteinizing hormone (LH)	prolactin (PRL)
calcitonin	glucagon	norepinephrine	steroid hormones
corticosteroids	glucocorticoids	pancreas	target cells
diabetes mellitus	gonads		thyroid gland

Crossword Puzzle

Use the Key Terms list from this chapter to fill in the crossword puzzle.

ACROSS

1. a family of steroid hormones, including progesterone, produced by the mammalian ovary
3. an amine hormone secreted by the adrenal medulla that prepares body organs for "fight or flight"
4. an endocrine gland that secretes thyroxine, triiodothyronine, and calcitonin
6. a regulatory chemical that travels in the blood from its production site, usually an endocrine gland, to other sites, where target cells respond to the regulatory signal
8. a protein hormone secreted by the anterior pituitary that stimulates milk production in mammals
9. an extension of the hypothalamus composed of nervous tissue that secretes hormones made in the hypothalamus
10. a protein hormone, secreted by islet cells in the pancreas that lowers the level of glucose in the blood
11. regulatory chemicals, lipids made from cholesterol that activates the transcription of specific genes in target cells
12. several chemically similar steroid hormones secreted by the gonads
16. four endocrine glands embedded in the surface of the thyroid gland that secrete parathyroid hormone
17. a pair of endocrine glands, located adjacent to a kidney in mammals, composed of an outer cortex and a central medulla
19. a peptide hormone secreted by the thyroid gland that lowers the blood calcium level
20. a human hormonal disease in which body cells cannot absorb enough glucose from the blood and become energy-starved
21. the central portion of an adrenal gland, controlled by nerve signals, secretes the fight or flight hormones epinephrine and norepinephrine
22. a family of hormones synthesized and secreted by the adrenal cortex, consisting of the mineralocorticoids and the glucocorticoids
23. sex organs in animals; ovaries and testes
24. an endocrine gland, adjacent to the hypothalamus and the posterior pituitary that synthesizes several hormones, including some that control the activity of other endocrine glands

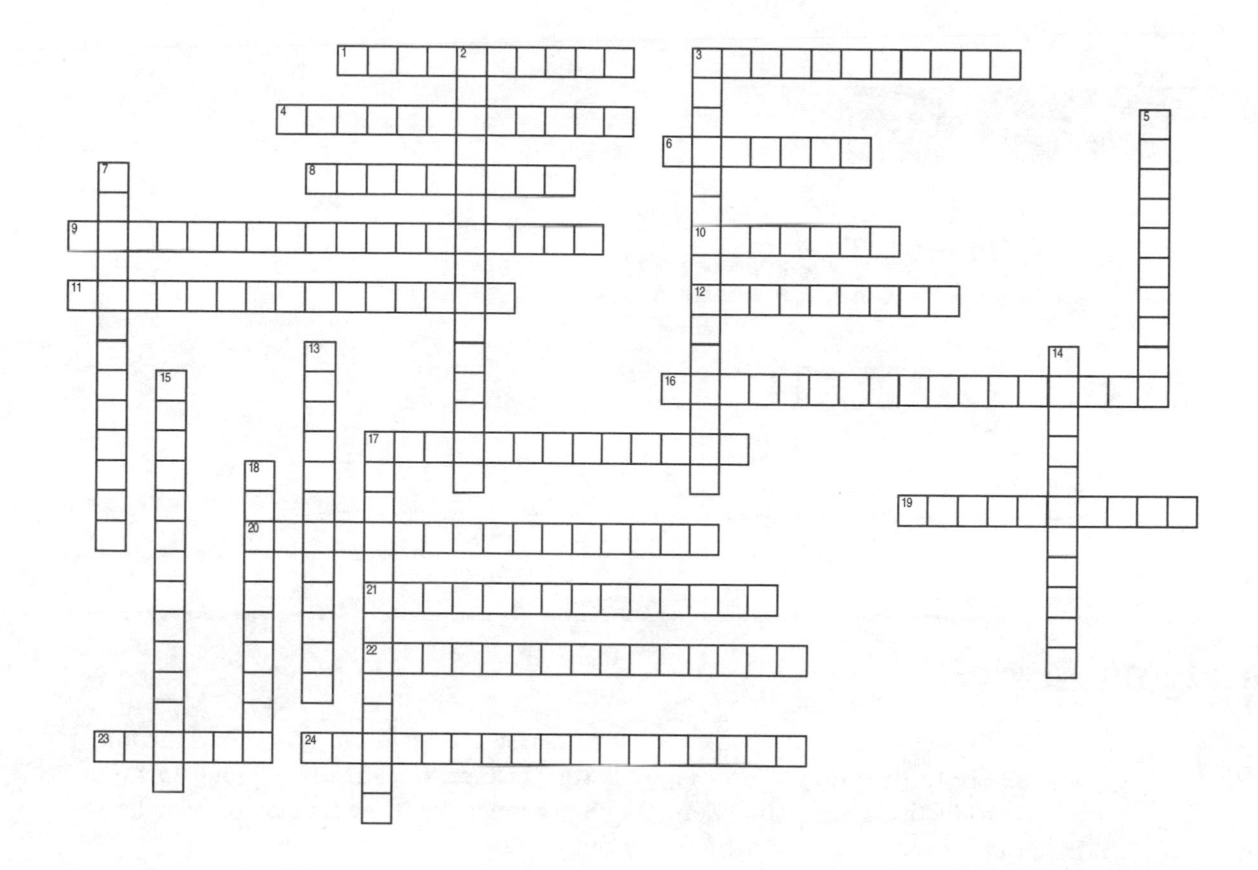

DOWN

2. the organ system consisting of ductless glands that secrete hormones and the molecular receptors on or in target cells that respond to the hormones
3. ductless glands that synthesize hormone molecules and secrete them directly into the bloodstream
5. clusters of endocrine cells in the pancreas that produce insulin and glucagon
7. a protein hormone secreted by the anterior pituitary that promotes development and growth and stimulates metabolism
13. the master control center of the endocrine system, located in the ventral portion of the vertebrate forebrain
14. cells that responds to a regulatory signal, such as a hormone
15. an amine hormone secreted by the adrenal medulla that prepares body organs for fight or flight
17. the outer portion of an adrenal gland, controlled by ACTH from the anterior pituitary
18. pain-inhibiting hormones produced by the brain and anterior pituitary

CHAPTER 26

Reproduction and Development

Studying Advice

a. Few systems of the body capture our interest and attention more than the one we use to reproduce. This chapter is your chance to learn more about how our reproductive systems function. The topics range from the tragedy of sexually transmitted disease to the joy of having children.

b. This chapter is filled with details of the structures and functions of the reproductive system. Create a system of note cards or other quick question-and-answer review. Create your note cards as you read through the chapter. Then review them daily to master the long list.

Student Media

Biology and Society on the Web

Learn what the National Cancer Institute has to say about fertility drugs and cancer.

Activities

26A Reproductive System of the Human Female
26B Reproductive System of the Human Male
26C Sea Urchin Embryonic Development
26D Frog Embryonic Development

Case Studies in the Process of Science

What Might Obstruct the Male Urethra?
What Determines Cell Differentiation in the Sea Urchin?

Evolution Connection on the Web

Examine research that proposes an alternative to the "grandmother hypothesis."

Organizing Tables

Compare the definitions of the forms of asexual reproduction in the table below. Provide at least one example of each.

TABLE 26.1

	Definition	Example
Binary fission		
Fission		
Fragmentation		
Regeneration		
Budding		

Compare oogenesis and spermatogenesis in the table below.

TABLE 26.2

	Oogenesis	Spermatogenesis
Where each process occurs		
Number of gametes produced from each parent cell		
The distribution of cytoplasm in the cells produced by this process		
The relative size, motility, and nutrient storage in the gametes produced		
Timing of completion of the process		

Identify the location where each of these processes occurs in humans.

TABLE 26.3

	Location
Sperm storage	
Fertilization	
Implantation	

Describe the structure and location of each of the developmental stages in the table below.

TABLE 26.4

	Structure	Location in the Mother's Reproductive Tract
Cleavage		
Blastocyst		
Gastrula		

Content Quiz

Directions: Identify the *one* best answer for the multiple-choice questions. For true/false questions, determine if the statement is true or false. If false, change the underlined word(s) to make the statement true. Finally, add the correct word(s) to the fill-in-the-blank questions to make the statements true.

Biology and Society: Rise of the Supertwins

1. Which one of the following statements about fertility drugs is *false*?
   A. Fertility drugs have allowed thousands of infertile couples to have a baby.
   B. About one out of every four women taking fertility drugs becomes pregnant with more than one embryo.
   C. Increased use of fertility drugs has decreased the multiple-birth rate.
   D. Fertility drugs stimulate the ovaries to release one or more eggs.

2. Which one of the following statements is *false*? Newborns from multiple-birth pregnancies
   A. are generally less healthy than newborns from single-birth pregnancies.
   B. have higher birth weights.
   C. are less likely to survive.
   D. are more likely to suffer life-long disabilities if they survive.

3. **True or False?** Fertility drugs stimulate the ovaries to release one or more eggs.
   ____

4. Increased use of ____________________ has caused the multiple-birth rate in the United States to soar.

Unifying Concepts of Animal Reproduction

Asexual Reproduction

Matching: Match the processes on the left to their best description on the right.

____	5. binary fission	A.	breaking up a parent body into pieces followed by regeneration
____	6. fission	B.	a single cell splits via mitosis into two genetically identical cells
____	7. fragmentation	C.	splitting off new individuals from existing ones
____	8. budding	D.	one individual splits into two or more about equal in size

9. Which one of the following statements is *false*? Asexual reproduction
   A. makes it easier to reproduce if an organism is sessile.
   B. makes it easier to reproduce if organisms are greatly isolated from one another.
   C. takes longer than sexual reproduction.
   D. allows an individual very well suited to its environment to quickly multiply and exploit available resources.
   E. removes the need to find a mate.

10. **True or False?** Asexual reproduction produces genetically <u>diverse</u> populations.

    ____

11. The creation of offspring that are genetically identical to a lone parent defines

    ________________ reproduction.

12. The regrowth of a whole animal from pieces is called ________________.

Sexual Reproduction

13. Compared to asexual reproduction, sexual reproduction is generally more adaptive when
    A. there is environmental stability.
    B. conditions favor the production of the greatest number of offspring.
    C. conditions favor the production of genetically identical offspring.
    D. the environment changes suddenly or drastically.
    E. more than one of the above is happening.

14. **True or False?** Sexual reproduction creates genetically <u>unique</u> offspring through the blending of the genotypes of two parents. ________________

15. **True or False?** The process of <u>external</u> fertilization occurs when sperm are deposited in or near the female reproductive tract and the gametes fuse within the female's body. ________________

16. The male gamete is the ________________ and the female gamete is the

    ________________.

17. Two haploid sex cells unite during fertilization to form a diploid

    ________________.

18. Some species are ________________, with both male and female reproductive systems.

Human Reproduction

Female Reproductive Anatomy

Matching: Match the structures on the left to their best description on the right.

_____ 19. cervix	A. narrow neck at the bottom of the uterus
_____ 20. follicles	B. the site of gamete production in females
_____ 21. oviduct	C. a short, sensitive shaft supporting a rounded glans
_____ 22. ovaries	D. a single egg surrounded by one or more layers of cells
_____ 23. vulva	E. birth canal
_____ 24. uterus	F. common site of fertilization
_____ 25. vagina	G. the actual site of pregnancy
_____ 26. clitoris	H. collective name for the female reproductive anatomy

27. Which one of the following statements is *false*?
    A. During a menstrual cycle, a woman typically releases one egg cell about every 28 days.
    B. The uterus is about the size and shape of a pear.
    C. After about the ninth week of development, the embryo is called a fetus.
    D. During sexual arousal, the vagina, labia minora, and clitoris engorge with blood.
    E. The uterus is lined with a blood-rich layer of tissue called the myometrium.

28. **True or False?** It is recommended that women have their cervix examined yearly via a Pap test. ____________________

29. The organs that produce gametes are called ____________________.

30. An egg cell is ejected from the follicle in the process of ____________________.

Male Reproductive Anatomy

Matching: Match the structures on the left to their best description on the right.

_____ 31. penis	A. a tube that stores sperm while they develop
_____ 32. testes	B. one of the glands producing the fluid portion of semen
_____ 33. urethra	C. the site of gamete production in males
_____ 34. epididymis	D. a sac containing the testes
_____ 35. scrotum	E. a shaft that supports a glans
_____ 36. prostate	F. a tube that conducts sperm and urine
_____ 37. vas deferens	G. a duct through which sperm travel during ejaculation

38. Which one of the following statements is *false*?
    A. About 95% of semen consists of sperm.
    B. The prostate gland is commonly diseased in men over 40.
    C. The testes function best at below-normal body temperature.
    D. The seminal vesicle adds part of the fluid that forms semen.
    E. Prostate cancer is the second most commonly diagnosed cancer in the United States.

39. **True or False?** Scientific studies suggest that circumcision has an overall positive impact on a man's health. ____________

40. The expulsion of sperm-containing fluid from the penis is called ____________.

Gametogenesis

41. Which one of the following statements is *false*?
    A. Meiosis in oogenesis yields cells of unequal size.
    B. Meiosis in spermatogenesis yields cells of equal size.
    C. Meiosis I of spermatogenesis produces four secondary spermatocytes.
    D. Meiosis in human males occurs only within seminiferous tubules.
    E. Meiosis in women ends after menopause.

42. Which one of the following statements is *false*?
    A. Polar bodies are produced during oogenesis but not spermatogenesis.
    B. Spermatogenesis produces four gametes, but oogenesis results in only one gamete from each parent cell.
    C. Sperm are small, motile, and contain few nutrients, while eggs are large, nonmotile, and well stocked with nutrients.
    D. Human females create mature ova only during fetal development.
    E. Human males create new sperm every day from puberty through old age.

43. **True or False?** A secondary oocyte completes meiosis II after fertilization.
    ____

44. The creation of gametes within the gonads defines ____________.

45. At birth, a baby girl has thousands of ____________, diploid cells that have paused in prophase of meiosis I.

46. Sperm develop within ____________ in the testes.

47. Prior to ejaculation, sperm are stored within the ____________.

The Female Reproductive Cycle

48. Which one of the following statements is *false*?
    A. The menstrual discharge consists of blood, clusters of endometrial cells, and mucus.
    B. The first day of menstruation is designated as the last day of the menstrual cycle.
    C. The endometrium will not be discharged if an embryo implants.
    D. The menstrual discharge leaves the body through the vagina.
    E. After menstruation, the endometrium regrows, reaching its maximum thickness in 20–25 days.

49. **True or False?** The menstrual cycle controls the growth and release of an ovum.

    ____

50. **True or False?** The growth of an ovarian follicle is stimulated by LH.

    ____

51. Uterine bleeding caused by the breakdown of the endometrium defines

    ____________________.

52. Ovulation typically takes place on day ____________________ of the 28-day cycle.

53. FSH and LH are produced by the ____________________.

54. A positive pregnancy test detects the hormone ____________________.

Reproductive Health

Contraception

Matching: Match the items on the left to their best description on the right.

____ 55. tubal ligation	A. removing the penis from the vagina before ejaculation
____ 56. vasectomy	B. removal of a short section from each vas deferens
____ 57. rhythm method	C. a thimble-shaped, small cap that tightly covers the cervix
____ 58. withdrawal	D. removal of a short section from each oviduct
____ 59. diaphragm	E. sperm-killing chemicals
____ 60. cervical cap	F. refraining from intercourse around the time of ovulation
____ 61. condom	G. a large dome-shaped rubber cap that covers the cervix
____ 62. spermicides	H. a barrier that fits over the penis or within the vagina

63. Which one of the following statements is *false*?
   A. The most widely used birth control pills contain synthetic estrogen and progestin.
   B. "The pill" prevents ovulation and keeps follicles from developing.
   C. Combined hormone contraceptives are also available as a shot, a ring inserted into the vagina, or a patch.
   D. Extensive evidence links the pill to cancers.
   E. Birth control pills require a physician's examination and prescription.

64. **True or False?** There is <u>no chemical</u> contraceptive currently available that prevents the production or release of sperm. ________________

65. Depo-Provera and Norplant use ________________ to prevent fertilization.

66. If pregnancy has already occurred, the drug ________________ can be used to induce an abortion during the first trimester of pregnancy.

67. Combination birth control pills can be used in high doses as emergency contraception, also called ________________.

Sexually Transmitted Diseases

68. Which one of the following statements is *false*?
   A. AIDS is caused by HIV.
   B. Very few STDs cause long-term problems or death if left untreated.
   C. STDs are most prevalent among teenagers and young adults.
   D. The best way to avoid both unwanted pregnancy and the spread of STDs is abstinence.
   E. Latex condoms provide the best dual protection for "safe sex."

69. **True or False?** Viral STDs <u>are not</u> curable. ________________

70. **True or False?** Many people infected with sexually transmitted diseases have <u>no apparent</u> symptoms. ________________

71. Chlamydia, gonorrhea, and syphilis are caused by ________________.

72. Yeast infections are caused by ________________.

73. Genital herpes and genital warts are caused by ________________.

Reproductive Technologies

Infertility

74. Infertility can be due to
    A. low sperm counts.
    B. inability of sperm to swim to an egg.
    C. a lack of eggs.
    D. failure to ovulate.
    E. all of the above.

75. Which one of the following statements is *false*?
    A. Low sperm counts may be due to a scrotum kept too warm by tight-fitting underwear.
    B. Drug therapies (including Viagra) and penile implants can be used to treat sterility.
    C. Hormone injections can induce ovulation but may result in multiple-embryo pregnancies.
    D. As with sperm, eggs can be obtained from a bank for injection into the uterus.
    E. A woman able to become pregnant but unable to support a growing fetus might hire a surrogate mother.

76. **True or False?** Infertility is usually due to problems with the woman.

    ____

77. The inability to maintain an erection defines ____________________.

78. Temporary ____________________ can result from alcohol or drug use, or because of psychological reasons.

In Vitro Fertilization

79. Which one of the following statements is *false*? In vitro fertilization
    A. costs around $10,000 for each attempt, successful or not.
    B. begins with the surgical removal of ova and the collection of sperm.
    C. was first performed in the United States in 1998.
    D. permits genetic testing of 8-cell embryos prior to implantation.
    E. is almost routine today.

80. **True or False?** By implanting only embryos of a certain sex, in vitro fertilization can be used by couples to select the sex of their baby. ____________________

81. "In vitro" literally means ____________________.

Human Development

Fertilization

82. Which one of the following does *not* occur during fertilization?
    A. Fusion of egg and sperm changes the egg so that other sperm cannot penetrate it.
    B. Fructose from the semen fuels movement of the sperm's tail.
    C. Fusion of egg and sperm activates the egg's metabolic machinery.
    D. The sperm penetrates the zona pellucida.
    E. Fusion of egg and sperm triggers the egg to undergo mitosis.

83. **True or False?** The sperm's thick head contains a diploid nucleus.

    ____

84. The enzymes that digest a hole in the zona pellucida are found in a sperm's

    ____________________.

85. The movement of the sperm tail is powered by ____________________, organelles clustered near the middle of the sperm.

Basic Concepts of Embryonic Development

Matching: Match the items on the left to their best description on the right.

_____	86. endoderm	A. gives rise to the nervous system
_____	87. ectoderm	B. results in an embryo shaped like a solid multicellular ball
_____	88. mesoderm	C. gives rise to the digestive system
_____	89. induction	D. an embryo shaped like a hollow ball with three main layers
_____	90. blastocyst	E. programmed cell death
_____	91. gastrula	F. gives rise to the heart, kidneys, and muscles
_____	92. cleavage	G. a way that cells influence an adjacent group of cells
_____	93. apoptosis	H. a fluid-filled hollow ball of about 100 cells

94. Which one of the following does *not* occur during cleavage?
    A. DNA replication
    B. cytokinesis
    C. an increase in the size of the embryo
    D. mitosis
    E. All of the above *do* occur during cleavage.

95. **True or False?** Changes in cell shape help to create embryonic structures.

    ____

Pregnancy and Early Development

Matching: Match the structures on the left to their best description on the right.

_____ 96. yolk sac — A. it produces the embryo's first blood cells and its first germ cells

_____ 97. allantois — B. a fluid-filled sac that encloses and protects the embryo

_____ 98. amnion — C. forms part of the umbilical cord

_____ 99. chorion — D. becomes part of the placenta

100. Which one of the following statements about the placenta is *false*?
    A. The embryonic and maternal blood supplies in the placenta flow into each other.
    B. Nutrients and oxygen are extracted from the mother's blood.
    C. Embryonic wastes are released into the mother's blood.
    D. Protective antibodies pass from the mother to the fetus.
    E. Most drugs can cross the placenta and harm the embryo.

101. **True or False?** Most viruses <u>cannot</u> cross the placenta and cause disease.

    ____

102. The outer cell layer of the gastrula, the ____________________, becomes part of the placenta.

103. The ____________________ is the organ that provides nourishment and oxygen to the embryo and helps dispose of its metabolic wastes.

104. The inner cell mass contains ____________________ cells, which have the potential to give rise to every type of cell in the body.

The Stages of Pregnancy

105. Which one of the following does *not* occur during the third trimester?
    A. The fetus gains the ability to maintain its own temperature.
    B. The fetus's bones begin to harden and muscles thicken.
    C. The fetus loses much of its fine body hair, except on its head.
    D. The fetus rotates so that its head points upward toward the mother's lungs.
    E. All of the above occur during the third trimester.

106. By the end of the ____________________ trimester the fetus's eyes are open and its teeth are forming.

107. During the ____________________ trimester, the fetus's circulatory system and respiratory system undergo changes that will allow the switch to air breathing.

108. All of a fetus's organs and major body parts are formed by the ____________________ trimester.

109. The most dramatic changes in a fetus occur during the ____________________ trimester.

Childbirth

110. Which one of the following statements about oxytocin is *false*?
   A. Oxytocin is a powerful stimulant for the smooth muscles in the wall of the uterus.
   B. Oxytocin stimulates the placenta to make prostaglandins.
   C. Oxytocin and prostaglandins cause uterine contractions.
   D. Estrogen triggers the formation of numerous oxytocin receptors on the uterus.
   E. All of the above statements are true.

111. **True or False?** The pituitary hormone <u>estrogen</u> promotes milk production by the mammary glands. ____________________

112. The ____________________ stage of labor is characterized by the birth of the child.

113. A child is born after a series of strong, rhythmic contractions of the uterus called ____________________.

114. During the final stage of labor, the ____________________ is delivered.

115. The longest stage of labor is ____________________, lasting 6–12 hours or longer.

Evolution Connection: Menopause and the Grandmother Hypothesis

116. Anthropologist Kristen Hawkes and her colleagues found that in a tribe of hunter-gatherers in northern Tanzania, children with caring grandmothers
   A. gained more weight and grew faster.
   B. answered math problems more quickly.
   C. had the best artistic talents.
   D. were generally overweight.
   E. usually did not have living parents.

117. **True or False?** Most species <u>lose</u> their reproductive capacity throughout life.
____

118. The cessation of ovulation and menstruation is called ____________________.

Word Roots

blasto = produce; **cyst** = sac, bladder (*blastocyst:* a hollow ball of cells produced one week after fertilization in humans)

contra = against (*contraception:* the prevention of pregnancy)

-ectomy = cut out (*vasectomy:* the cutting of each vas deferens to prevent sperm from entering the urethra)

endo = inside (*endometrium:* the inner lining of the uterus, which is richly supplied with blood vessels)

epi = above, over (*epididymis:* a coiled tubule located adjacent to the testes where sperm are stored)

fertil = fruitful (*fertilization:* the union of haploid gametes to produce a diploid zygote)

gastro = stomach, belly (*gastrulation:* the formation of a gastrula from a blastula)

labi = lip; **major** = larger (*labia majora:* a pair of thick, fatty ridges that enclose and protect the labia minora and vestibule)

oo = egg; **genesis** = producing (*oogenesis:* the process in the ovary that results in the production of female gametes)

tri = three (*trimester:* a three-month period)

tropho = nourish (*trophoblast:* the outer epithelium of the blastocyst, which forms the fetal part of the placenta)

Key Terms

acrosome
allantois
amnion
asexual reproduction
barrier methods
binary fission
birth control pills
blastocyst
budding
cervix
chorion
chorionic villi
cleavage
clitoris
contraception
copulation
ectoderm
ejaculation
embryo
emergency contraception
endoderm
endometrium
epididymis
external fertilization
fertilization
fetus
fission
follicles
fragmentation
gametes
gametogenesis
gastrula
gastrulation
gestation
glans
gonads
hermaphrodites
hymen
impotence
in vitro fertilization (IVF)
induction
infertility
internal fertilization
labia majora
labia minora
labor
menstrual cycle
menstruation
mesoderm
morning after pills (MAPs)
oogenesis
ovarian cycle
ovaries
oviduct
ovulation
ovum
oxytocin
penis
placenta
polar body
prepuce
primary oocyte
primary spermatocytes
programmed cell death (apoptosis)
prostate
regeneration
reproduction
reproductive cycle

rhythm method (natural family planning)
scrotum
secondary oocyte
secondary spermatocytes
semen
seminal vesicle
seminiferous tubules
sexual reproduction
sexually transmitted diseases (STDs)
sperm
spermatids
spermatogenesis
spermicides
stem cells
testes
trimesters
trophoblast
tubal ligation
umbilical cord
urethra
uterus
vagina
vas deferens
vasectomy
vulva
withdrawal
yolk sac
zygote

Crossword Puzzle

Use the Key Terms list from this chapter to fill in the crossword puzzle.

ACROSS

4. the hormonally synchronized cyclic buildup and breakdown of the endometrium of some primates, including humans
7. uniting sperm and egg in a laboratory container, followed by the placement of a resulting early embryo in the mother's uterus
10. a pouch of skin outside the abdomen that houses a testis
12. in most mammals, the organ that provides nutrients and oxygen to the embryo and helps dispose of its metabolic wastes
13. a pair of outer thickened folds of skin that protect the female genital region
14. relatively unspecialized cells that can give rise to one or more types of specialized cells
16. the copulatory structure of male mammals
17. a then membrane that partly covers the vaginal opening in the human female
19. in vertebrate animals, the extra embryonic membrane that encloses the fluid-filled amniotic sac containing the embryo
24. an extra embryonic membrane that develops from endoderm
25. a means of asexual reproduction in which a parent organism, often a single cell, divides into two individuals of about equal size
26. the creation of offspring by a single parent, without the participation of a sperm and egg
29. in animal development, the succession of rapid cell divisions without cell growth, that converts the animal zygote into a ball of cells
30. in mammalian development, the outer portion of a blastocyst
31. the neck of the uterus, which opens into the vagina
33. the middle layer of the three embryonic cell layers in a gastrula
34. a gland in human males that secretes an acid-neutralizing component of semen
35. female gonads that produces egg cells and reproductive hormones
36. a duct that conveys urine from the urinary bladder to the outside
37. surgical removal of a section of the two sperm ducts to prevent sperm from reaching the urethra

DOWN

1. the sperm-containing fluid that is ejaculated by the male during orgasm
2. the fertilized egg
3. in human development, a trio of three month long periods of pregnancy
5. discharge of semen from the penis
6. sexual intercourse, necessary for internal fertilization to occur
8. part of the female reproductive system between the uterus and the outside opening
9. the creation of new individuals from existing ones
11. the state of carrying developing young within the female reproductive tract
15. the union of the nucleus of a sperm cell with the nucleus of an egg cell, producing a zygote
18. clusters of cells that surround, protect, and nourish a developing egg cell in the ovary
19. a membrane-enclosed sac at the tip of a sperm
20. an organ in the female that engorges with blood and becomes erect during sexual arousal
21. a mammalian embryo made up of a hollow ball of cells that results from cleavage and that implants in the mother's endometrium
22. a developing human from the ninth week of gestation until birth
23. the innermost of three embryonic cell layers in a gastrula
27. a developing stage of a multicellular organism
28. sec cells; haploid eggs or sperm
32. during embryonic development, the influence of one group of cells on another group of cells

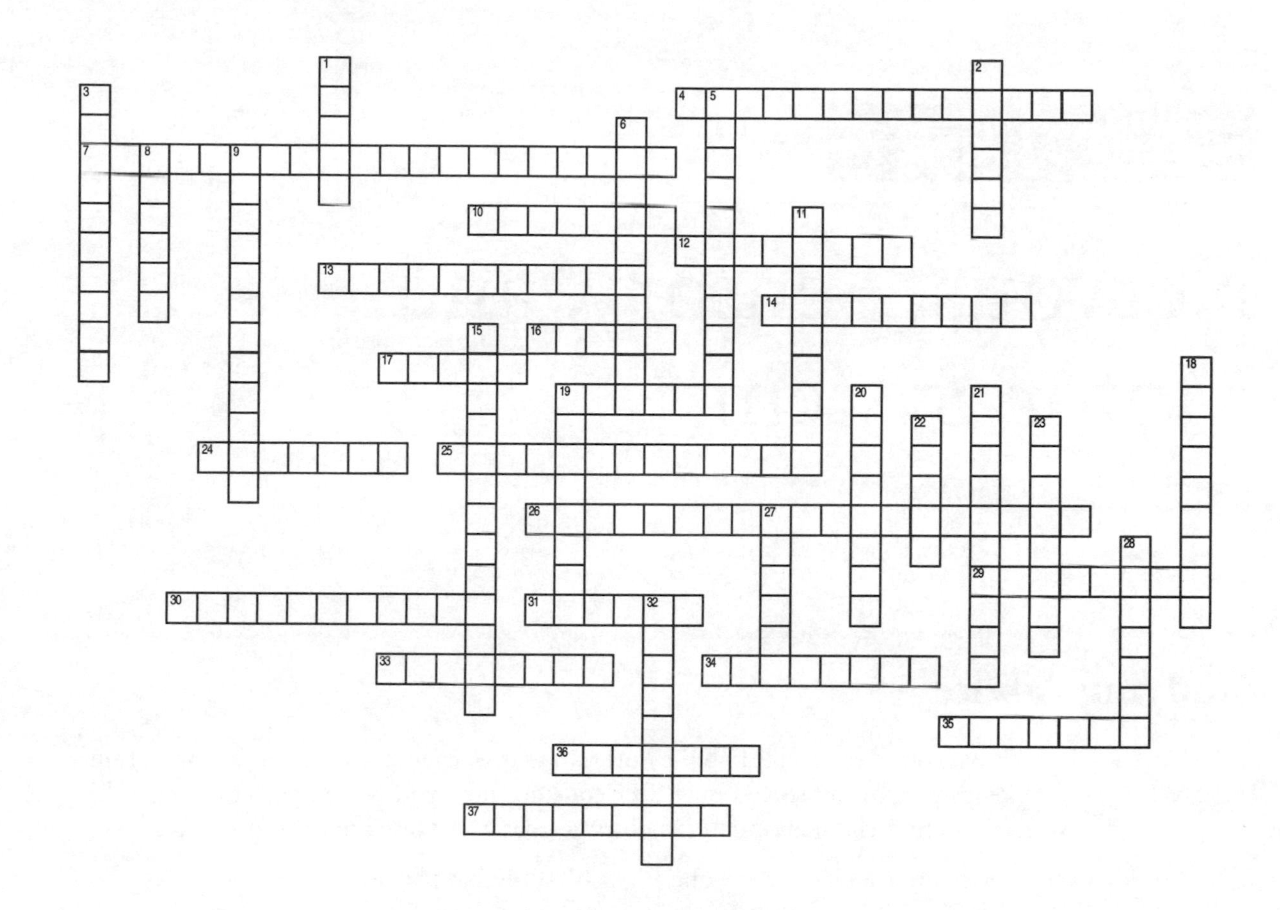

CHAPTER 27

Nervous, Sensory, and Motor Systems

Studying Advice

a. Have you ever thought about how you think? Ever wondered how your brain keeps everything sorted out? Ever consider how you remember anything? This fascinating chapter reveals insight into many of these amazing processes.

b. This is not a chapter for one night of studying. It is long and it contains a long list of vocabulary. The organizing tables on the following pages help you to sort out some of the information, but you will need to create other systems to learn the vocabulary. Consider using the matching questions in this chapter to create note cards to learn and review the brief definitions of the structures.

Student Media

Biology and Society on the Web

Learn more about the symptoms, possible causes, and treatment of depression.

Activities

27A Neuron Structure
27B Nerve Signals: Action Potentials
27C Neuron Communication
27D Structure and Function of the Eye
27E The Human Skeleton
27F Skeletal Muscle Structure
27G Muscle Contraction

Case Studies in the Process of Science

What Triggers Nerve Impulses?

How Do Electrical Stimuli Affect Muscle Contraction?

Evolution Connection on the Web

Discover more about planarians and their relationship to other organisms.

Organizing Tables

Compare the two main subdivisions of the nervous system in the table below. In the function(s) category, indicate which systems are responsible for sensory input, integration, and motor output.

TABLE 27.1

	Structural Components	Function(s)
CNS		
PNS		

Compare the structures and functions of the parts of a neuron in the table below.

TABLE 27.2

	Structural Components	Function(s)
Cell body		
Dendrite		
Axon		

Compare the structures and functions of the components of the peripheral nervous system in the table below.

TABLE 27.3

	Structural Components	Function(s)
Sensory Division (two types of neurons)		
Motor Division Somatic nervous system		
Motor Division Autonomic nervous system Sympathetic division		
Motor Division Autonomic nervous system Parasympathetic division		

Compare the five general categories of sensory receptors in the table below.

TABLE 27.4

	Location	Function(s)
Pain receptors		
Thermoreceptors		
Mechanoreceptors		

continued

TABLE 27.4, *continued*

	Location	Function(s)
Chemoreceptors		
Electromagnetic receptors		

Content Quiz

Directions: Identify the *one* best answer for the multiple-choice questions. For true/false questions, determine if the statement is true or false. If false, change the underlined word(s) to make the statement true. Finally, add the correct word(s) to the fill-in-the-blank questions to make the statements true.

Biology and Society: Battling Depression

1. Which one of the following is *not* a characteristic of depression?
   A. persistent sadness
   B. loss of interest in pleasurable activities
   C. changes in sleeping patterns
   D. changes in body weight
   E. increased energy

2. **True or False?** Once thought to be a purely psychological condition, depression is now known to be a disorder of the brain. ____________________

3. Depression is frequently associated with a lower level of one particular neurotransmitter, called ____________________

An Overview of Animal Nervous Systems

Organization of Nervous Systems; Neurons

4. If you touch something sharp and pull your finger back to avoid injury, the signals travel from
   A. sensory input to integration to motor output.
   B. sensory input to motor output to integration.
   C. integration to sensory input to motor output.
   D. motor output to integration to sensory input.
   E. motor output to sensory input to integration.

5. **True or False?** Signals from the central nervous system are sent out to effectors by <u>sensory</u> neurons. ____________________

6. **True or False?** The <u>central</u> nervous system is mostly composed of nerves that carry signals into and out of the CNS. ____________________

7. A(n) ____________________ is a communication line made from cable-like bundles of neuron fibers tightly wrapped in connective tissue.

8. The conveyance of signals to the CNS from sensory receptors is called ____________________.

9. The part of a neuron that houses the nucleus is called the ____________________.

10. Outnumbering neurons are ____________________ cells that protect, insulate, and reinforce the neurons.

11. Axons that convey signals very rapidly are enclosed along most of their length by an insulating material called the ____________________.

12. Signals are carried toward another neuron or toward an effector by ____________________.

13. Gaps in the myelin sheath form ____________________, the only points on the axon where signals can be transmitted.

Sending a Signal Through a Neuron

14. Stimulating a neuron's plasma membrane to generate a nerve signal is most like
    A. hitting a baseball with a bat.
    B. turning on a flashlight to create light.
    C. bouncing a ball against the ground.
    D. cutting up paper with scissors.
    E. constructing a dog house.

15. Which one of the following statements about membrane potentials is *false*?
   A. The potential energy of a cell is in an electrical charge difference across a neuron's plasma membrane.
   B. During a resting membrane potential, the cytoplasm just inside the membrane is negative in charge.
   C. A membrane stores energy by joining opposite charges together.
   D. A cell's membrane has channels and pumps that regulate the passage of positive ions, contributing to the resting membrane potential.
   E. A cell's membrane has channels and pumps that regulate the passage of positive ions.

16. How do the events of an action potential cause the "domino effect" of a nerve signal?
   A. No ions are allowed to pass through the membrane where an action potential occurs, so ions cross at the next available point.
   B. Inflowing negative ions trigger the opening of channels in the membrane next to the action potential.
   C. Inflowing positive ions trigger the opening of channels in the membrane next to the action potential.
   D. Changes in the myelin sheath create new openings in regions next to the place where an action potential is occurring.

17. How do action potentials relay different intensities of information to the central nervous system?
   A. A stronger action potential is sent when a signal is stronger.
   B. A weaker action potential is sent when a signal is stronger.
   C. Stronger signals cause a greater surge of ions across the membrane.
   D. The frequency of action potentials changes with the intensity of stimuli.

18. **True or False?** Action potentials are <u>all-or-none</u> events. ____________________

19. **True or False?** Most of the dissolved proteins and other large organic molecules inside a neuron are <u>positively</u> charged. ____________________

20. The voltage across a plasma membrane of a resting neuron is called the ____________________.

21. Any factor that causes a nerve signal to be generated is called a(n) ____________________.

22. If a stimulus is strong enough, a sufficient number of channels open to reach the ____________________, the minimum change in a membrane's voltage that must occur to trigger an action potential.

Passing a Signal from a Neuron to a Receiving Cell

Matching: Match each item on the left to its best description on the right.

_____ 23. caffeine	A. decrease our perception of pain
_____ 24. nicotine	B. chemically similar to dopamine and norepinephrine
_____ 25. alcohol	C. a strong depressant, it seems to increase the effects of GABA
_____ 26. opiates	D. a disease associated with a lack of dopamine
_____ 27. antipsychotic drugs	E. binds to and activates receptors for acetylcholine
_____ 28. Ritalin	F. counters the effects of inhibitory neurotransmitters
_____ 29. endorphins	G. a disease associated with an excess of dopamine
_____ 30. Parkinson disease	H. drugs that bind to endorphin receptors
_____ 31. schizophrenia	I. block dopamine receptors

32. Where are neurotransmitters located before an action potential travels down a nerve?
    A. in the synaptic cleft
    B. in the synaptic knob
    C. in the membrane of the neuron
    D. in the dendrites
    E. in the nerve cell body

33. What happens to neurotransmitters after they bind to a receiving neuron's plasma membrane?
    A. They are absorbed by the receiving neuron's plasma membrane.
    B. They are sent to the nerve cell body.
    C. They travel down the receiving neuron's plasma membrane.
    D. They are broken down or transported back to the sending neuron for recycling.

34. **True or False?** Synapses <u>can be</u> either electrical or chemical.

    ____

35. **True or False?** In a <u>chemical</u> synapse, an action potential jumps directly from one neuron to the next. ____________________

36. The narrow gap separating a synaptic knob of the sending neuron from the receiving neuron is called a(n) ________________.

37. A relay point between two neurons or between a neuron and an effector cell is a(n) ________________.

38. In a(n) ________________ synapse, an action potential is converted to chemical signals.

39. A(n) ________________ carries information from one nerve cell to another in a chemical synapse.

The Human Nervous System: A Closer Look

The Central Nervous System

40. Which one of the following statements about the spinal cord is *false*?
    A. The spinal cord functions like a telephone cable jam-packed with wires.
    B. The spinal cord is often severed.
    C. The spinal cord contains white matter.
    D. The spinal cord contains gray matter.
    E. Meninges protect the brain and spinal cord.

41. **True or False?** White matter mainly consists of axons with light-colored myelin sheaths. ________________

42. **True or False?** The concentration of the nervous system at the head end of the body is called centralization. ________________

43. **True or False?** Paralysis of the lower half of the body is called quadriplegia. ________________

44. The presence of a central nervous system distinct from a peripheral nervous system is called ________________.

45. The ________________ is the master control center of the nervous system.

46. An infection of the meninges is called ________________.

The Peripheral Nervous System

47. Which one of the following does *not* occur when the sympathetic division of the autonomic nervous system is activated?
   A. pupils dilate
   B. heart accelerates
   C. glucose is released into the bloodstream
   D. epinephrine and norepinephrine are released into the bloodstream
   E. salivary production is increased

48. **True or False?** The sensory division of the PNS consists of neurons that bring in information about the outside environment and neurons that provide information about the body itself. ________________

49. **True or False?** The sympathetic division primes the body for digesting food and resting. ________________

50. **True or False?** Neurons of the autonomic nervous system carry signals to skeletal muscles. ________________

51. Feeling pain from an internal organ on the surface of the body is called ________________.

The Human Brain

Matching: Match each structure on the left to its best description on the right.

____ 52. thalamus	A.	the highly folded outer surface of the cerebrum
____ 53. hypothalamus	B.	consists of the medulla oblongata and pons
____ 54. brainstem	C.	regulates body temperature, blood pressure, hunger, and thirst
____ 55. cerebellum	D.	consists of the thalamus, hypothalamus, and cerebral cortex
____ 56. corpus callosum	E.	provides coordination of movement and balance
____ 57. cerebral cortex	F.	a band that connects the cerebral hemispheres
____ 58. limbic system	G.	sorts data into categories; suppresses or enhances other signals

59. Which one of the following statements is *false*?
   A. The cerebral cortex is divided into right and left sides.
   B. The corpus callosum restricts communication between the two hemispheres.
   C. Each side of the cerebral cortex has four lobes.
   D. The association areas are the sites of higher mental activities.
   E. Language results from extremely complex interactions among several association areas.

60. **True or False?** The medulla oblongata is a planning center for body movements.

____

61. Areas in the two cerebral hemispheres become specialized for different functions in the process of ___________________.

62. One part of the hypothalamus functions as a timing mechanism, a sort of ___________________ clock.

The Senses

Sensory Input

Matching: Match each structure on the left to its best description on the right.

_____ 63.	pain receptors	A.	detect various forms of mechanical energy
_____ 64.	thermoreceptors	B.	respond to chemicals
_____ 65.	mechanoreceptors	C.	respond to energy of various wavelengths
_____ 66.	chemoreceptors	D.	detect either heat or cold
_____ 67.	electromagnetic receptors	E.	respond to excess heat or pressure

68. Which one of the following receptors is *not* found in skin?
    A. thermoreceptors
    B. electromagnetic receptors
    C. pain receptors
    D. mechanoreceptors
    E. All of the above are found in the skin.

69. **True or False?** Receptor cells convert one type of signal (the stimulus) into an electrical signal in a process called sensory adaptation. ___________________

70. **True or False?** The stronger the stimulus, the larger the receptor potential.

____

71. The change in membrane potential in a receptor cell is called the ___________________.

72. Local regulators that increase pain by sensitizing pain receptors are called ___________________.

Vision

Matching: Match each structure on the left to its best description on the right.

_____ 73. astigmatism	A. at the retina's center of focus
_____ 74. farsightedness	B. a pigmented layer
_____ 75. presbyopia	C. a mucous membrane that helps keep the eye moist
_____ 76. nearsightedness	D. a tough, whitish layer of connective tissue
_____ 77. cones	E. secretes a dilute salt solution
_____ 78. rods	F. blurred vision caused by a misshapen lens or cornea
_____ 79. lacrimal gland	G. the opening in the center of the iris
_____ 80. conjunctiva	H. it lets light into the eye and also helps focus light
_____ 81. glaucoma	I. a layer that contains photoreceptor cells
_____ 82. fovea	J. a type of farsightedness due to inflexibility of the lens
_____ 83. retina	K. it gives the eye its color
_____ 84. choroid	L. detect shades of gray in the retina
_____ 85. iris	M. ability to see far but not near
_____ 86. pupil	N. caused by increased pressure inside the eye
_____ 87. sclera	O. ability to see near but not far
_____ 88. cornea	P. detect color in the retina

89. Which one of the following statements is *false*?
    - A. The lens of the eye focuses light onto the retina.
    - B. The shape of the lens is controlled by the muscles attached to the choroid.
    - C. When the eye focuses on a nearby object the lens becomes thinner and flatter.
    - D. There are no photoreceptor cells in the part of the retina where the optic nerve passes through the back of the eye.
    - E. An infection or allergic reaction may cause inflammation of the conjuctiva, a condition called conjunctivitis.

90. **True or False?** The much smaller chamber in front of the lens contains a thin fluid, the <u>vitreous</u> humor. ____________________

91. Rods contain a visual pigment called ____________________, which can absorb dim light. Cones contain a visual pigment called ____________________.

Hearing

Matching: Match each structure on the left to its best description on the right.

_____ 92. pinna	A. a tube extending from the pinna to the eardrum
_____ 93. auditory canal	B. the bendable structure we commonly call our "ear"
_____ 94. eustachian tube	C. one of the channels in the inner ear
_____ 95. cochlea	D. an array of hair cells embedded in a basilar membrane
_____ 96. organ of Corti	E. conducts air between the middle ear and the back of the throat
_____ 97. eardrum	F. a sheet of tissue that separates the outer ear from the middle ear

Sequencing: Indicate the correct sequence of the following nine events that occur in the process of hearing. Place a "1" in front of the first event, a "2" in front of the second, etc.

_____ 98. Hair cells develop a receptor potential and release more neurotransmitter molecules at its synapse with a sensory neuron.

_____ 99. As a pressure wave passes through the cochlea, it pushes downward and makes the basilar membrane vibrate.

_____ 100. When a hair cell's projections are bent, ion channels in its plasma membrane open, and positive ions enter the cell.

_____ 101. A vibrating object creates pressure waves in the surrounding air.

_____ 102. Vibrations pass through the hammer, anvil, and stirrup in the middle ear.

_____ 103. The sensory neuron sends more action potentials to the brain through the auditory nerve.

_____ 104. Waves make the eardrum vibrate with the same frequency as the sound.

_____ 105. Vibration of the basilar membrane makes the hairlike projections on the hair cells alternately brush against and draw away from the overlying membrane.

_____ 106. The stirrup transmits the vibrations to the inner ear, producing pressure waves in the fluid within the cochlea.

107. The actual sensory transduction converting sound into an action potential occurs in

A. the outer ear.

B. the middle ear.

C. the inner ear.

D. all of the above.

E. none of the above.

108. Deafness can be caused by
   A. the inability to conduct sounds.
   B. a ruptured eardrum.
   C. stiffening of the middle-ear bones.
   D. damage to receptor cells or neurons.
   E. all of the above.

109. **True or False?** The basilar membrane is <u>not</u> uniform along its spiraling length. ________________

110. A ringing or buzzing sound in your ears is a condition called ________________.

Motor Systems

The Skeletal System

111. Which one of the following statements about osteoporosis is *false*?
   A. Prevention of osteoporosis begins with sufficient calcium intake while bones are still increasing in density.
   B. Walking, jogging, and lifting weights builds bone mass and is beneficial throughout life.
   C. Increased levels of estrogen contribute to osteoporosis.
   D. Insufficient exercise, an inadequate intake of protein and calcium, smoking, and diabetes mellitus may contribute to osteoporosis.
   E. Treatments include calcium and vitamin supplements, hormone replacement therapy, and drugs that slow bone loss or increase bone formation.

112. Which one of the following is an autoimmune disease in which the joints become highly inflamed and their tissues possibly destroyed?
   A. rheumatoid arthritis
   B. Parkinson disease
   C. tinnitus
   D. osteoporosis
   E. multiple sclerosis

113. **True or False?** The <u>axial</u> skeleton is made up of the bones of the limbs, shoulders, and pelvis. ________________

114. **True or False?** Humans, like all vertebrates, have an <u>endoskeleton</u>, hard supporting elements situated among soft tissues. ________________

115. **True or False?** The shaft of a long bone surrounds a central cavity that contains <u>red</u> bone marrow, mostly stored fat brought into the bone by the blood.

   ____

116. A(n) ________________ joint allows us to rotate the forearm at the elbow.

117. A freely moving ________________ joint joins the humerus to the scapula and the femur to the pelvis.

118. A(n) ________________ joint between the humerus and the head of the ulna permits movement in a single plane.

The Muscular System; Stimulus and Response: Putting It All Together

119. Which one of the following does *not* occur during a muscle contraction?
   A. A myosin head gains energy from the breakdown of NADH.
   B. An energized myosin head binds to an exposed binding site on actin.
   C. The molecular event that actually causes sliding is called the power stroke.
   D. During the power stroke, the myosin head bends back to its low-energy position, pulling the thin filament toward the center of the sarcomere.
   E. On the next power stroke, the myosin head attaches to another binding site ahead of the previous one on the thin filament.

120. Which one of the following summarizes the activities of actin and myosin during a muscle contraction?
   A. cock, detach, attach, bend
   B. detach, cock, attach, bend
   C. bend, cock, detach, attach
   D. attach, cock, bend, detach
   E. bend, detach, attach, cock

121. A motor unit functions most like
   A. blowing into a tuba to make music.
   B. breaking a stick in half.
   C. burning gasoline to make a truck engine function.
   D. a switch that controls a set of lights.
   E. assembling pieces of a puzzle.

122. Which one of the following does *not* occur when a muscle relaxes?
   A. Motor neurons stop sending action potentials to the muscle fibers.
   B. Actin binding sites are blocked again.
   C. The ER releases Ca^{2+} into the cytoplasm.
   D. Sarcomeres stop contracting.

123. Which one of the following is an example of a motor response by a baseball player?
   A. seeing a ball
   B. hearing the crack of the bat against the ball
   C. determining where the ball will go
   D. reaching out to catch the ball
   E. feeling joy in success

124. **True or False?** A muscle <u>cannot</u> relengthen on its own. ____________________

125. **True or False?** <u>Thin</u> filaments are composed of myosin. ____________________

126. **True or False?** A sarcomere contracts when its thin filaments <u>slide across</u> its thick filaments. ____________________

127. Muscles attach to bone by ____________________

128. A myofibril consists of repeating units called ____________________.

129. Each muscle fiber consists of a bundle of smaller ____________________.

130. A motor neuron and all the muscle fibers it controls define a(n) ____________________.

131. A motor neuron forms synapses with the muscle fibers at ____________________ junctions.

Evolution Connection: Evolution of the Eye

132. Which one of the following statements about planarians vision is *false*?
    A. The brain of planarians forms an image based upon information from the eyes.
    B. The eyespots contain photoreceptors.
    C. The openings of the two eyes face opporite directions.
    D. The brain compares the rate of nerve impulses coming from the two eyespots.

133. **True or False?** Planarians tend to move <u>toward</u> a light source.

____

134. Planarians have ____________________ that provide information about light intensity and direction.

Word Roots

auto = self (*autonomic nervous system:* a subdivision of the motor nervous system of vertebrates that regulates the internal environment)

cephalo = head (*cephalization:* the clustering of sensory neurons and other nerve cells to form a small brain near the anterior end and mouth of animals with elongated, bilaterally symmetrical bodies)

dendro = tree (*dendrite:* one of usually numerous, short, highly branched processes of a neuron that conveys nerve impulses toward the cell body)

inter = between (*interneurons:* an association neuron; a nerve cell within the central nervous system that forms synapses with sensory and motor neurons and integrates sensory input and motor output)

neuro = nerve; **trans** = across (*neurotransmitter:* a chemical messenger released from the synaptic terminal of a neuron at a chemical synapse that diffuses across the synaptic cleft and binds to and stimulates the postsynaptic cell)

para = near (*parasympathetic division:* one of two divisions of the autonomic nervous system)

soma = body (*somatic nervous system:* the branch of the motor division of the vertebrate peripheral nervous system composed of motor neurons that carry signals to skeletal muscles in response to external stimuli)

syn = together (*synapse:* the locus where a neuron communicates with a postsynaptic cell in a neural pathway)

Key Terms

action potential
appendicular skeleton
aqueous humor
arthritis
astigmatism
auditory canal
autonomic nervous system
axial skeleton
axon
ball-and-socket joints
basal ganglia
basilar membrane
biological clock
brain
brainstem
cell body
central nervous system (CNS)
centralization
cephalization
cerebellum
cerebral cortex
cerebrospinal fluid
cerebrum
chemoreceptors
choroid
cochlea
cones
conjunctiva
cornea
corpus callosum
dendrites
eardrum
effectors
electromagnetic receptors
endoskeleton
Eustachian tube
farsightedness
fovea
glaucoma
gray matter
hinge joint
hypothalamus
inner ear
integration
interneurons
iris
lacrimal gland
lateralization
lens
limbic system
mechanoreceptors
medulla oblongata
meninges
middle ear
motor division
motor neurons
motor output
motor units
myelin sheath
myofibrils
nearsightedness
nerve
nervous system
neurons
neurotransmitter
nodes of Ranvier
organ of Corti
osteoporosis
outer ear
pain receptors
parasympathetic division
peripheral nervous system (PNS)
photopsins
photoreceptors
pinna
pivot joint
pons
pupil
receptor potential
red bone marrow
referred pain
resting potential
retina
rheumatoid arthritis
rhodopsin
rods
sarcomere
sclera
sensory adaptation
sensory division
sensory input
sensory neurons
sensory transduction
skeletal muscle
somatic nervous system
spinal cord
stimulus
supporting cells
sympathetic division
synapse
synaptic cleft
synaptic knobs
tendons
thalamus
thermoreceptors
thick filaments
thin filaments
threshold potential
vitreous humor
white matter
yellow bone marrow

Crossword Puzzle

Use the Key Terms list from this chapter to fill in the crossword puzzle.

ACROSS

1. the largest, most sophisticated, and most dominant part of the vertebrate forebrain, made up of right and left cerebral hemispheres
4. one of three main regions of the vertebrate ear; a chamber containing three small bones (the hammer, anvil, and stirrup)
6. a hard skeleton located within the soft tissues of an animal; includes spicules of sponges, the hard plates of echinoderms, and the cartilage and bony skeletons of many vertebrates
9. cells, tissues, or organs capable of carrying out some action in response to a command from the nervous system
13. the integration and command center of the nervous system; the brain and, in vertebrates, the spinal cord
17. nerve cells; the fundamental structural and functional units of the nervous system, specialized for carrying signals from one location in the body to another
18. a plasmalike liquid in the space between the lens and the cornea in the vertebrate eye
20. blurred vision caused by a misshapen lens or cornea
21. part of the vertebrate hindbrain; mainly a planning center that interacts closely with the cerebrum in coordinating body movement
23. a neuron fiber that conducts signals to another neuron or to an effector cell
29. the opening in the iris that admits light into the interior of the vertebrate eye
30. a transparent frontal portion of the sclera that admits light into the vertebrate eye
31. part of the vertebrate outer ear that channels sound waves from the pinna or outer body surface to the eardrum
38. components of the skeletal system that support the central trunk of the body; the skull, backbone, and rib cage in a vertebrate
39. a cablelike bundle of neuron fibers (axons and dendrites) tightly wrapped in connective tissue
41. a motor neuron and all the muscle fibers it controls
42. a skeletal disorder characterized by inflamed joints and deterioration of the cartilage between bones
43. nerve cells that convey command signals from the central nervous system to effector cells, such as muscle cells or gland cells
44. photoreceptor cells in the vertebrate retina, enabling vision in dim light
45. the colored part of the vertebrate eye, formed by the anterior portion of the choroid
46. a junction, or relay point, between two neurons, or between a neuron and an effector cell
47. neuron fibers that convey signals from their tips in-ward, toward the rest of the neuron
48. a layer of connective tissue forming the outer surface of the vertebrate eye

DOWN

2. joints that allows rotation and movement in several planes
3. a disorder of the vertebrate eye in which increased pressure may lead to blindness
5. nerve cells, entirely within the central nervous system, that integrate sensory signals and may relay command signals to motor neurons
7. one of three main regions of the vertebrate ear; includes the cochlea, organ of Corti, and semicircular canals
8. an inability to focus on close objects; occurs when the eyeball is shorter than normal and the focal point of the lens is behind the retina
10. a coiled tube in the inner ear of birds and mammals that contains the hearing organ, the organ of Corti
11. in vertebrates, the photoreceptor cells in the retina, stimulated by bright light and enabling color vision
12. the master control center of the endocrine system, located in the ventral portion of the vertebrate forebrain
14. the structure in an eye that focuses light rays onto the retina

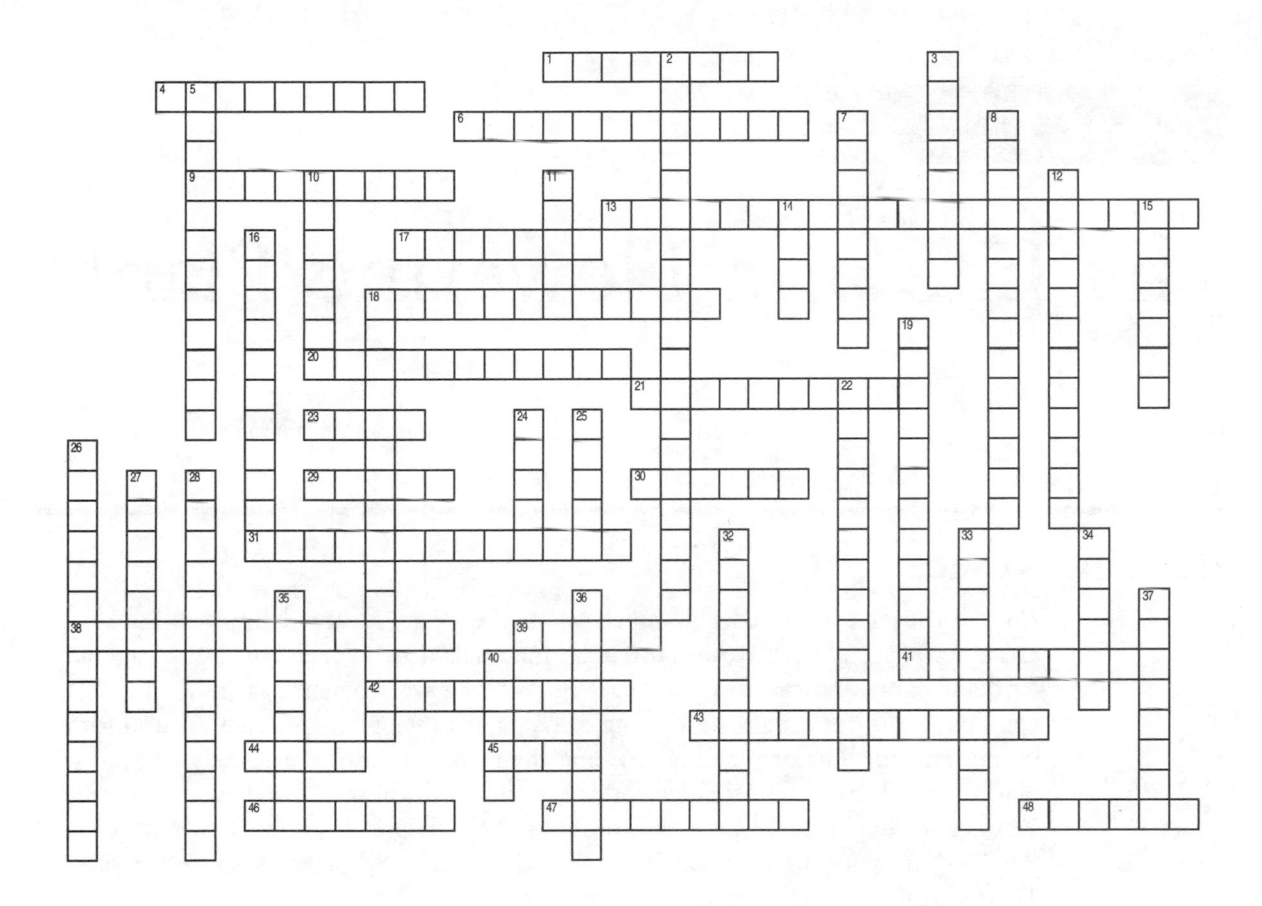

15. a sheet of connective tissue separating the outer ear from the middle ear that vibrates when stimulated by sound waves and passes the waves to the middle ear
16. a mucous membrane that helps keep the eye moist; lines the inner surface of the eyelids and covers the front of the eyeball, except the cornea
18. a self-propagating change in the voltage across a neuron's plasma membrane; a nerve signal
19. a functional unit of several integrating and relay centers located deep in the human forebrain; interacts with the cerebral cortex in creating emotions and storing memories
22. a gland above the eye that secretes tears
24. an eye's center of focus and the place on the retina where photoreceptors are highly concentrated
25. the flaplike part of the outer ear, projecting from the body surface of many birds and mammals
26. a folded sheet of gray matter forming the surface of the cerebrum
27. a thin, pigmented layer in the vertebrate eye, surrounded by the sclera
28. the concentration of the nervous system at the anterior end
32. a joint that allows movement in only one plane; in humans, examples include the elbow and knee
33. a joint that allows precise rotations in multiple planes. An example in humans is the wrist
34. the light-sensitive layer in an eye, made up of photoreceptor cells and sensory neurons
35. the part of a cell, such as a neuron, that houses the nucleus
36. a functional unit of the vertebrate brain, composed of the midbrain, medulla oblongata, and the pons; serves mainly as a sensory filter, selecting which information reaches higher brain centers
37. one of three main regions of the ear in reptiles, birds, and mammals; made up of the auditory canal and, in many birds and mammals
40. the part of the central nervous system involved in regulating and controlling body activity and interpreting information from the senses transmitted through the nervous system

CHAPTER 28

The Life of a Flowering Plant

Studying Advice

Have you ever considered how much plants are a part of our daily lives? Look around you right now. How many plant materials do you see? This book page is made of paper, derived from wood. Perhaps you have a pencil nearby with a wooden shaft? You might have a bag of chips or other plant product around for a snack. The curtains or material covering the furniture nearby are likely covered with plant-derived materials (such as cotton), as is the clothing you are currently wearing. Plants and their products are all around us. This chapter is about their parts, what they do, and what we do with them. This chapter should be read before Chapter 29. The figures in this chapter, and the organizing tables on the following pages, are especially helpful in understanding and learning this information.

Student Media

Biology and Society on the Web

Discover how the organism that caused the Irish potato famine continues to threaten potato crops today.

Activities

28A Roots, Stems, and Leaves

28B The Life Cycle of a Flowering Plant

28C Seed and Fruit Formation and Germination

28D Primary and Secondary Growth

Case Studies in the Process of Science

How Do the Tissues in Monocot and Dicot Stems Compare?

What Tells Desert Seeds When to Germinate?

Evolution Connection on the Web

Learn about the efforts to address the decline in populations of migratory pollinators.

Organizing Tables

Compare the structure of monocots and dicots in the table below.

TABLE 28.1

	Monocots	**Dicots**
The number of seed leaves		
Pattern of veins in the leaves		
The number of flower petals and other parts		
Root systems		
Examples		

Compare the structures and functions of the following types of plant cells.

TABLE 28.2

	Structure	**Function**
Parenchyma cells		
Collenchyma cells		
Sclerenchyma cells		
Water-conducting cells		
Food-conducting cells		

Compare annuals, biennials, and perennials in the table below.

TABLE 28.3

	Definition	Examples
Annuals		
Biennials		
Perennials		

Content Quiz

Directions: Identify the *one* best answer for the multiple-choice questions. For true/false questions, determine if the statement is true or false. If false, change the underlined word(s) to make the statement true. Finally, add the correct word(s) to the fill-in-the-blank questions to make the statements true.

Biology and Society: Plant Cloning—Feast and Famine

1. Which one of the following statements is *false*?
   A. Potatoes were introduced to Europe in the sixteenth century.
   B. Potatoes have been grown in South America for thousands of years.
   C. A potato famine caused Ireland's population to drop by about 25% in just a few years.
   D. A mature potato can be split and planted and each part can develop into a new potato plant that is genetically identical to the parent plant.
   E. A genetically homogenous population is more likely to survive when an environment changes.

2. Which one, if any, of the following does *not* come from plants?
   A. lumber
   B. fabric
   C. paper
   D. industrial chemicals
   E. medicines
   F. All of the above *do* come from plants.

3. **True or False?** Plant <u>roots</u> prevent soil erosion. ____________________

4. A late blight is a(n) ____________________-like organism that causes potato rot.

The Structure and Function of a Flowering Plant

Monocots and Dicots

5. Which one of the following statements is *false*?
   A. Monocots have leaves with parallel veins.
   B. Monocots have flowers with petals and other parts in multiples of three.
   C. Monocots have a large, vertical root.
   D. Monocots have stems with vascular tissues arranged in a scattered array of bundles.
   E. Monocots include the orchids, palms, lilies, and grains and other grasses.

6. Which one of the following statements is *false*?
   A. Dicots have leaves with a multibranched network of veins.
   B. Dicots have a large, vertical root.
   C. Dicots have flowers with petals and other parts in multiples of four or five.
   D. Dicots have stems that do not have vascular bundles.
   E. Dicots include shrubs, trees, and many of our food crops.

7. **True or False?** Dicot embryos have <u>one leaf</u>. ____________________

8. The first leaves that appear on a plant embryo are ____________________.

Plant Organs: Roots, Stems, and Leaves

9. Which one of the following is *not* an evolutionary adaptation that made it possible for plants to move onto land? The ability to
   A. use photosynthesis.
   B. absorb light and take in carbon dioxide from the air.
   C. take up water and minerals from the soil.
   D. survive dry conditions.

10. Which one of the following is *not* a function of a plant's root system?
    A. anchor the plant in the soil
    B. produce sugar
    C. transport minerals and water
    D. absorb minerals and water
    E. store food

11. **True or False?** In a tree, the trunk and branches are its roots.

    ____

12. **True or False?** Grasses and most other monocots have long leaves without petioles. ____________________

13. **True or False?** The spines of a barrel cactus are modified stems.

    ____

14. Carrots and turnips are examples of ____________________, which store food.

15. Stems, leaves, and adaptations for reproduction are all part of the ____________________ system of a plant.

16. In a plant stem the points at which leaves are attached are called ____________________, and the portions of the stem in between are called ____________________.

17. In the process called ____________________, the terminal bud of a plant produces hormones that inhibit growth of its axillary buds.

18. Removing the terminal buds of some plants stimulates the growth of ____________________ buds.

19. Using asexual reproduction, a strawberry plant produces a new plant at the tip of its ____________________.

20. A sweet potato is a modified ____________________, but a white potato is a modified ____________________.

21. Stems and leaves depend on the water and minerals absorbed by ____________________.

22. Located near root tips, tiny projections called ____________________ increase the surface area for absorption of water and minerals.

23. The primary site of photosynthesis in plants is the ____________________.

24. A sweet-pea plant uses a modified leaf called a(n) ________________ to climb up its supports.

Plant Cells

Matching: Match each item on the left to its best description on the right.

_____ 25. chloroplasts	A. provide support in growing parts of plant
_____ 26. central vacuole	B. cell part made mainly of cellulose
_____ 27. cell walls	C. form chains with overlapping ends; are dead when mature
_____ 28. parenchyma cells	D. helps maintain turgor in a cell
_____ 29. collenchyma cells	E. form chains with overlapping ends; are alive when mature
_____ 30. food-conducting cells	F. organelles that contain photosynthetic pigments
_____ 31. water-conducting cells	G. their rigid cell walls support the plant similar to steel beams in a building
_____ 32. sclerenchyma cells	H. most abundant type of cell in most plants

33. **True or False?** The single most distinctive feature of plant cells is their vacuole surrounding the plasma membrane. ________________

34. Plant cell walls are made mainly of the structural carbohydrate ________________.

35. Long water-conducting cells with tapered ends are called ________________; wider, shorter, and less tapered cells are called ________________.

Plant Tissues and Tissue Systems

Matching: Match each item on the left to its best description on the right.

_____ 36. phloem	A. the first defense against physical damage and infections
_____ 37. xylem	B. a waxy coating
_____ 38. epidermis	C. site of food storage and water and mineral absorption in a root
_____ 39. cuticle	D. regulate the size of the stomata
_____ 40. endodermis	E. cells that conduct water and dissolved minerals
_____ 41. pith	F. fills the center of the stem in dicots
_____ 42. guard cells	G. tiny pores between two specialized epidermal cells
_____ 43. mesophyll	H. the ground tissue system of a leaf
_____ 44. cortex	I. food-conducting cells
_____ 45. stomata	J. the innermost layer of cortex

46. **True or False?** All plant stems have vascular tissue systems arranged in numerous vascular bundles. ____________________

47. **True or False?** Each vein in a leaf is a vascular bundle composed of xylem and phloem surrounded by a sheath of parenchyma cells. ____________________

48. Made up of xylem and phloem, the ____________________ tissue system provides support and transports water and nutrients throughout the plant.

49. Filling the spaces between the epidermis and vascular tissue system, the ____________________ system makes up the bulk of a young plant.

50. A plant tissue composed of more than one type of cell is called a ____________________ tissue.

The Life Cycle of a Flowering Plant

The Flower

Matching: Match each item on the left to its best description on the right.

_____ 51.	petals	A. the female organ of the flower
_____ 52.	stamens	B. the receiving surface for pollen grains
_____ 53.	ovule	C. the male organs
_____ 54.	sepals	D. it leads to the ovary at the base of the stigma
_____ 55.	carpel	E. often bright and colorful, they advertise a flower to pollinators
_____ 56.	ovary	F. contains the developing egg and cells that support it
_____ 57.	style	G. a terminal sac on the stamen
_____ 58.	anther	H. enclose and protect the flower bud
_____ 59.	stigma	I. houses reproductive structures called the ovules

60. **True or False?** A flower's main parts—the sepals, petals, stamens, and carpels—are modified stems. ____________________

61. **True or False?** Many plants can reproduce sexually and asexually. ____________________

62. **True or False?** Using sexual reproduction, a single plant can produce many identical plants quickly and efficiently. ____________________

63. In angiosperms, the organ specific to sexual reproduction is the ____________________.

64. Fertilization occurs in the ____________________.

Pollination and Fertilization

Matching: Match each item on the left to its best description on the right.

____ 65. sporophyte	A.	a plant's haploid generation
____ 66. gametophyte	B.	the female gametophyte of angiosperms
____ 67. pollen grain	C.	structure that contains immature male gametophytes
____ 68. embryo sac	D.	the diploid plant body
____ 69. pollination	E.	the delivery of pollen to the stigma of a carpel
____ 70. fertilization	F.	the union of male and female gametes

71. **True or False?** The pollen grain germinates on the <u>stamen</u>.

    ____

72. The formation of a zygote and a cell with a triploid nucleus is called

    ____________________.

Seed Formation

73. After a mature seed is produced,
    A. the endosperm disintegrates.
    B. the seed coat splits open.
    C. the zygote divides by meiosis.
    D. the embryo stops developing.
    E. an embryonic root emerges.

74. **True or False?** The <u>stigma</u> encloses and protects the endosperm.

    ____

75. The ____________________ is a multicellular mass that nourishes the embryo until it becomes a self-supporting seedling.

Fruit Formation; Seed Germination

76. Which one of the following statements about fruits is *false*?
    A. Fruits are highly variable.
    B. A corn kernel is a fruit.
    C. A pea pod is a fruit.
    D. Fruits develop after the seeds develop.
    E. A fruit houses and protects seeds and helps disperse them from the parent plant.

77. Germination is typically triggered by
   A. mechanical damage to the seed coat.
   B. fertilization.
   C. destruction of the endosperm.
   D. ingestion by an animal.
   E. the absorption of water by the seed.

78. **True or False?** The first thing to emerge from a germinating seed is the embryonic shoot. ________________

79. **True or False?** In the wild, only a few seedlings endure long enough to reproduce. ________________

80. Fleshy, edible fruits entice animals that help spread ________________.

Plant Growth

81. A plant that lives for two years, typically flowering in the second year, is
   A. an annual.
   B. a biennial.
   C. a biannual.
   D. a perennial.
   E. a centennial.

82. **True or False?** Organisms that grow throughout their lives show determinate growth. ________________

83. Plants that live and reproduce for many years are called ________________.

84. Plants that live and reproduce in just one year or growing season are called ________________.

Primary Growth: Lengthening

85. Which one of the following statements about meristems is *false*?
   A. Meristems consists of unspecialized cells that divide and generate new cells and tissues.
   B. Meristems are present only during the embryonic stages of a plant's life.
   C. Cell division in the apical meristems of roots and shoots contributes to primary growth.
   D. Growth in all plants is made possible by meristems.

86. **True or False?** Cells produced by secondary growth form tissues that develop into the epidermis, cortex, and vascular tissue. ________________

87. At the tip of a root is a(n) ________________, a thimblelike cone of cells that protects the apical meristem.

88. Meristems at the tips of roots and in the terminal and axillary buds of shoots are called ________________.

89. Cell division in the apical meristems of roots and shoots produces ________________ growth.

Secondary Growth: Thickening

90. Secondary growth involves cell division in two meristems:
    A. the vascular cambium and the cork cambium.
    B. the vascular cambium and the apical meristems.
    C. the cork cambium and the apical meristems.
    D. wood rays and apical meristems.
    E. the cork cambium and rays.

91. Which one of the following statements about growth rings is *false*?
    A. Annual growth rings result from layers of uneven secondary xylem growth due to seasonal variations.
    B. Spring wood cells are usually smaller and thicker walled than those produced in summer.
    C. Each tree ring consists of a cylinder of spring wood surrounded by a cylinder of summer wood.
    D. The vascular cambium becomes dormant each year during winter.

92. After several decades of growth by a tree, which one of the following is typically dead?
    A. the vascular cambium
    B. the cork cambium
    C. the youngest secondary phloem
    D. mature cork cells
    E. cells in the wood rays

93. **True or False?** Stems and roots often thicken as a result of primary growth. ________________

94. **True or False?** After several decades of secondary growth, the bulk of a tree trunk is dead tissue. ________________

95. During secondary growth, the ________________ first appears as a cylinder of actively dividing cells between the primary xylem and primary phloem.

96. Yearly production of a new layer of secondary ________________ accounts for most of the growth in thickness of a perennial plant.

97. Secondary xylem consists of xylem cells and fibers that have thick walls rich in ________________.

98. Cork is produced by a meristem called ________________.

99. Everything external to the vascular cambium (the secondary phloem, cork cambium, and cork) is called ________________.

Evolution Connection: The Interdependence of Angiosperms and Animals

100. Which one of the following do angiosperms *not* rely upon for pollination and seed dispersal?
    A. insects
    B. fish
    C. mammals
    D. birds

101. **True or False?** Flowers that are pollinated by <u>birds</u> often have markings that reflect ultraviolet light. ________________

102. A high-energy fluid that plants use only for attracting pollinators is called ________________.

Word Roots

apic = the tip; **meristo** = divided (*apical meristems:* embryonic plant tissue on the tips of roots and in the buds of shoots that supplies cells for the plant to grow)

bienn = every two years (*biennial:* a plant that requires two years to complete its life cycle)

coll = glue; **enchyma** = an infusion (*collenchyma cells:* a flexible plant cell type that occurs in strands or cylinders that support young parts of the plant without restraining growth)

endo = inner; **derm** = skin (*endodermis:* the innermost layer of the cortex in plant roots)

epi = over (*epidermis:* the dermal tissue system in plants; the outer covering of animals)

gamet = a wife or husband (*gametophyte:* the multicellular haploid form in organisms undergoing alternation of generations, which mitotically produces haploid gametes that unite and grow into the sporophyte generation)

inter = between (*internode:* the segment of a plant stem between the points where leaves are attached)

meso = middle; **phyll** = a leaf (*mesophyll:* the ground tissue of a leaf, sandwiched between the upper and lower epidermis and specialized for photosynthesis)

perenni = through the year (*perennial:* a plant that lives for many years)

phloe = the bark of a tree (*phloem:* the portion of the vascular system in plants consisting of living cells arranged into elongated tubes that transport sugar and other organic nutrients throughout the plant)

sporo = a seed; **phyto** = a plant (*sporophyte:* the multicellular diploid form in organisms undergoing alternation of generations that results from a union of gametes and that meiotically produces haploid spores that grow into the gametophyte generation)

stam = standing upright (*stamen:* the pollen-producing male reproductive organ of a flower, consisting of an anther and filament)

xyl = wood (*xylem:* the tube-shaped, nonliving portion of the vascular system in plants that carries water and minerals from the roots to the rest of the plant)

Key Terms

annuals
anther
apical dominance
apical meristem
bark
biennials
carpel
collenchyma cells
cork
cork cambium
cortex
cotyledons
cuticle
determinate growth
dicot
double fertilization
embryo sac
endodermis
endosperm
epidermis
fertilization
filament
food-conducting cells
fruit
gametophyte
germinates
ground tissue system
guard cells
indeterminate growth
internodes
leaves
meristem
mesophyll
monocot
nodes
ovary
ovule
parenchyma cells
perennials
petals
phloem
pith
pollen grain
pollination
primary growth
root cap
root hairs
root system
sclerenchyma cells
secondary growth
seed coat
sepals
shoot system
sporophyte
stamens
stems
stigma
stomata
style
tissue systems
vascular cambium
vascular tissue system
water-conducting cells
wood
xylem

Crossword Puzzle

Use the Key Terms list from this chapter to fill in the crossword puzzle.

ACROSS

2. the green tissue in the interior of a leaf; a leaf's ground tissue system, the main site of photosynthesis
4. in seed plants, the delivery, by wind or animals, of pollen from the male parts of a plant to the stigma of a carpel on the female
6. a tissue of mostly parenchyma cells that makes up the bulk of a young plant and is continuous throughout its body
10. specialized, living plant cells with a thin primary wall; also called sieve-tube member
14. a flowering plant whose embryos have a single seed leaf, or cotyledon
15. a sac in which pollen grains develop, located at the tip of a flower's stamen
17. a ripened, thickened ovary of a flower, which protects dormant seeds and aids in their dispersal
18. parts of a plant's shoot system that supports the leaves and reproductive structures
22. in a plant, the hormonal inhibition of axillary buds by a terminal bud
26. a meristem at the tip of a plant root or in the terminal or axillary bud of a shoot
28. in flowering plants, the basal portion of a carpel in which the egg-containing ovules develop
29. plants that complete their life cycle in two years
30. the portion of a plant's vascular system that conveys sugars, nutrients, and hormones throughout a plant; made up of food-conducting cells
32. in flowering plants, a nutrient-rich mass formed by the union of a sperm cell with two polar nuclei during double fertilization
35. in a flowering plant, the stalk of a stamen
37. part of the ground tissue system of a dicot plant
42. in flowering plants, the formation of both a zygote and a cell with a triploid nucleus, which develops into the endosperm
43. secondary xylem of a plant
44. the innermost layer of the cortex of a plant root; forms a selective barrier, determining which substances pass from the cortex into the vascular tissue
45. plants that live for many years
46. modified leaves of a flowering plant
47. growth that continues throughout life, as in most plants

DOWN

1. termination of growth after reaching a certain size, as in most animals
3. the female gametophyte contained in the ovule of a flowering plant
5. a reproductive structure in a seed plant, containing the female gametophyte and the developing egg; an ovule develops into a seed
7. a flowering plant whose embryo has two seed leaves, or cotyledons
8. the first leaves that appear on an embryo of a flowering plant; seed leaves
9. all the tissues external to the vascular cambium in a plant that is growing in thickness
11. meristematic tissue that produces cork cells during secondary growth of a plant
12. portions of a plant stem between two nodes
13. in plants, supportive cells with a rigid secondary wall hardened with lignin
15. plants that complete their life cycle in a single year or growing season
16. in plants, cells with a thick primary wall and no secondary wall, functioning mainly in supporting growing parts
19. the outermost protective layer of a plant's bark, produced by the cork cambium
20. in plants, a waxy coating on the surface of stems and leaves that helps retain water
21. in plants, the ground tissue system of a root, made up mostly of parenchyma cells, which store food and absorb minerals that have passed through the epidermis
23. modified leaves of a flowering plant

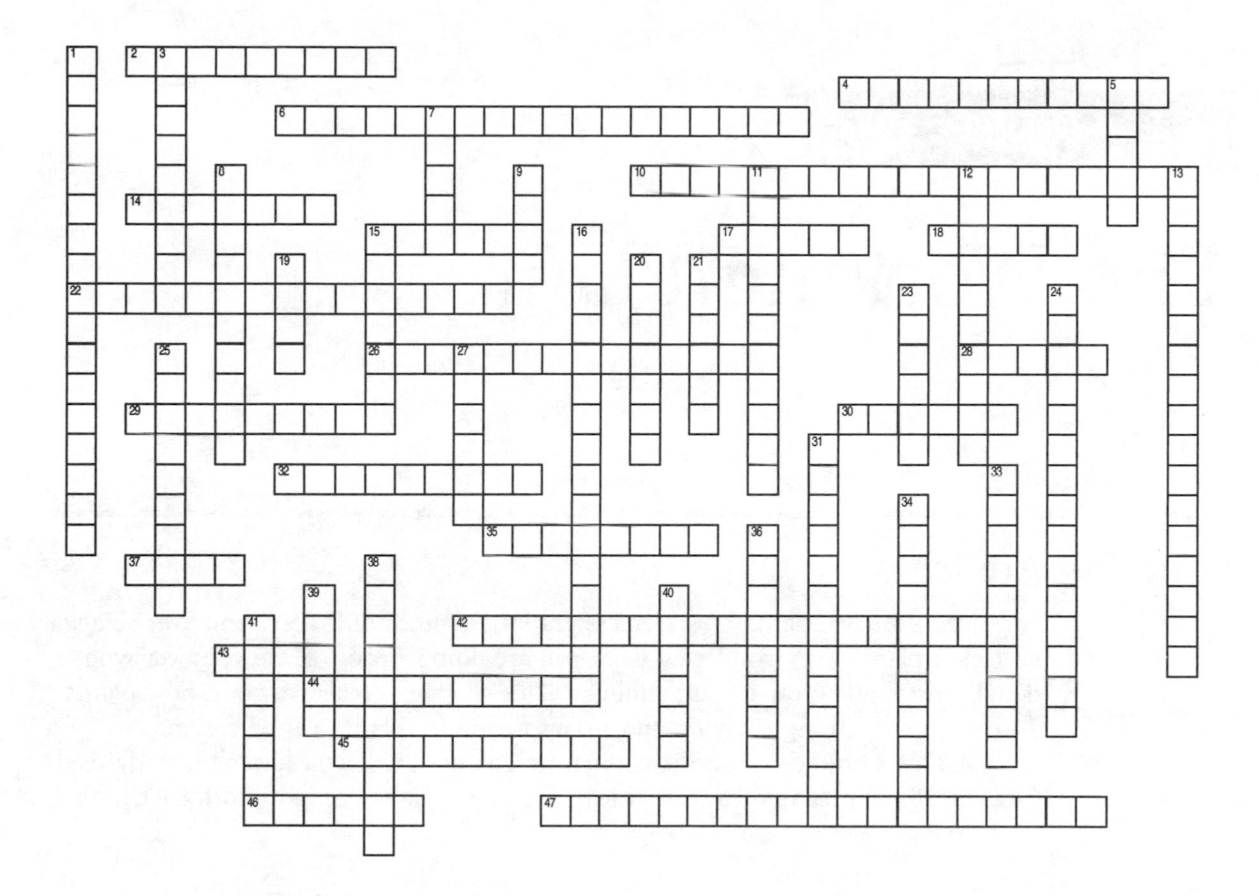

24. in plants, relatively unspecialized cells with a thin primary wall and no secondary wall; functions in photosynthesis, food storage, and aerobic respiration and may differentiate into other cell types
25. in plants, the tissue system forming the protective outer covering of leaves, young stems, and young roots
27. the female part of a flower, consisting of a stalk with an ovary at the base and a stigma, which traps pollen, at the tip
31. the union of the nucleus of a sperm cell with the nucleus of an egg cell, producing a zygote
33. to initiate growth, as in a plant seed
34. the multicellular haploid form in the life cycle of organisms undergoing alternation of generations
36. plant tissue consisting of undifferentiated cells that divide and generate new cells and tissues
38. specialized epidermal cells in plants that regulate the size of a stoma, allowing gas exchange between the surrounding air and the photosynthetic cells in the leaf
39. points of attachment of a leaf on a stem
40. main sites of photosynthesis in a plant; consist of a flattened blade and a stalk (petiole) that join the leaves to the stem
41. a cone of cells at the tip of a plant root that protects the root's apical meristem

CHAPTER 29

The Working Plant

Studying Advice

Have you ever wondered how seedless grapes are produced? Have you ever struggled to keep a plant alive, wondering what you are doing wrong? Have you ever wondered how plants grow toward sunlight? This chapter reveals why and how plants do many of the things they do and makes caring for plants a little easier.

Chapter 29 builds upon the content in Chapter 28. If you haven't already read Chapter 28, be prepared to read at least sections of it as background for Chapter 29.

Student Media

Biology and Society on the Web

Examine the similarities—and differences—in transport systems in plants and animals.

Activities

29A Absorption of Nutrients from Soil

29B Transpiration

29C Transport in Phloem

29D Leaf Drop

29E Flowering Lab

Case Studies in the Process of Science

How Does Acid Precipitation Affect Mineral Deficiency?

What Determines if Water Moves Into or Out of a Plant Cell?

How Is the Rate of Transpiration Calculated?

What Plant Hormones Affect Organ Formation?

Evolution Connection on the Web

Discover the interrelatedness of honeybees, mites, and fruit and vegetable crops.

Organizing Tables

Distinguish between micronutrients and macronutrients in the table below.

TABLE 29.1

	Definition	Examples
Micronutrients		
Macronutrients		

Distinguish between the three types of tropisms in the table below.

TABLE 29.2

	Definition	Examples
Phototropism		
Thigmotropism		
Gravitropism		

Content Quiz

Directions: Identify the *one* best answer for the multiple-choice questions. For true/false questions, determine if the statement is true or false. If false, change the underlined word(s) to make the statement true. Finally, add the correct word(s) to the fill-in-the-blank questions to make the statements true.

Biology and Society: Pass the Plant Sap, Please

1. Which one of the following statements is *false*?
   A. Maple syrup is made from the sap of a maple tree.
   B. Maple sap is clear and nearly tasteless.
   C. Maple sap is mostly sugar, with about 2–3% water.
   D. Maple sap is boiled down to make maple syrup.
   E. In early spring, sap is transported upward in the xylem to nourish the developing leaf buds.

2. **True or False?** Sap is collected in early fall. ________________

3. Watery solutions that move through the vascular system of a plant are called ________________.

How Plants Acquire and Transport Nutrients

Plant Nutrition

4. Which one of the following is a micronutrient?
   A. carbon
   B. oxygen
   C. hydrogen
   D. nitrogen
   E. iron

5. Which one of the following is a macronutrient?
   A. manganese
   B. copper
   C. sulfur
   D. chlorine
   E. zinc

6. A plant needs only minute quantities of micronutrients because it
   A. can easily get them from the soil.
   B. uses micronutrients over and over.
   C. can produce them from macronutrients.
   D. really needs them only a few days of the year.
   E. needs them only to reproduce.

7. **True or False?** Plants require relatively large amounts of <u>micronutrients</u>.

   ____

8. **True or False?** It is <u>relatively easy</u> to diagnose nutrient deficiencies in plants.

   ____

9. **True or False?** A deficiency of any micronutrient <u>can kill</u> a plant.

   ____

10. An element is considered a(n) ____________________ plant nutrient if the plant must obtain it to complete its life cycle.

From the Soil into the Roots

11. Which one of the following statements about how plants obtain nutrients is *false*?
    A. A plant's source of carbon is atmospheric CO_2.
    B. A plant's source of oxygen is atmospheric CO_2.
    C. A plant's source of hydrogen is water.
    D. To reach the xylem, soil solutions must pass through the plasma membranes of root cells.
    E. A plant absorbs all of its essential nutrients from the soil.

12. **True or False?** <u>All</u> substances that enter a plant root are dissolved in water.

    ____

13. Roots efficiently extract nutrients from soil due to their ____________________, extensions of epidermal cells that dramatically increase the surface available for absorption.

14. Most plants gain absorptive surface through mutually beneficial symbiotic associations with fungi in an association called ____________________.

15. The selectively permeable plasma membrane of ____________________ cells regulates what solutes reach the xylem.

The Role of Bacteria in Nitrogen Nutrition

Matching: Match each item on the left to its best description on the right.

____ 16. nitrogen-fixing bacteria	A.	add ammonium to soil by decomposing organic matter
____ 17. ammonifying bacteria	B.	bacteria that convert atmospheric N_2 to ammonium
____ 18. nitrifying bacteria	C.	plants that produce their seeds in pods
____ 19. legumes	D.	the process of converting atmospheric N_2 to ammonium
____ 20. root nodules	E.	convert soil ammonium to nitrate
____ 21. nitrogen fixation	F.	plant root cells containing vesicles filled with bacteria

22. **True or False?** Plants <u>can</u> use the form of nitrogen found in air.

    ____

23. For plants to absorb ____________________ from the soil, it must first be converted to ammonium ions (NH_4^+) or nitrate ions (NO_3^-).

The Transport of Water

24. Transpiration exerts a pull on a tense string of water molecules that is held together by ____________________ and helped upward by ____________________.
    A. cohesion; adhesion
    B. adhesion; cohesion
    C. cohesion; gravity
    D. adhesion; gravity
    E. gravity; cohesion

25. Which one, if any, of the following environmental conditions would *not* increase the amount of transpiration in a plant?
    A. intense sunlight
    B. warm weather
    C. high humidity
    D. windy conditions
    E. All of the above *would* increase the amount of transpiration in a plant.

26. Which one of the following statements is *false*?
    A. Stomata are usually open during the day.
    B. Stomata are usually open at night to allow oxygen to enter from the atmosphere.
    C. During the day, CO_2 can enter the leaf from the atmosphere.
    D. Stomata may also close during the day if a plant is losing water too fast.
    E. Two guard cells flanking each stoma control its opening by changing shape.

27. **True or False?** The sticking together of molecules of the same kind defines adhesion. ____________________

28. **True or False?** The transport of xylem sap requires no energy expenditure by the plant. ____________________

29. Mature water-conducting cells of ____________________ conduct xylem sap from a plant's roots to the tips of its leaves.

The Transport of Sugars

30. Which one of the following statements about the pressure-flow mechanism is *false*?
    A. Each food-conducting tube in phloem tissue has a source end and a sink end.
    B. At source, sugar is loaded from a photosynthetic cell into a phloem tube.
    C. At the sugar sink, water and sugar enter the phloem tube.
    D. At the sugar sink, the exit of water lowers the water pressure in the tube.
    E. Phloem sap always flows from a sugar source to a sugar sink.

31. **True or False?** A sugar sink is a location in a plant where sugar is being produced. ____________________

32. **True or False?** Phloem sap moves freely from the cytoplasm of one cell to the next. ____________________

33. The main function of ____________________ is to transport sugars produced by a plant using photosynthesis.

Plant Hormones

Auxin; Ethylene; Cytokinins; Gibberellins; Abscisic Acid

Matching Match each hormone on the left to its one best description on the right.

_____ 34. auxin	A. promotes seed germination
_____ 35. ethylene	B. promotes cell elongation in stems
_____ 36. cytokinin	C. promotes fruit ripening and dropping of leaves
_____ 37. gibberellin	D. inhibits seed germination
_____ 38. abscisic acid	E. promotes cytokinesis

39. Which one of the following statements about plant hormones is *false*?
   A. Plants produce hormones in very large amounts.
   B. Each type of hormone can produce a variety of effects.
   C. Small amounts of hormones can have profound effects on target cells.
   D. Hormones play critical roles in the development of plants.
   E. Hormones play critical roles in the growth of plants.

40. Which one of the following statements about auxin is *false*?
   A. Auxin promotes cell elongation in stems only within a certain concentration range.
   B. Auxin is produced by developing seeds and promotes the growth of fruit.
   C. Auxin is produced by phloem cells inside of a shoot.
   D. Auxin will migrate from the bright side to the dark side of a shoot.
   E. Auxin promotes growth in stem diameter.

41. Which one of the following hormones enters the shoot system from the roots and counters the inhibitory effects of auxin coming down from the terminal buds?
   A. gibberellins
   B. auxin
   C. ethylene
   D. cytokinins
   E. none of the above

42. Which of the following pairs stimulate cell elongation and cell division in stems and influence fruit development?
   A. gibberellins and auxin
   B. cytokinins and ethylene
   C. abscisic acid and ethylene
   D. auxin and abscisic acid
   E. cytokinins and abscisic acid

43. **True or False?** An uneven distribution of auxin causes the dark side of a shoot to grow <u>slower</u> than the light side. ____________________

44. **True or False?** The same hormone <u>may have different</u> effects at different concentrations in the same target cell. ____________________

45. **True or False?** Animals <u>and plants</u> use hormones to regulate their internal activities. ____________________

46. **True or False?** Some growers retard ripening by storing apples in containers flushed with <u>ethylene</u>. ____________________

47. Regulatory chemicals that travel from their production sites to affect other parts of the body are called ____________________.

48. The term ____________________ is used for any chemical substance that promotes seedling elongation.

49. The primary internal signal that enables plants to withstand drought is ____________________.

50. Spraying unfertilized plants with synthetic ____________________ can produce seedless tomatoes, cucumbers, and eggplants.

51. The growth of a shoot toward light is called ____________________.

Response to Stimuli

Tropisms: Photoperiods

52. Short-day plants are actually responding to
    A. warmer daytime temperatures.
    B. stronger intensities of light.
    C. longer nights.
    D. phases of the moon.
    E. the total amount of sunlight through the day.

53. **True or False?** Long-day plants usually flower in late <u>summer or early fall</u>.
____

54. Growth of a plant in response to touch is ____________________.

55. Growth of a plant in response to gravity is ____________________.

Evolution Connection: The Interdependence of Organisms

56. Most flowering plants depend on
    A. insects or other animals for pollination and seed dispersal.
    B. nitrogen-fixing bacteria.
    C. the fungi of mycorrhizae.
    D. the sun for photosynthesis.
    E. all of the above.

57. **True or False?** The fossil record shows that mycorrhizae have existed since plants first evolved. ________________

58. A needlelike mouthpart is used by ________________ to feed on plant sap.

59. The garden praying mantis picured on this book's cover feeds on ________________.

Word Roots

aux = grow, enlarge (*auxins:* a class of plant hormones, including indoleacetic acid, having a variety of effects, such as phototropic response through the stimulation of cell elongation, stimulation of secondary growth, and the development of leaf traces and fruit)

cyto = cell; **kine** = moving (*cytokinins:* a class of related plant hormones that retard aging and act in concert with auxins to stimulate cell division, influence the pathway of differentiation, and control apical dominance)

gibb = humped (*gibberellins:* a class of related plant hormones that stimulate growth in the stem and leaves, trigger the germination of seeds and breaking of bud dormancy, and stimulate fruit development with auxin)

macro = large (*macronutrients:* elements required by plants and animals in relatively large amounts)

micro = small (*micronutrients:* elements required by plants and animals in very small amounts)

myco = a fungus (*mycorrrhizae:* mutualistic associations of plant roots and fungi)

photo = light; **trop** = turn, change (*phototropism:* growth of a plant shoot toward or away from light)

Key Terms

abscisic acid (ABA)
adhesion
auxin
cohesion
cytokinins
essential plant nutrient
ethylene
gibberellins
gravitropism
macronutrients
micronutrients
mycorrhiza
nitrogen fixation
phloem sap
photoperiod
phototropism
pressure-flow mechanism
root nodules
sugar sink
sugar source
thigmotropism
transpiration
transpiration-cohesion-tension mechanism
tropisms
xylem sap

Crossword Puzzle

Use the Key Terms list from this chapter to fill in the crossword puzzle.

ACROSS

1. a plant organ in which sugar is being produced by either photosynthesis or the breakdown of starch
6. a plant organ that is a net consumer or storer of sugar
7. the conversion of atmospheric nitrogen into nitrogen compounds that plants can absorb and use
8. a plant hormone that inhibits cell division and promotes dormancy
11. the length of the day relative to the length of the night; an environmental stimulus that plants use to detect the time of year
13. the solution of sugars, other nutrients, and hormones conveyed throughout a plant via phloem tissue
14. a plant's growth response to gravity
15. chemical substances that an organism must obtain in relatively large amounts
17. a gas that functions as a hormone in plants, triggering aging responses such as fruit ripening and leaf drop
19. the growth of a plant shoot in response to light
20. the method by which phloem sap is transported through a plant from a sugar source, where sugars are produced, to a sugar sink, where sugars are used
21. elements that an organism needs in very small amounts and that functions as a component or cofactor of enzymes
22. swellings on a plant root consisting of plant cells that contain nitrogen-fixing bacteria

DOWN

2. a family of plant hormones that trigger the germination of seeds and interact with auxins in regulating growth and fruit development
3. a chemical element that is required for a plant to grow from a seed and complete the life cycle, producing another generation of seeds
4. growth movement of a plant in response to touch
5. a family of plant hormones that promote cell division, retard aging in flowers and fruits, and may interact antagonistically with auxins in regulating plant growth and development
8. the attraction between different kinds of molecules
9. the attraction between molecules of the same kind
10. a plant hormone, whose chief effect is to promote the growth and development of shoots
12. the evaporative loss of water from a plant
16. the solution of inorganic nutrients conveyed in xylem tissue from a plant's roots to its shoots
18. growth responses that make a plant grow toward or away from a stimulus

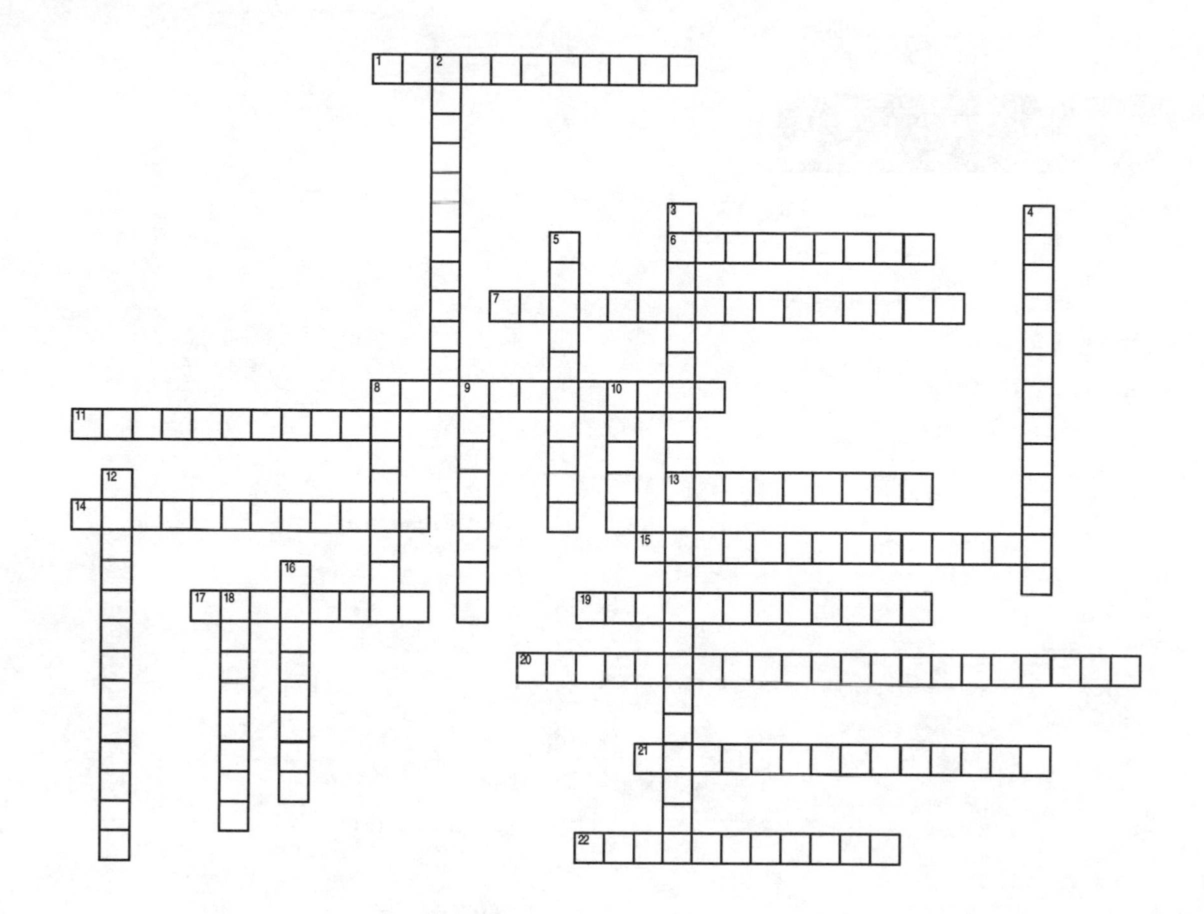

1
2
3
4
5
6
7
8
9
10
11
12
13
14
15
16
17
18
19
20
21
22

ANSWER KEY

Chapter 1

Content Quiz Answers

1. A
2. C
3. True
4. biosphere
5. ecology
6. cell
7. genome
8. B
9. D
10. C
11. A
12. D
13. B
14. A
15. False, three
16. taxonomy
17. domains
18. E
19. D
20. B
21. True
22. Evolution
23. natural selection
24. B
25. D
26. D
27. C
28. A
29. True
30. natural selection
31. C
32. discovery
33. science
34. D
35. True
36. hypothesis
37. E
38. mimicry
39. variable
40. D
41. B
42. False, great
43. evolution
44. technology

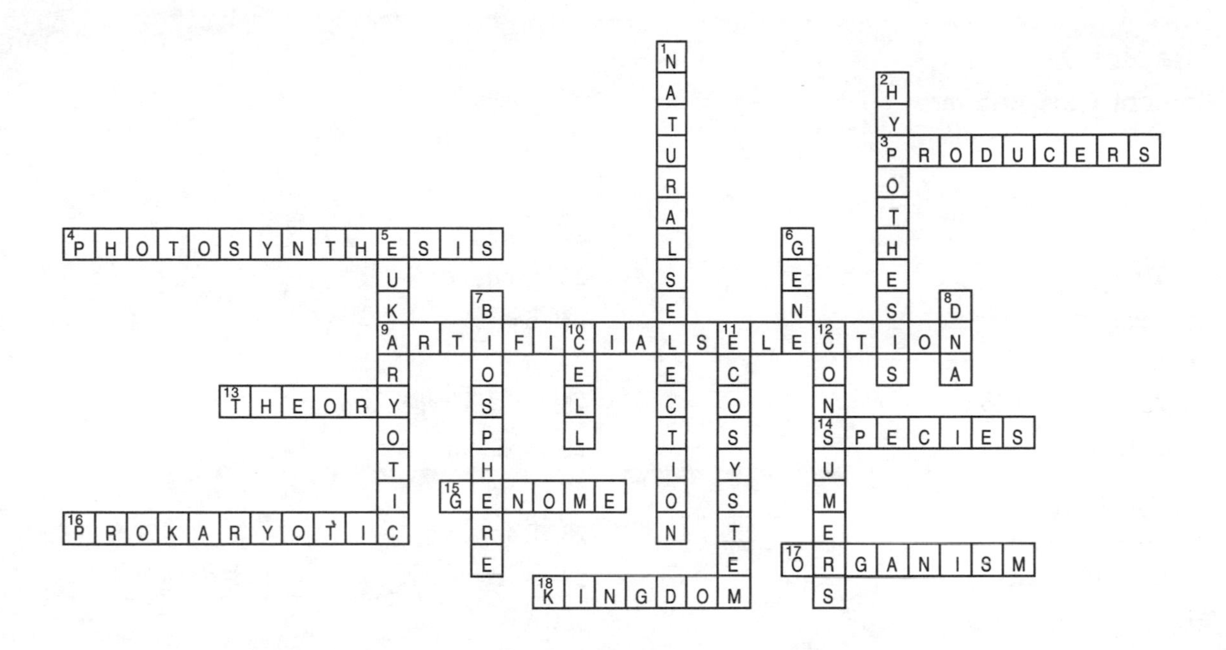
1 NATURALSELECTION
2 HYPOTHESIS
3 PRODUCERS
4 PHOTOSYNTHESIS
5 EUKARYOTIC
6 GENE
7 BIOSPHERE
8 DNA
9 ARTIFICIALSELECTION
10 CELL
11 ECOSYSTEM
12 CONSUMERS
13 THEORY
14 SPECIES
15 GENOME
16 PROKARYOTIC
17 ORGANISM
18 KINGDOM

Chapter 2

Content Quiz Answers

1. E
2. True
3. chemical
4. A
5. False, population
6. cells, atoms
7. C
8. True
9. trace elements
10. D
11. C
12. D
13. True
14. protons, neutrons
15. E
16. B
17. B
18. True
19. ions
20. D
21. True
22. products
23. B
24. B
25. False, oxygen atom
26. covalent
27. A
28. B
29. A
30. A
31. True
32. solute, solution
33. C
34. E
35. D
36. False, 10 times
37. buffers
38. C
39. False, center
40. volcanoes

1 CHEMICALELEMENT
2 ATOMICNUMBER
3 ECOSYSTEM
4 MASSNUMBER
5 CHEMICALREACTION
6 HYDROGENBOND
7 ION
8 ISOTOPE
9 ACID
AQUEOUSSOLUTION
10 POLARMOLECULE
11 HEAT
12 SOLUTE
13 PROTON
14 COHESION
15 SOLVENT
16 BASE
17 MOLECULE
18 PHSCALE
19 PRODUCTS
20 BUFFER
21 TEMPERATURE
22 MASS
23 POPULATION
24 IONICBOND
25 NUCLEUS
26 COVALENTBOND
27 CELL
28 NEUTRON
29 REACTANT
30 TISSUE
31 MATTER
32 ATOM
33 ELECTRON
34 CHEMICALBOND
COMMUNITY
35 COMPOUND
36 SOLUTION
37 TRACEELEMENT
38 ORGANISM
39 RADIOACTIVEISOTOPE
40 ORGAN
41 EVAPORATIVECOOLING
42 ORGANSYSTEM

Chapter 3
Content Quiz Answers

1. D
2. True
3. lactose
4. C
5. A
6. True
7. covalent
8. A
9. False, water
10. polymers, monomers
11. C
12. A
13. B
14. E
15. E
16. False, cellulose
17. True
18. isomers
19. monosaccharides, dehydration synthesis
20. glucose
21. E
22. C
23. D
24. A
25. True
26. anabolic steroids
27. hydrophobic
28. E
29. C
30. D
31. B
32. True
33. primary
34. denaturation
35. C
36. B
37. A
38. False, DNA
39. nucleotides
40. True
41. amino acid

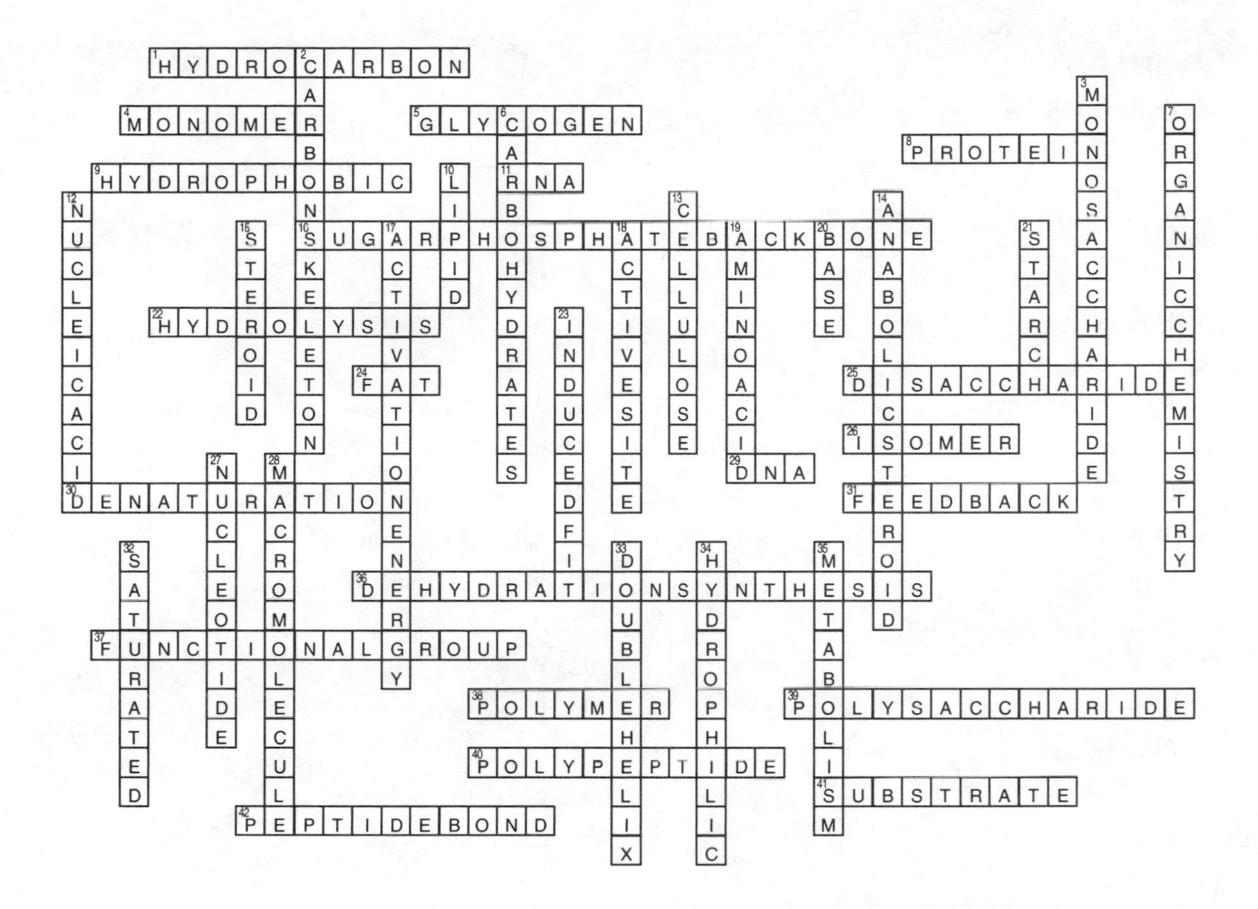
1 HYDROCARBON
2 CARBONSKELETON
3 MONOSACCHARIDE
4 MONOMER
5 GLYCOGEN
6 CARBOHYDRATES
7 ORGANICCHEMISTRY
8 PROTEIN
9 HYDROPHOBIC
10 LIPID
11 RNA
12 NUCLEICACID
13 CELLULOSE
14 ANABOLICSTEROID
15 STEROID
16 SUGARPHOSPHATEBACKBONE
17 ACTIVATIONENERGY
18 ACTIVESITE
19 AMINOACID
20 BASE
21 STARCH
22 HYDROLYSIS
23 INDUCEDFIT
24 FAT
25 DISACCHARIDE
26 ISOMER
27 NUCLEOTIDE
28 MACROMOLECULE
29 DNA
30 DENATURATION
31 FEEDBACK
32 SATURATED
33 DOUBLEHELIX
34 HYDROPHILIC
35 METABOLISM
36 DEHYDRATIONSYNTHESIS
37 FUNCTIONALGROUP
38 POLYMER
39 POLYSACCHARIDE
40 POLYPEPTIDE
41 SUBSTRATE
42 PEPTIDEBOND

Chapter 4

Content Quiz Answers

1. D
2. A
3. False, cell walls
4. antibiotics
5. B
6. C
7. C
8. True
9. scanning
10. eukaryotic, prokaryotic
11. B
12. True
13. fluid mosaic
14. D
15. A
16. True
17. messenger RNA
18. C
19. A
20. False, Golgi apparatus
21. Golgi apparatus
22. lysosomes
23. E
24. E
25. True
26. ATP
27. E
28. D
29. False, microtubules
30. microtubules
31. A
32. B
33. B
34. False
35. plasma membrane
36. communicating
37. tight
38. anchoring
39. A
40. True
41. phospholipids

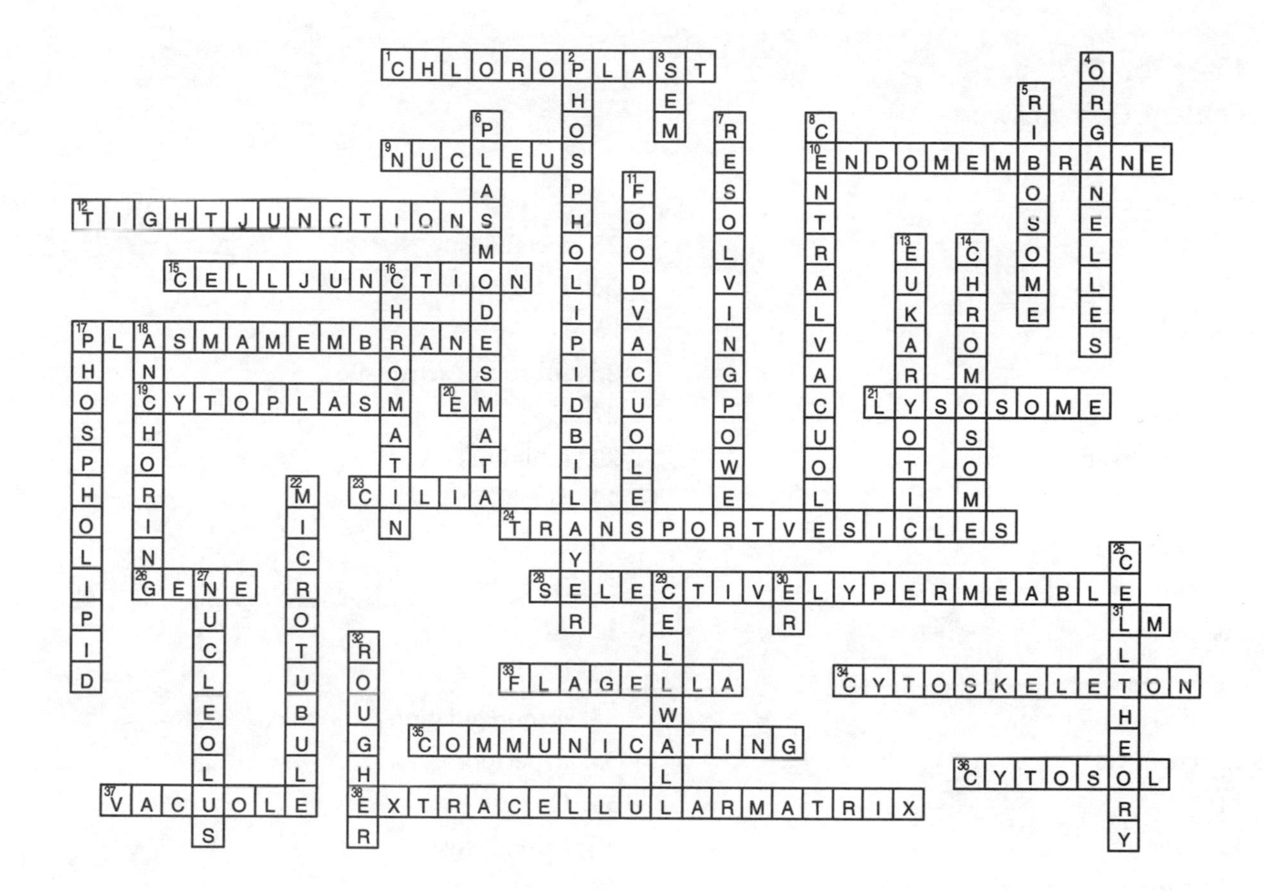
1 CHLOROPLAST
2 PHOSPHOLIPIDBILAYER
3 SEM
4 ORGANELLES
5 RIBOSOME
6 PLASMODESMATA
7 RESOLVINGPOWER
8 CENTRALVACUOLE
9 NUCLEUS
10 ENDOMEMBRANE
11 FOODVACUOLE
12 TIGHTJUNCTIONS
13 EUKARYOTIC
14 CHROMOSOME
15 CELLJUNCTION
16 CHROMATIN
17 PLASMAMEMBRANE
17 PHOSPHOLIPID
18 ANCHORING
19 CYTOPLASM
20 EM
21 LYSOSOME
22 MICROTUBULE
23 CILIA
24 TRANSPORTVESICLES
25 CELLTHEORY
26 GENE
27 NUCLEOLUS
28 SELECTIVELYPERMEABLE
29 CELLWALL
30 ER
31 LM
32 ROUGHER
33 FLAGELLA
34 CYTOSKELETON
35 COMMUNICATING
36 CYTOSOL
37 VACUOLE
38 EXTRACELLULARMATRIX

Chapter 5

Content Quiz Answers

1. A
2. False, less
3. cellulose
4. B
5. True
6. conservation of energy
7. C
8. True
9. entropy
10. A
11. C
12. False, potential
13. chemical
14. B
15. True
16. calorie, kilocalorie
17. False, more
18. triphosphate tail
19. B
20. D
21. A
22. ADP
23. metabolism
24. False, lower
25. D
26. substrate, active site
27. C
28. inhibitors
29. A
30. C
31. B
32. A
33. C
34. osmoregulation
35. hypertonic
36. C
37. False, active
38. D
39. pinocytosis
40. E
41. True
42. C
43. True
44. molecular

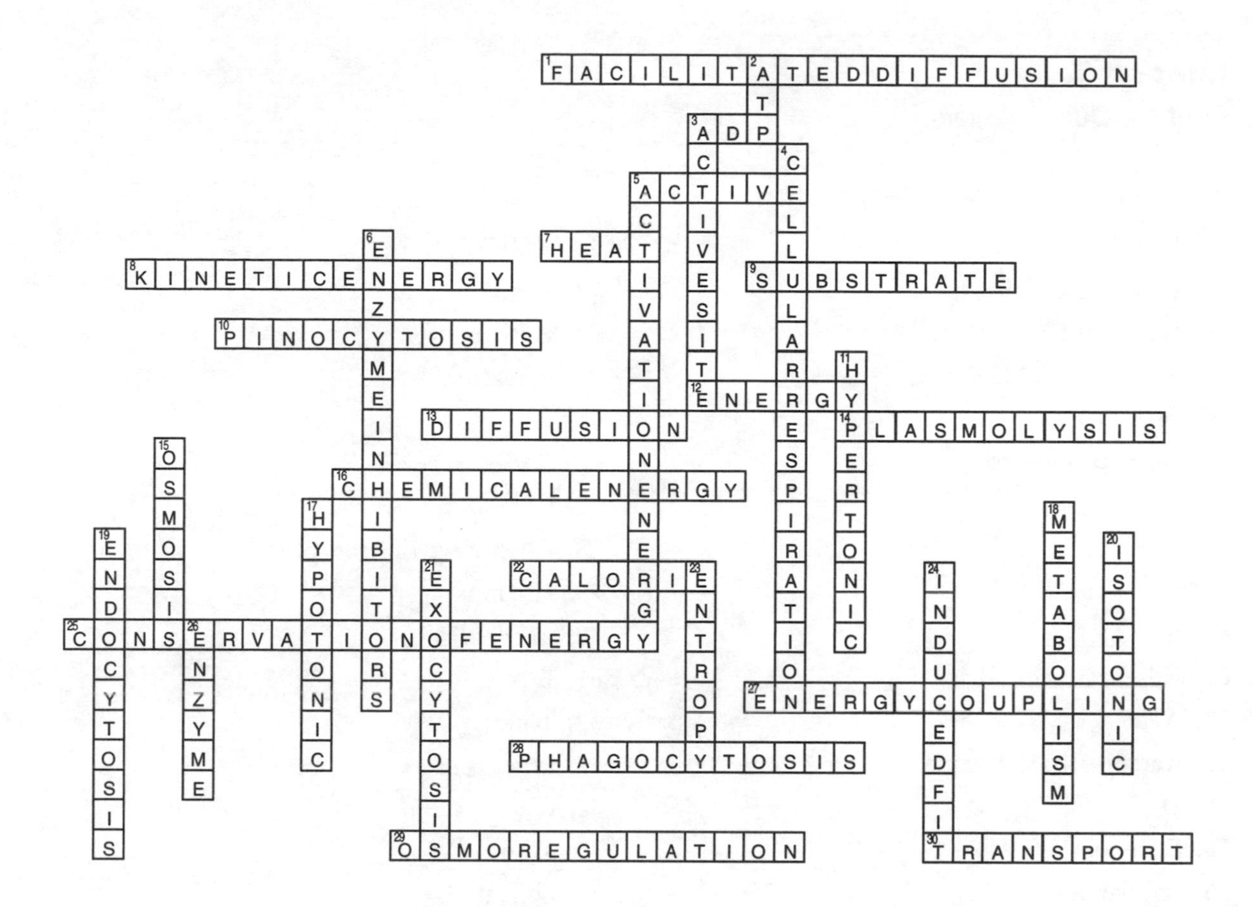

FACILITATEDDIFFUSION
ATP
ADP
ACTIVESITE
CELLULARRESPIRATION
ACTIVE
ACTIVATIONENERGY
ENZYMEINHIBITORS
HEAT
KINETICENERGY
SUBSTRATE
PINOCYTOSIS
HYPERTONIC
ENERGY
DIFFUSION
PLASMOLYSIS
OSMOSIS
CHEMICALENERGY
HYPOTONIC
METABOLISM
ENDOCYTOSIS
ISOTONIC
EXOCYTOSIS
CALORIE
ENTROPY
INDUCEDFIT
CONSERVATIONOFENERGY
ENZYME
ENERGYCOUPLING
PHAGOCYTOSIS
OSMOREGULATION
TRANSPORT

Chapter 6

Content Quiz Answers

1. D
2. False, anaerobic
3. lactic acid
4. A
5. True
6. False, producers
7. heterotrophs
8. A
9. D
10. E
11. False, mitochondria
12. ATP
13. water, carbon dioxide
14. D
15. False, oxygen
16. cellular respiration
17. C
18. True
19. hydrogen
20. B
21. A
22. B
23. False, oxygen
24. oxidation
25. C
26. D
27. C
28. False, electron transport
29. H^+ ions
30. metabolism
31. B
32. True
33. lactic acid
34. C
35. True
36. obligate
37. C
38. True
39. oxygen

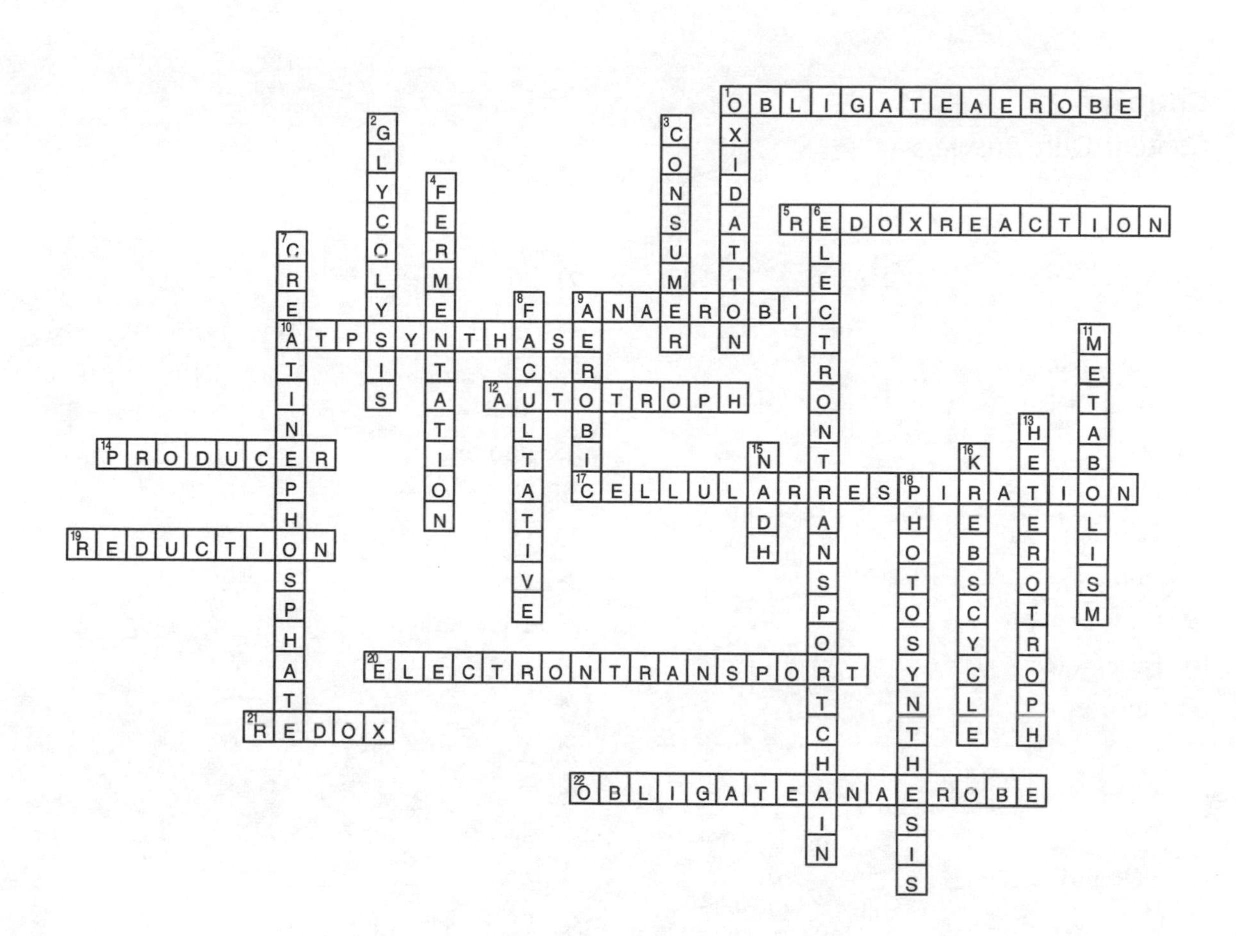

1 OBLIGATEAEROBE
OXIDATION
2 GLYCOLYSIS
3 CONSUMER
4 FERMENTATION
5 REDOXREACTION
6 ELECTRONTRANSPORTCHAIN
7 CREATINEPHOSPHATE
8 FACULTATIVE
9 ANAEROBIC
AEROBIC
10 ATPSYNTHASE
11 METABOLISM
12 AUTOTROPH
13 HETEROTROPH
14 PRODUCER
15 NADH
16 KREBSCYCLE
17 CELLULARRESPIRATION
18 PHOTOSYNTHESIS
19 REDUCTION
20 ELECTRONTRANSPORT
21 REDOX
22 OBLIGATEANAEROBE

Chapter 7

Content Quiz Answers

1. A
2. C
3. False, plants
4. fossil fuels
5. D
6. D
7. False, stroma
8. stomata
9. B
10. False, water
11. glucose
12. A
13. C
14. True
15. ATP and NADPH
16. C
17. True
18. more
19. D
20. False, thylakoid
21. Chlorophyll *a*
22. C
23. D
24. True
25. photon
26. C
27. D
28. B
29. True
30. NADPH
31. A
32. A
33. True
34. oxygen
35. A
36. False, decrease
37. greenhouse effect
38. C
39. False, aerobic
40. oxygen

1 GREENHOUSE EFFECT
2 REACTIONCENTER
3 PRIMARYELECTRONACCEPTOR
4 THYLAKOIDS
5 STROMA
6 WAVELENGTH
7 GRANA
8 PHOTON
9 CAMPLANTS
CALVINCYCLE
10 PHOTOSYSTEM
11 GLOBALWARMING
12 LIGHTREACTION
13 CHLOROPHYLLA
14 NADPH
15 CHLOROPLAST
16 MESOPHYLL
17 ELECTROMAGNETICSPECTRUM

Chapter 8
Content Quiz Answers

1. A
2. False, ten
3. infertility
4. A
5. True
6. chromosomes
7. D
8. False, half
9. asexual
10. C
11. False, proteins
12. centromere
13. D
14. True
15. True
16. mitosis, cytokinesis
17. S
18. A
19. B
20. D
21. B
22. centrosomes
23. B
24. C
25. True
26. benign
27. chemotherapy
28. D
29. False, autosomes
30. homologous
31. B
32. True
33. 25
34. E
35. B
36. True
37. haploid
38. A
39. False, do
40. meiosis II
41. C
42. False, meiosis
43. chiasma
44. E
45. E
46. False, sex chromosomes
47. Down syndrome
48. C
49. True
50. tetraploid

CROSSINGOVER
LIFECYCLE
POLYPLOID
SISTERCHROMATIDS
INTERPHASE
CYTOKINESIS
KARYOTYPE
TRISOMY
FERTILIZATION
HAPLOID
TELOPHASE
MEIOSIS
CANCER
BENIGNTUMOR
MITOTICPHASE
MALIGNANTTUMOR
MITOSIS
ANAPHASE
ASEXUALREPRODUCTION
AUTOSOME
CELLCYCLE
CLEAVAGEFURROW
MITOTICSPINDLE
GAMETE
GENOME
SARCOMA
METASTASIS
LEUKEMIA
LYMPHOMA
CHROMOSOME
PROPHASE
METAPHASE
CELLPLATE
HOMOLOGOUSCHROMOSOMES
DIPLOID
CENTROSOME
DOWNSYNDROME
ZYGOTE
CARCINOMAS
CHIASMA
GENETICRECOMBINATION
CENTROMERE

Chapter 9

Content Quiz Answers

1. D
2. True
3. fetal cells
4. A
5. False, pollinated the plants by hand
6. F_2
7. C
8. D
9. A
10. True
11. locus
12. D
13. False, all
14. monohybrid
15. C
16. True
17. genotype
18. C
19. True
20. D
21. B
22. False
23. True
24. carriers
25. E
26. False, more
27. heterozygotes
28. Huntington disease
29. cystic fibrosis
30. B
31. False, did not show
32. hypercholesterolemia
33. A
34. C
35. False, dominant
36. AB
37. B
38. True
39. pleiotropy
40. E
41. True
42. polygenic inheritance
43. A
44. True
45. genetic
46. D
47. B
48. True
49. chromosome theory of inheritance
50. C
51. False, incorrect
52. fruit flies
53. D
54. True
55. sex
56. A
57. False, occur
58. False, several
59. hemophilia
60. Duchenne muscular dystrophy
61. D
62. True
63. Y

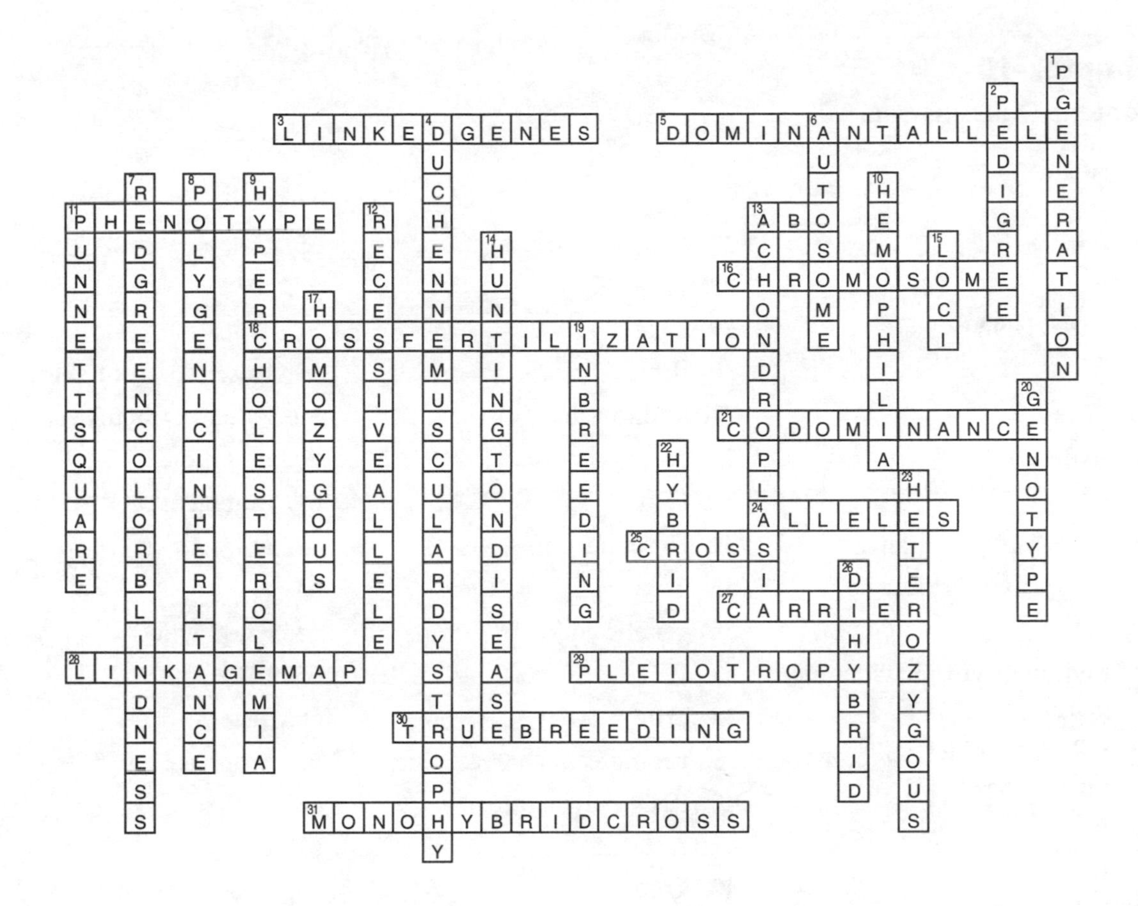

1 PGENERATION
2 PEDIGREE
3 LINKEDGENES
4 DUCHENNEMUSCULARDYSTROPHY
5 DOMINANTALLELE
6 AUTOSOME
7 REDGREENCOLORBLINDNESS
8 POLYGENICINHERITANCE
9 HYPERCHOLESTEROLEMIA
10 HEMOPHILIA
11 PHENOTYPE
11 PUNNETTSQUARE
12 RECESSIVEALLELE
13 ABO
13 ACHONDROPLASIA
14 HUNTINGTONSDISEASE
15 LOCI
16 CHROMOSOME
17 HOMOZYGOUS
18 CROSSFERTILIZATION
19 INBREEDING
20 GENOTYPE
21 CODOMINANCE
22 HYBRID
23 HETEROZYGOUS
24 ALLELES
25 CROSS
26 DIHYBRID
27 CARRIER
28 LINKAGEMAP
29 PLEIOTROPY
30 TRUEBREEDING
31 MONOHYBRIDCROSS

Chapter 10

Content Quiz Answers

1. E
2. B
3. False, DNA
4. B
5. True
6. structure
7. C
8. False, is missing an
9. deoxyribose, ribose
10. uracil
11. thymine, cytosine, adenine, guanine
12. polynucleotides, nucleotides
13. B
14. A
15. False, outside
16. GCTA
17. C
18. B
19. False, many replication origins
20. DNA polymerase
21. DNA polymerase
22. D
23. B
24. False, proteins
25. genotype, phenotype
26. transcription, translation
27. C
28. True
29. True
30. three
31. amino acids
32. A
33. C
34. False, no gaps
35. True
36. genetic code
37. C
38. True
39. promoter
40. RNA polymerase
41. A
42. True
43. introns, exons
44. B
45. E
46. D
47. False, anticodon
48. mRNA, tRNA
49. A
50. True
51. amino acid
52. A, P, translocation
53. stop codon
54. A
55. C
56. A
57. True
58. False, does not shift
59. False, not beneficial
60. mutation
61. mutagens
62. D
63. A
64. False, lytic
65. False, can
66. True
67. prophage
68. E
69. E
70. False, are not
71. True
72. provirus
73. plasma membrane, nuclear membrane
74. B
75. plasma membrane
76. retrovirus
77. reverse transcriptase
78. D
79. True
80. True
81. True

Chapter 11

Content Quiz Answers

1. B
2. True
3. stem
4. E
5. False, some
6. cellular differentiation
7. A
8. False, DNA
9. genes
10. B
11. True
12. True
13. regeneration
14. A
15. E
16. False, nucleus
17. False, 1950s
18. E
19. False, Adult
20. therapeutic
21. D
22. D
23. A
24. B
25. False, RNA
26. operon
27. C
28. B
29. D
30. D
31. B
32. False, prevent
33. True
34. False, DNA
35. enhancers
36. transcription factors
37. silencers
38. True
39. signal-transduction
40. A
41. C
42. D
43. True
44. False, DNA
45. True
46. False, not caused
47. growth factors
48. E
49. True
50. True
51. carcinogens
52. E
53. False, similar
54. nucleotides

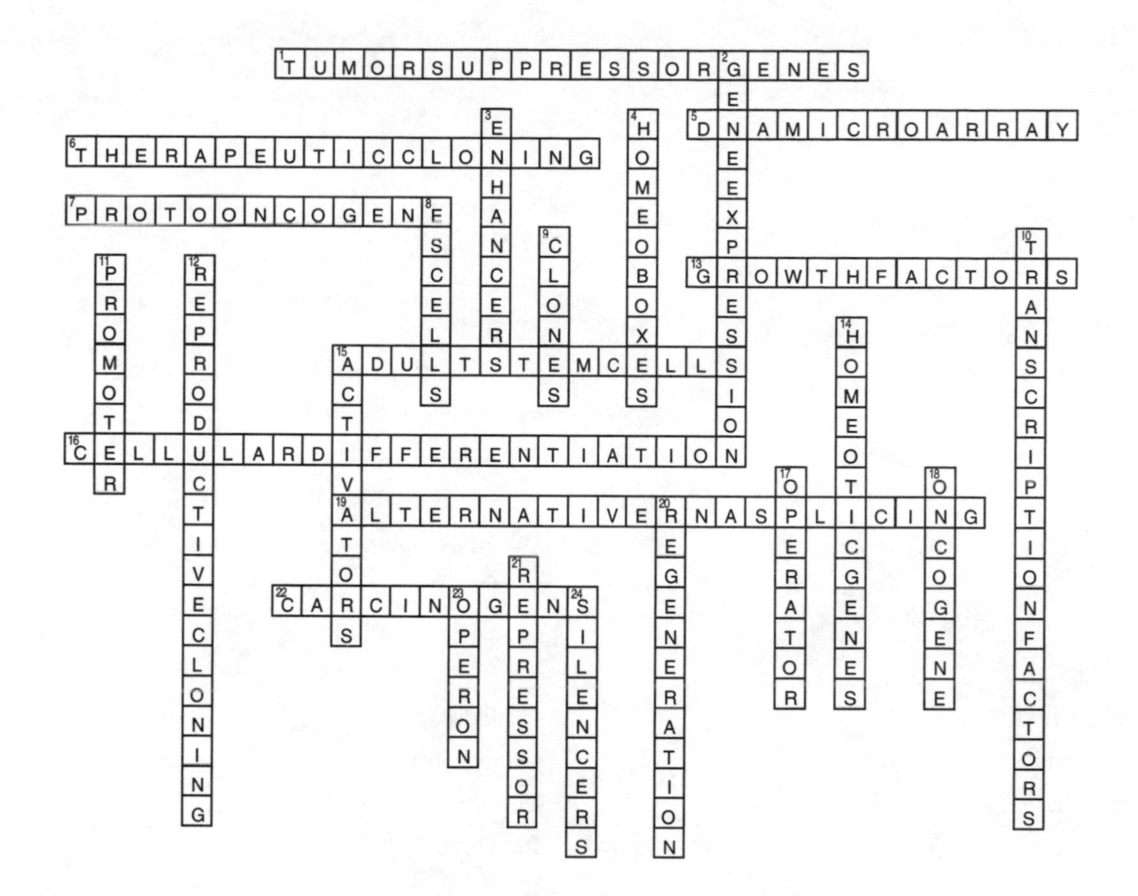
1 TUMORSUPPRESSORGENES
2 GENEEXPRESSION
3 ENHANCER
4 HOMEOBOXES
5 DNAMICROARRAY
6 THERAPEUTICCLONING
7 PROTOONCOGENE
8 ESCELLS
9 CLONES
10 TRANSCRIPTIONFACTORS
11 PROMOTER
12 REPRODUCTIVECLONING
13 GROWTHFACTORS
14 HOMEOTICGENES
15 ADULTSTEMCELLS
15 ACTIVATORS
16 CELLULARDIFFERENTIATION
17 OPERATOR
18 ONCOGENE
19 ALTERNATIVERNASPLICING
20 REGENERATION
21 REPRESSOR
22 CARCINOGENS
23 OPERON
24 SILENCERS

Chapter 12

Content Quiz Answers

1. D
2. E
3. False, gene
4. genetic material
5. A
6. True
7. True
8. recombinant DNA technology
9. biotechnology
10. E
11. True
12. vitamin A
13. bacteria
14. cholera
15. 4
16. 2
17. 7
18. 1
19. 6
20. 3
21. 5
22. D
23. False, separate from
24. True
25. False, the same
26. True
27. nucleic acid probe
28. restriction enzymes
29. genomic library
30. vectors
31. restriction sites
32. DNA ligase
33. C
34. True
35. DNA fingerprinting
36. A
37. polymerase chain reaction
38. RFLP analysis
39. DNA polymerase
40. forensics
41. D
42. A
43. False, telomeres
44. genomics
45. repetitive DNA
46. D
47. False, 98.5%
48. C
49. D
50. False, have not
51. sequenced
52. E
53. False, artificial
54. stem
55. E
56. True
57. C
58. True
59. C
60. False, do
61. True

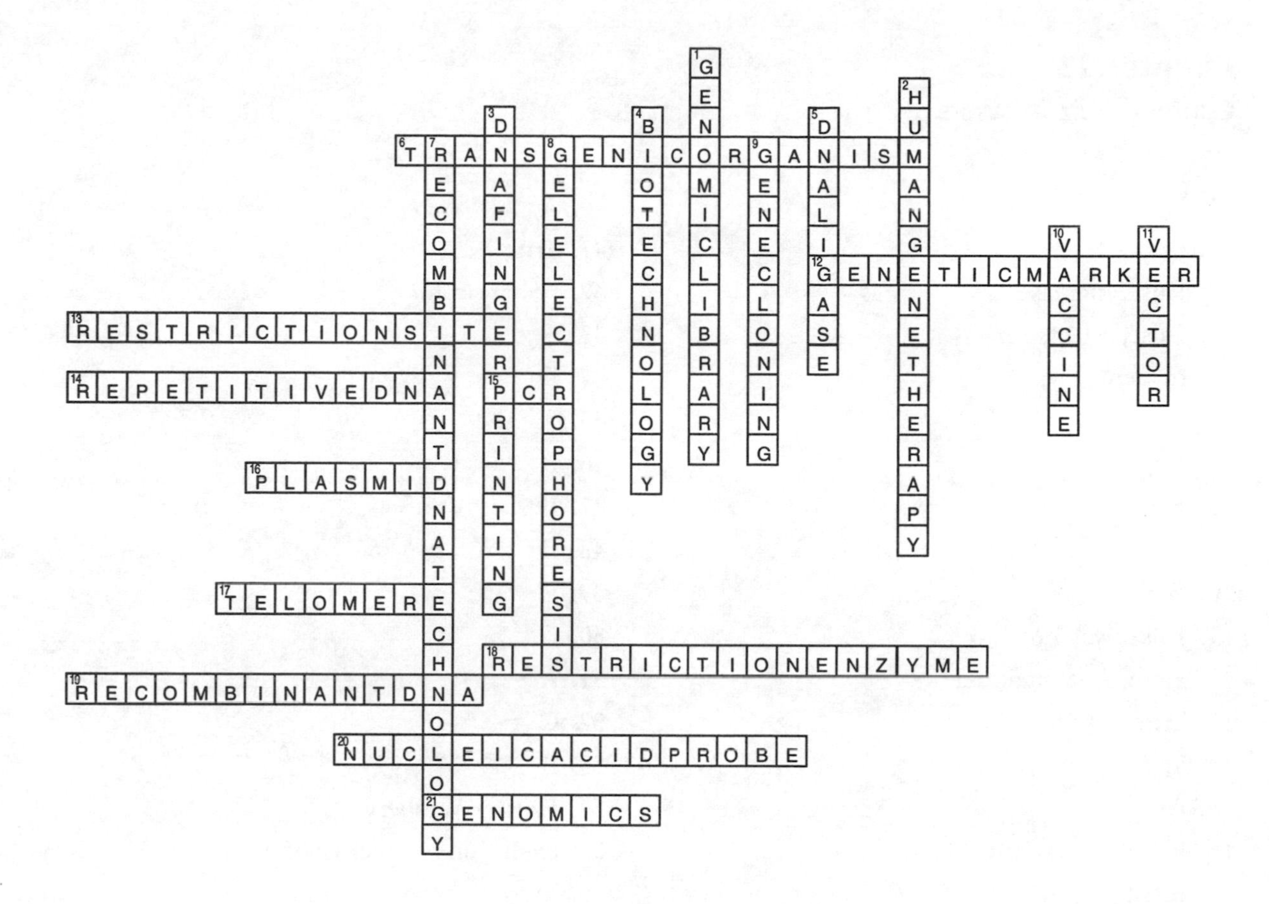

1 GENOMICLIBRARY
2 HUMANGENETHERAPY
3 DNAFINGERPRINTING
4 BIOTECHNOLOGY
5 DNALIGASE
6 TRANSGENICORGANISM
7 RECOMBINANTDNATECHNOLOGY
8 GELELECTROPHORESIS
9 GENECLONING
10 VACCINE
11 VECTOR
12 GENETICMARKER
13 RESTRICTIONSITE
14 REPETITIVEDNA
15 PCR
16 PLASMID
17 TELOMERE
18 RESTRICTIONENZYME
19 RECOMBINANTDNA
20 NUCLEICACIDPROBE
21 GENOMICS

Chapter 13

Content Quiz Answers

1. D
2. C
3. False, increase
4. False, selects for
5. descent
6. C
7. D
8. B
9. D
10. B
11. False, was not
12. False, Anaximander
13. True
14. True
15. A
16. False, prokaryotes
17. fossil
18. D
19. C
20. A
21. False, does not support
22. A
23. E
24. True
25. common ancestor
26. B
27. D
28. True
29. False, most
30. South America
31. environment
32. A
33. True
34. True
35. C
36. False, can
37. False, populations
38. population genetics
39. modern synthesis
40. A
41. True
42. polymorphic
43. C
44. D
45. A
46. B
47. B
48. D
49. E
50. C
51. A
52. C
53. B
54. Hardy-Weinberg
55. evolution or microevolution
56. D
57. A
58. D
59. D
60. False, decrease
61. True
62. True
63. False, does not have
64. genetic drift
65. mutation
66. E
67. C
68. B
69. C, D
70. A
71. True
72. stabilizing
73. extinction
74. C
75. A
76. True
77. carriers

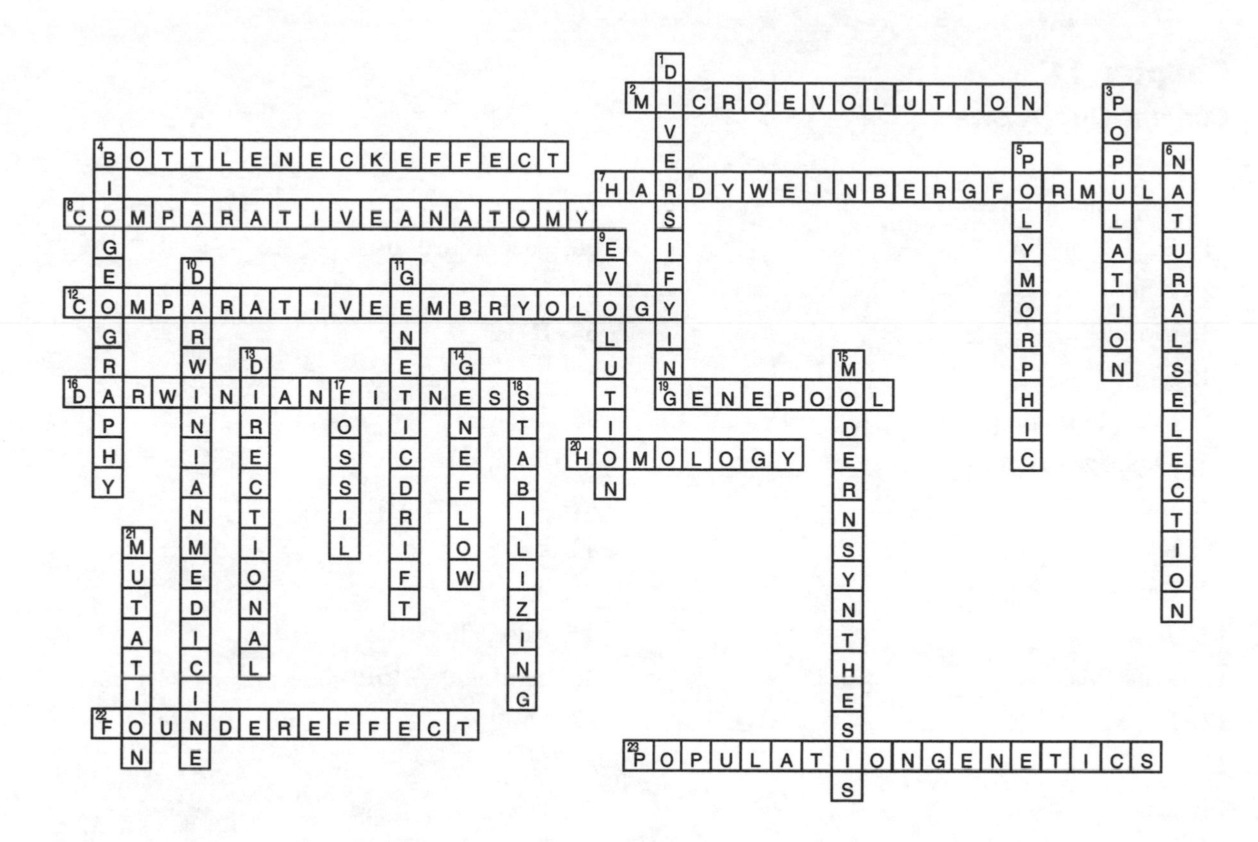

D
MICROEVOLUTION
POPULATION
BOTTLENECKEFFECT
POLYMORPHIC
NATURALSELECTION
HARDYWEINBERGFORMULA
COMPARATIVEANATOMY
BIOGEOGRAPHY
DIVERSIFYING
EVOLUTION
DARWINIANMEDICINE
GENETICDRIFT
COMPARATIVEEMBRYOLOGY
DIRECTIONAL
GENEFLOW
MODERNSYNTHESIS
DARWINIANFITNESS
FOSSIL
STABILIZING
GENEPOOL
HOMOLOGY
MUTATION
FOUNDEREFFECT
POPULATIONGENETICS

Chapter 14
Content Quiz Answers

1. E
2. True
3. iridium
4. E
5. False, branching
6. Galápagos Islands
7. speciation
8. D
9. True
10. True
11. Ernst Mayr
12. D
13. A
14. True
15. hybrid sterility
16. D
17. D
18. False, two
19. sympatric
20. allopatric
21. sympatric
22. allopatric
23. allopatric
24. sympatric
25. A
26. C
27. True
28. False, little
29. B
30. False, could not have
31. exaptation
32. C
33. A
34. D
35. True
36. paedomorphosis
37. C
38. D
39. B
40. D
41. False, relative
42. radiometric dating
43. B
44. A
45. False, allopatric
46. continental drift
47. D
48. D
49. False, six
50. A
51. E
52. A
53. True
54. binomial
55. A
56. D
57. E
58. D
59. False, analogous, homologous
60. the same structure
61. analogous
62. homologous
63. convergent, homology
64. cladistic analysis
65. E
66. False, Protista
67. cladistic analysis
68. B
69. True

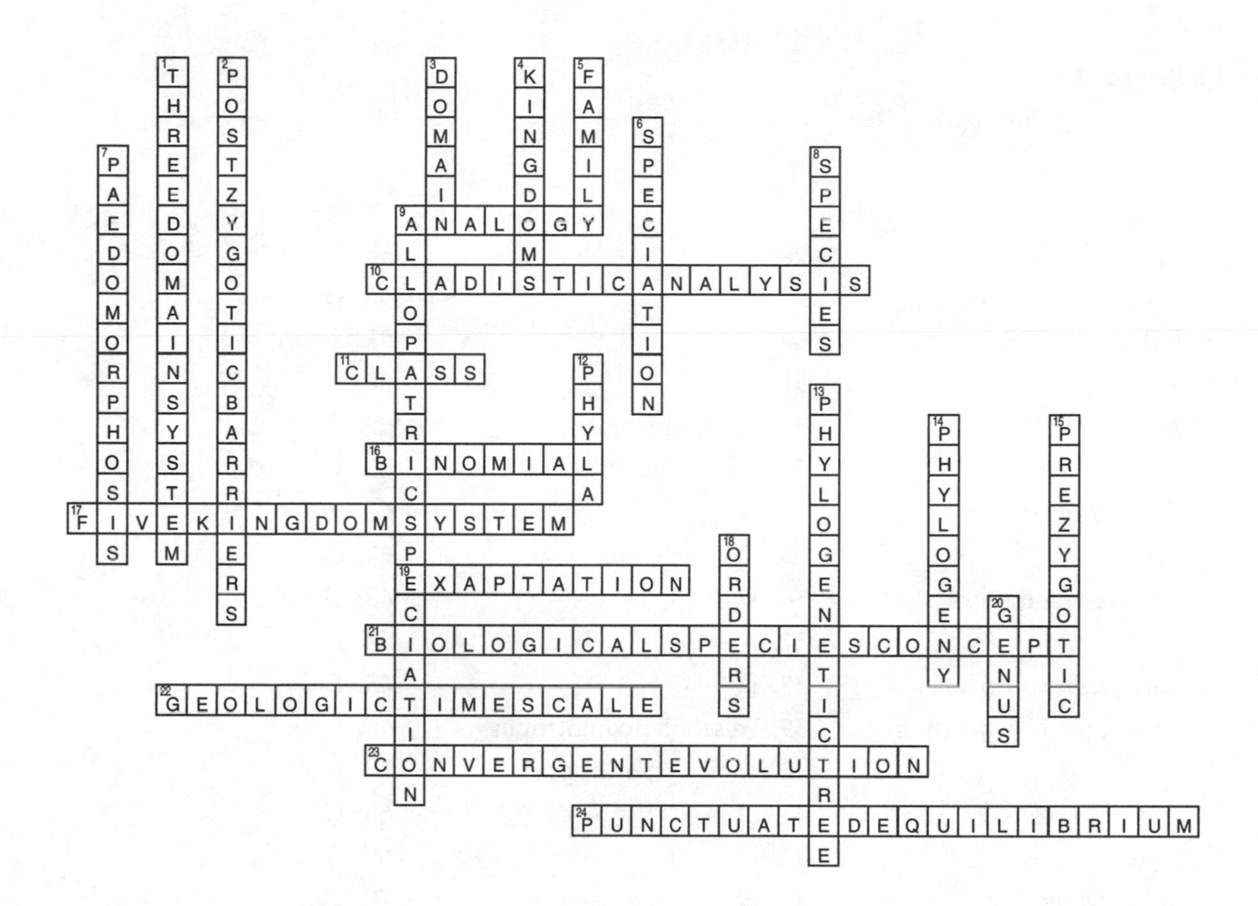
1 THREEDOMAINSYSTEM
2 POSTZYGOTICBARRIERS
3 DOMAIN
4 KINGDOM
5 FAMILY
6 SPECIATION
7 PAEDOMORPHOSIS
8 SPECIES
9 ANALOGY
9 ALLOPATRICSPECIATION
10 CLADISTICANALYSIS
11 CLASS
12 PHYLA
13 PHYLOGENETICTREE
14 PHYLOGENY
15 PREZYGOTIC
16 BINOMIAL
17 FIVEKINGDOMSYSTEM
18 ORDERS
19 EXAPTATION
20 GENUS
21 BIOLOGICALSPECIESCONCEPT
22 GEOLOGICTIMESCALE
23 CONVERGENTEVOLUTION
24 PUNCTUATEDEQUILIBRIUM

Chapter 15
Content Quiz Answers

1. E
2. True
3. bubonic plague
4. A
5. B
6. D
7. False, prokaryotes
8. True
9. cellular respiration
10. E
11. False, different from
12. spontaneous generation
13. D
14. E
15. True
16. False, would have
17. ribozymes/other RNA
18. E
19. True
20. False, would survive
21. bacteria
22. C
23. False, anaerobic
24. archaea
25. F
26. C
27. E
28. A
29. B
30. D
31. A
32. False, different
33. binary fission
34. B
35. C
36. D
37. A
38. D
39. False, photoautotrophs
40. chemoheterotrophs
41. D
42. B
43. False, are not
44. bioremediation
45. B
46. E
47. C
48. C
49. False, aerobic
50. mitochondria
51. protists
52. H
53. K
54. G
55. I
56. D
57. B
58. F
59. E
60. J
61. A
62. C
63. G
64. D
65. E
66. B
67. F
68. C
69. A
70. D
71. B
72. D
73. B
74. True
75. green algae
76. pigments
77. D
78. True
79. Volvox

1 CHEMOHETEROTROPHS
2 SPIROCHETES
3 CHEMOAUTOTROPHS
4 ENDOSPORES
5 PHOTOHETEROTROPHS
6 DIATOMS
7 PROTOZOANS
8 PHOTOAUTOTROPHS
9 CILIATES
10 RIBOZYMES
11 ALGAE
11 AMOEBAS
12 APICOMPLEXANS
13 PLASMODIALSLIMEMOLDS
14 BINARYFISSION
15 ARCHAEA
16 PROTISTS
17 BACTERIA
17 BIOREMEDIATION
18 PATHOGENS
19 PLANKTON
20 BIOGENESIS
21 FLAGELLATES
22 BACILLI
23 ENDOTOXINS
24 GREENALGAE
25 FORAMS
26 CELLULARSLIMEMOLDS
27 EXOTOXIN
28 PSEUDOPODIA
29 DINOFLAGELLATES
30 ENDOSYMBIOSIS
31 SEAWEEDS

Chapter 16
Content Quiz Answers

1. D
2. False, more
3. coniferous
4. C
5. E
6. D
7. A
8. A
9. B
10. A
11. D
12. True
13. mycorrhizae
14. stomata
15. cuticle
16. roots; shoots
17. lignin
18. D
19. False, would
20. charophyceans
21. B
22. D
23. C
24. A
25. False, angiosperms
26. seed
27. D
28. B
29. False, diploid
30. False, gametophyte
31. sporophyte; gametophyte
32. alternation of generations
33. B
34. True
35. coal
36. C
37. E
38. A
39. C
40. True
41. False, do not have
42. ionizing radiation
43. pollen
44. female gametophytes
45. D
46. F
47. B
48. E
49. C
50. A
51. A
52. C
53. True
54. False, angiosperms
55. False, flower
56. fruit
57. plants
58. C
59. True
60. True
61. C
62. A
63. D
64. False, chitin
65. spores
66. hydrolytic enzymes
67. hyphae
68. mycelium
69. H
70. False, less
71. decomposers
72. D
73. False, mutualistic
74. lichen

1 VASCULARTISSUE
2 LICHENS
3 STIGMA
4 GYMNOSPERMS
5 FRUIT
6 MUTUALISM
7 SHOOTS
8 OVULES
9 CUTICLE
10 FLOWER
11 DOUBLEFERTILIZATION
12 FERN
13 CARPEL
14 FOSSILFUELS
15 MYCORRHIZAE
16 ALTERNATIONOFGENERATIONS
16 ABSORPTION
17 ENDOSPERM
18 MYCELIUM
19 CHAROPHYCEANS
20 OVARY
21 SPORES
22 XYLEM
23 SEED
24 BRYOPHYTES
25 ANTHER
26 GAMETOPHYTE
27 ROOTS
28 MOSSES
29 GERMINATE
30 PHLOEM
31 ANGIOSPERMS
32 GAMETANGIA
33 LIGNIN
34 STOMATA
35 SPOROPHYTE
36 PETALS
37 STYLE
38 STAMEN
39 POLLEN
40 SEPALS

Chapter 17

Content Quiz Answers

1. E
2. True
3. quolls
4. D
5. A
6. True
7. False, zygote, blastula, gastrula
8. metamorphosis
9. nervous
10. C
11. True
12. True
13. E
14. C
15. A
16. bilateral
17. pseudocoelom
18. D
19. D
20. False, are marine
21. choanocytes
22. B
23. A
24. False, cnidocytes
25. polyp
26. medusa
27. A
28. True
29. False, more than one
30. rare/not well cooked
31. B
32. B
33. True
34. pseudocoelom
35. D
36. False, bivalves
37. cephalopods
38. radula
39. mantle
40. B
41. D
42. False, free-living
43. soil; castings
44. anticoagulant; blood clots
45. D
46. E
47. C
48. A
49. C
50. B
51. D
52. True
53. False, complete
54. exoskeleton
55. crustaceans
56. entomology
57. segments
58. insects
59. A
60. True
61. radial; bilateral
62. L
63. J
64. F
65. M
66. N
67. B
68. C
69. A
70. K
71. H
72. G
73. I
74. E
75. D
76. A
77. B
78. False, includes
79. lancelets; tunicates
80. C
81. B
82. D
83. E
84. A
85. D
86. True
87. lateral line
88. operculum
89. swim bladder
90. C
91. C
92. E
93. skin
94. tetrapods
95. A
96. E
97. False, 10% of
98. amniotic
99. ectotherms
100. B
101. D
102. False, is not
103. keratin
104. A
105. C
106. False, marsupial
107. monotremes
108. Primates
109. A
110. A

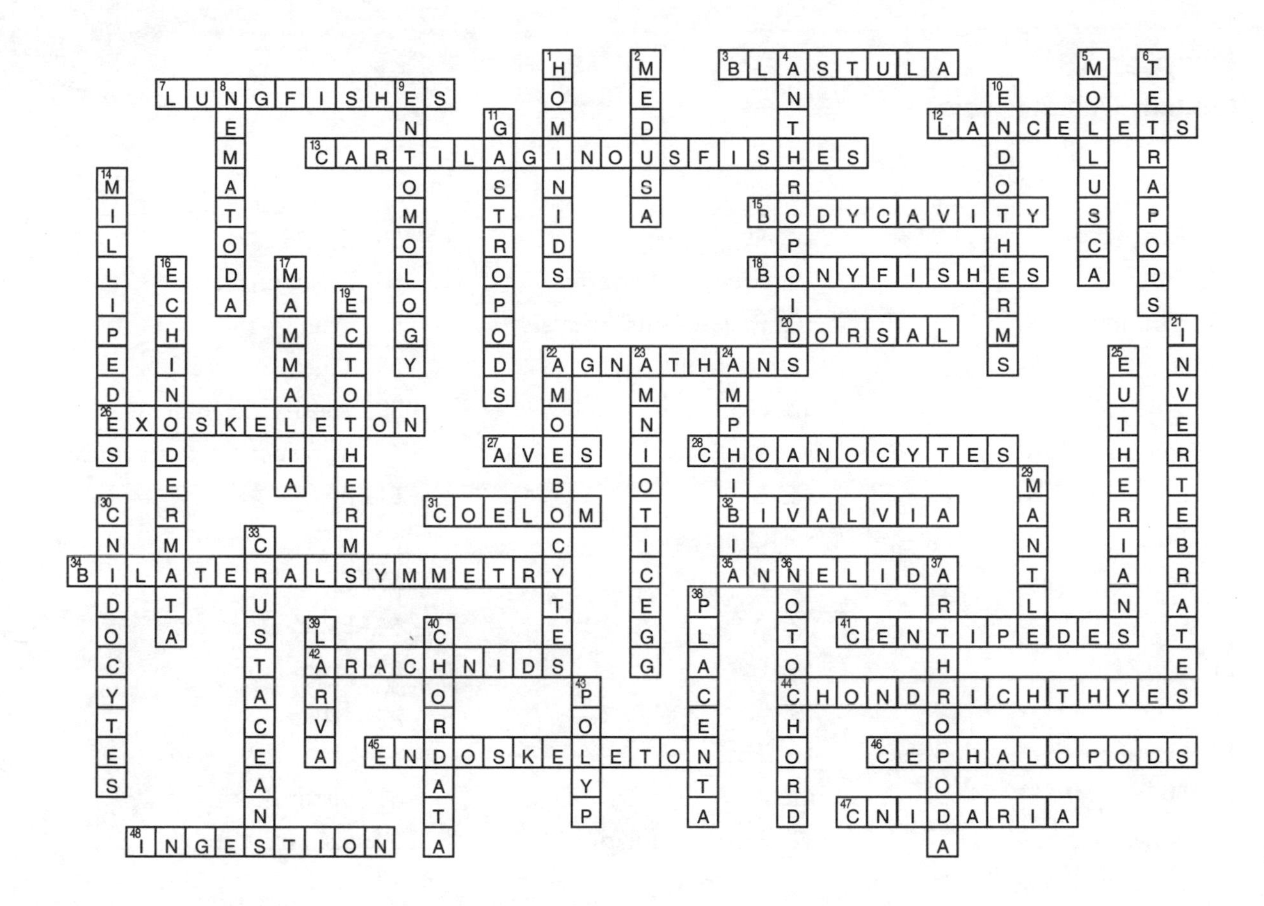

111. True
112. False, Old World
113. prosimians
114. C
115. E
116. D
117. A
118. B
119. A
120. C
121. E
122. False, are not
123. False, at different rates
124. True
125. True
126. Four
127. culture
128. A
129. True

Chapter 18

Content Quiz Answers

1. F
2. True
3. 6 billion
4. A
5. True
6. abiotic
7. B
8. D
9. A
10. C
11. E
12. True
13. biosphere
14. D
15. False, increases
16. ecology
17. B
18. D
19. D
20. True
21. True
22. False, surface
23. True
24. winds
25. C
26. A
27. B
28. D
29. False, Endotherms
30. False, plants, animals
31. acclimation
32. ecological
33. anatomical
34. B
35. C
36. D
37. A
38. C
39. C
40. True
41. A
42. False, uniform
43. random
44. B
45. B
46. False, intermediate
47. True
48. exponential growth
49. logistic growth
50. population-limiting
51. C
52. B
53. D
54. D
55. E
56. False, increases
57. True
58. True
59. density-dependent
60. intraspecific competition
61. A
62. D
63. C
64. False, a decrease
65. True
66. True
67. True
68. birth; death
69. age structure
70. A
71. True
72. life table
73. survivorship curve
74. life history
75. C
76. True
77. opportunistic
78. C
79. True

1 ACCLIMATION
1 ABIOTICCOMPONENT
2 DENSITYDEPENDENTFACTOR
3 BIOSPHERE
4 LIFETABLE
5 ECOSYSTEM
6 BIOTICCOMPONENT
7 DISPERSIONPATTERN
8 CARRYINGCAPACITY
9 CLUMPED
10 ECOSYSTEMECOLOGY
11 RANDOM
12 COMMUNITY
13 ORGANISMALECOLOGY
14 AGESTRUCTURE
15 MARKRECAPTUREMETHOD
16 COMMUNITYECOLOGY
17 ECOLOGY
18 LIFEHISTORY
19 LOGISTICGROWTHMODEL
20 POPULATION
21 HABITATS
22 EXPONENTIALGROWTHMODEL
23 EQUILIBRIALLIFEHISTORY
24 UNIFORM
25 OPPORTUNISTICLIFEHISTORIES

Chapter 19

Content Quiz Answers

1. B
2. False, shallow
3. predators
4. ecosystems
5. D
6. C
7. A
8. B
9. B
10. True
11. species diversity
12. environment
13. community
14. C
15. C
16. D
17. True
18. interspecific
19. interspecific competition
20. niche
21. resource partitioning
22. C
23. A
24. D
25. D
26. B
27. False, pursue
28. False, cryptic
29. True
30. False, Müllerian
31. warning
32. keystone
33. C
34. D
35. B
36. False, parasitism
37. True
38. False, few
39. symbiont, host
40. C
41. B
42. True
43. ecological succession
44. C
45. True
46. intermediate disturbance
47. C
48. False, sunlight
49. True
50. ecosystem
51. abiotic
52. air; soil
53. trophic structure
54. C
55. D
56. A
57. B
58. B
59. D
60. C
61. False, producers
62. trophic levels
63. food chain
64. prokaryotes; fungi
65. webs
66. D
67. E
68. True
69. True
70. biomass
71. primary productivity
72. C
73. True
74. omnivores
75. A
76. C
77. False, ammonium or nitrate
78. True
79. True
80. biogeochemical
81. B
82. True
83. False, does not necessarily have
84. convergence/convergent evolution
85. D
86. C
87. A
88. B
89. E
90. C
91. C
92. False, high
93. True
94. C
95. C
96. True
97. True
98. True
99. False, the same factors as
100. False, phytoplankton
101. estuary
102. pelagic
103. aphotic
104. hydrothermal vent
105. C
106. A
107. True
108. coevolution

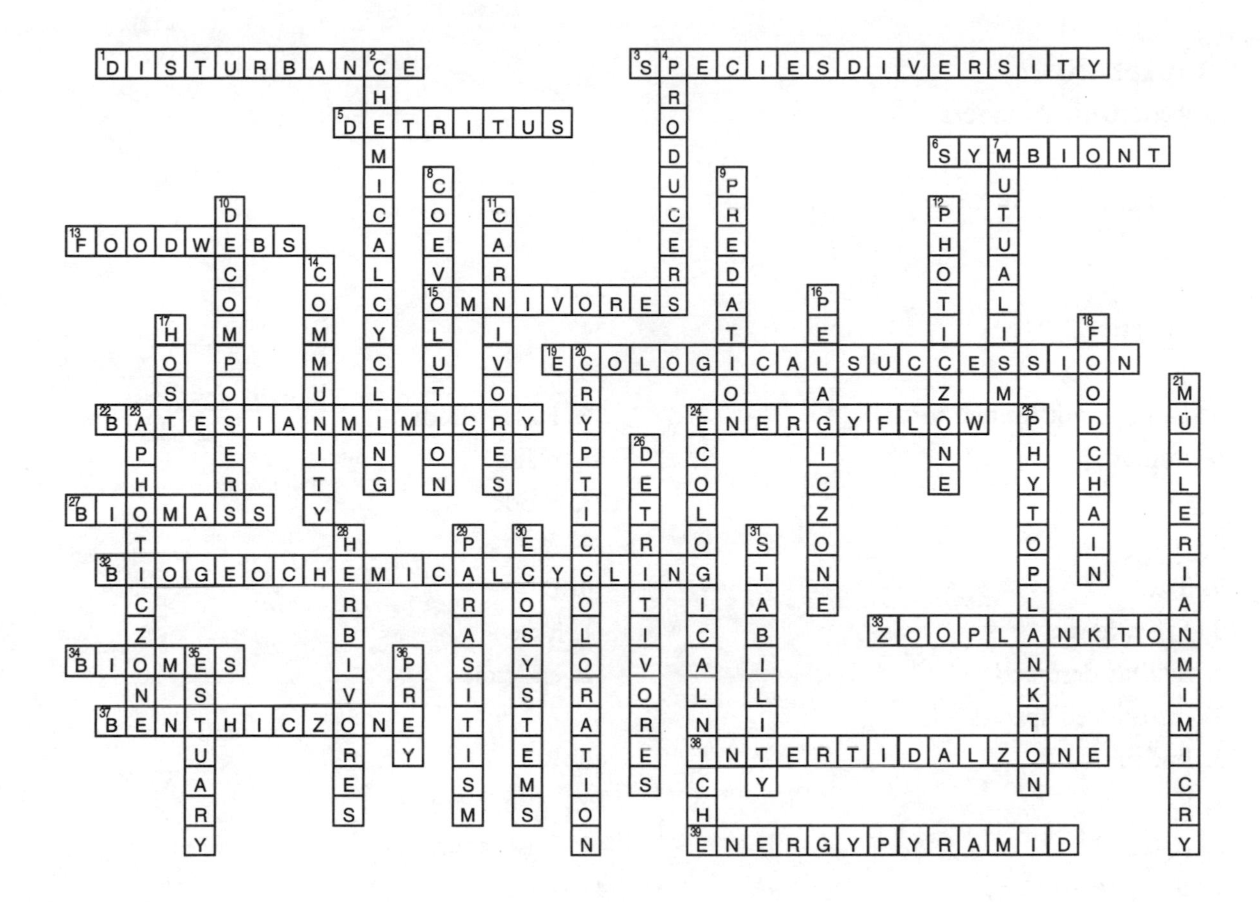
DISTURBANCE
SPECIESDIVERSITY
DETRITUS
SYMBIONT
FOODWEBS
OMNIVORES
ECOLOGICALSUCCESSION
BATESIANMIMICRY
ENERGYFLOW
BIOMASS
BIOGEOCHEMICALCYCLING
ZOOPLANKTON
BIOMES
BENTHICZONE
INTERTIDALZONE
ENERGYPYRAMID
CHEMICALCYCLING
PRODUCERS
MUTUALISM
COEVOLUTION
CARNIVORES
PREDATION
DECOMPOSERS
PHOTICZONE
COMMUNITY
PELAGICZONE
HOST
FOODCHAIN
CRYPTICCOLORATION
MÜLLERIANMIMICRY
APHOTICZONE
ECOLOGICALNICHE
PHYTOPLANKTON
DETRITIVORES
HERBIVORES
PARASITISM
ECOSYSTEMS
STABILITY
ESTUARY
PREY

Chapter 20

Content Quiz Answers

1. A
2. True
3. diversity
4. C
5. False, is not limited to
6. humans
7. D
8. A
9. True
10. False, fewer
11. habitat destruction
12. introduced species
13. kudzu
14. C
15. A
16. B
17. C
18. True
19. eutrophication
20. E
21. True
22. DDT
23. C
24. D
25. False, fossil fuels
26. True
27. greenhouse effect
28. Antarctica
29. D
30. A
31. B
32. True
33. C
34. C
35. A
36. False, trillion
37. plants
38. exotic
39. A
40. D
41. True
42. endemic
43. A
44. B
45. False, before
46. False, source
47. True
48. True
49. True
50. endangered
51. threatened
52. E
53. B
54. False, increased
55. True
56. False, can
57. landscape ecology
58. movement corridor
59. zoned reserve
60. sustainable development
61. biophilia

Across

1. BIODIVERSITYHOTSPOT
2. EUTROPHICATION
4. BIOLOGICALMAGNIFICATION
5. ENDANGEREDSPECIES
6. MOVEMENTCORRIDOR
7. ENDEMICSPECIES
8. LANDSCAPEECOLOGY
12. CONSERVATIONBIOLOGY
13. OZONELAYER
14. GREENHOUSEEFFECT
15. INTRODUCEDSPECIES
16. SUSTAINABLEDEVELOPMENT

Down

1. BIODIVERSITYCRISIS
3. POPULATIONFRAGMENTATION
9. SINKHABITAT
10. SOURCEHABITAT
11. ZONEDRESERVE

Chapter 21
Content Quiz Answers

1. A
2. False, heat exhaustion
3. heat stroke
4. True
5. anatomy, physiology
6. D
7. A
8. B
9. C
10. B
11. A
12. False, neuron
13. loose connective tissue
14. B
15. True
16. organ systems
17. B
18. False, an open
19. circulatory
20. D
21. True
22. homeostasis
23. A
24. A
25. False, inhibit
26. negative
27. D
28. A
29. False, endotherms
30. thermoregulation
31. E
32. True
33. False, osmoregulators
34. osmoregulation
35. C
36. A
37. B
38. C
39. A
40. A
41. D
42. True
43. dialysis
44. A
45. False, convergent
46. fusiform

1 CONVERGENT EVOLUTION
2 FILTRATION
3 POSITIVE FEEDBACK
4 ORGAN SYSTEM
5 FIBROUS CONNECTIVE TISSUE
6 EXCRETION
7 NEURON
8 OSMOREGULATOR
9 NEPHRON
10 OSMOCONFORMERS
11 URETER
12 ENDOTHERMS
13 PHYSIOLOGY
14 ORGAN
15 EPITHELIAL TISSUE
16 SECRETION
17 ADIPOSE TISSUE
18 MUSCLE TISSUE
19 INTERSTITIAL FLUID
20 CARDIAC MUSCLE
21 HOMEOSTASIS
22 BLOOD
23 LOOSE CONNECTIVE TISSUE
24 BONE
25 FILTRATE
26 REABSORPTION
27 CARTILAGE
28 URETHRA
29 TISSUE
30 DIALYSIS
31 OSMOREGULATION
32 CONNECTIVE TISSUES
33 NEGATIVE FEEDBACK

Chapter 22
Content Quiz Answers

1. B
2. True
3. anorexia
4. D
5. C
6. A
7. B
8. B
9. A
10. False, Chemical
11. monomers
12. hydrolases
13. food vacuoles
14. I
15. F
16. A
17. H
18. K
19. J
20. C
21. L
22. B
23. E
24. G
25. D
26. N
27. M
28. D
29. C
30. False, heartburn
31. amylase
32. peristalsis
33. sphincters
34. rectum
35. A
36. True
37. False, plant
38. True
39. False, inorganic
40. basal metabolic rate
41. diet
42. enzymes
43. recommended daily allowances
44. energy
45. heat
46. D
47. False, Undernutrition
48. protein
49. B
50. True
51. fatty

1 ESSENTIALNUTRIENTS
6 GASTRICJUICE
8 ACIDCHYME
11 BASALMETABOLICRATE
14 SALIVARYAMYLASE
15 CALORIE
16 JEJUNUM
17 OMNIVORES
20 ALIMENTARYCANAL
22 ANUS
23 HERBIVORES
24 MINERALS
25 INGESTION
26 DUODENUM
28 PEPSIN
29 ESOPHAGUS
31 PERISTALSIS
32 KILOCALORIE
33 LARGEINTESTINE
1 ELIMINATION
2 RECTUM
3 STOMACH
4 BILE
5 FOODVACUOLES
7 DIGESTION
8 ABSORPTION
9 FECES
10 ILEUM
12 ESSENTIALAMINOACID
13 GALLBLADDER
18 VITAMINS
19 SMALLINTESTINE
21 CARNIVORES
27 MOUTH
30 PHARYNX

Chapter 23

Content Quiz Answers

1. B
2. False, the heart
3. circulatory
4. C
5. B
6. True
7. False, carbon dioxide
8. cardiovascular
9. D
10. True
11. double
12. D
13. True
14. False, speed up
15. diastole
16. heart murmur
17. C
18. A
19. True
20. True
21. hypertension
22. E
23. A
24. E
25. True
26. True
27. thrombus
28. plasma
29. red blood
30. B
31. False, heart attack
32. False, heart attack
33. interstitial fluid
34. atherosclerosis
35. B
36. False, more
37. tracheae
38. gills
39. oxygen, ATP
40. A
41. B
42. False, alveolus
43. vocal cords
44. C
45. E
46. False, diaphragm
47. negative
48. carbon dioxide
49. A
50. True
51. higher, oxygen

1 LUNGS
2 PLATELETS
3 FIBRINOGEN
4 PLASMA
5 TRACHEA
6 NEGATIVEPRESSUREBREATHING
7 CAPILLARYBEDS
8 HEARTRATE
9 DIAPHRAGM
10 CIRCULATORYSYSTEM
10 CARDIOVASCULARSYSTEM
11 HEMOGLOBIN
12 BRONCHI
13 DIASTOLE
14 SYSTOLE
15 VEINS
16 PACEMAKER
17 PULMONARYCIRCUIT
18 HEARTATTACK
19 ARTERIES
19 ALVEOLI
20 CLOSEDCIRCULATORYSYSTEM
21 LEUKOCYTES
22 BLOODPRESSURE
23 EMPHYSEMA
24 RESPIRATORYSYSTEM
25 LARYNX
26 ATRIUM
27 PHARYNX
28 VENTRICLE
29 ELECTROCARDIOGRAM
30 OPENCIRCULATORYSYSTEM

Chapter 24

Content Quiz Answers

1. B
2. False, helper T
3. True
4. immune
5. D
6. True
7. mucus
8. lysozyme
9. C
10. True
11. interferons
12. complement
13. A
14. B
15. False, do not address
16. pyrogens
17. C
18. lymph
19. lymph nodes
20. E
21. B
22. False, ineffective
23. False, Passive
24. antigen
25. antibodies
26. immunity
27. vaccination or immunization
28. B
29. E
30. A
31. E
32. D
33. False, T cells
34. False, B cells
35. B, T
36. Y
37. D
38. E
39. A
40. True
41. False, memory cells
42. monoclonal antibodies
43. memory cells
44. E
45. False, histamine
46. True
47. allergens
48. anaphylactic shock
49. B
50. D
51. C
52. A
53. C
54. True
55. autoimmune
56. A
57. True
58. lymphocytes
59. Immunodeficiency
60. immune
61. C
62. True
63. influenza

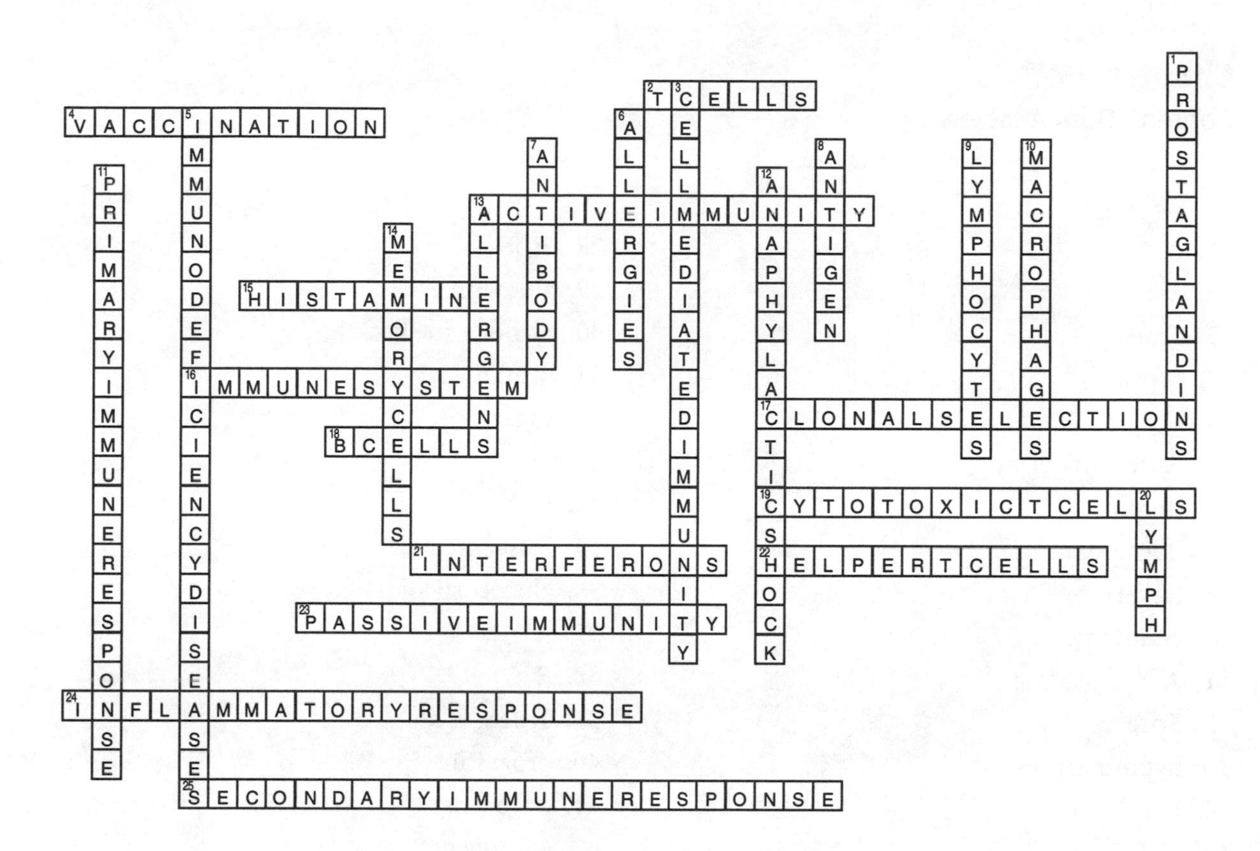

1 PROSTAGLANDINS
2 TCELLS
3 CELLMEDIATEDIMMUNITY
4 VACCINATION
5 IMMUNODEFICIENCYDISEASES
6 ALLERGIES
7 ANTIBODY
8 ANTIGEN
9 LYMPHOCYTES
10 MACROPHAGES
11 PRIMARYIMMUNERESPONSE
12 ANAPHYLACTICSHOCK
13 ACTIVEIMMUNITY
13 ALLERGENS
14 MEMORYCELLS
15 HISTAMINE
16 IMMUNESYSTEM
17 CLONALSELECTION
18 BCELLS
19 CYTOTOXICTCELLS
20 LYMPH
21 INTERFERONS
22 HELPERTCELLS
23 PASSIVEIMMUNITY
24 INFLAMMATORYRESPONSE
25 SECONDARYIMMUNERESPONSE

Chapter 25

Content Quiz Answers

1. B
2. A
3. True
4. breast
5. C
6. False, endocrine
7. steroid
8. hormone
9. target
10. endocrine
11. A
12. True
13. hypothalamus
14. E
15. C
16. False, prolactin
17. antidiuretic
18. endorphins
19. dwarfism
20. E
21. B
22. E
23. True
24. goiter
25. antagonistic
26. calcitonin
27. D
28. True
29. glucagon
30. diabetes mellitus
31. insulin
32. sugar
33. D
34. E
35. False, medulla
36. glucocorticoids
37. A
38. E
39. True
40. progestins
41. estrogens
42. androgens
43. B
44. F
45. G
46. A
47. H
48. E
49. C
50. D
51. B
52. D
53. True

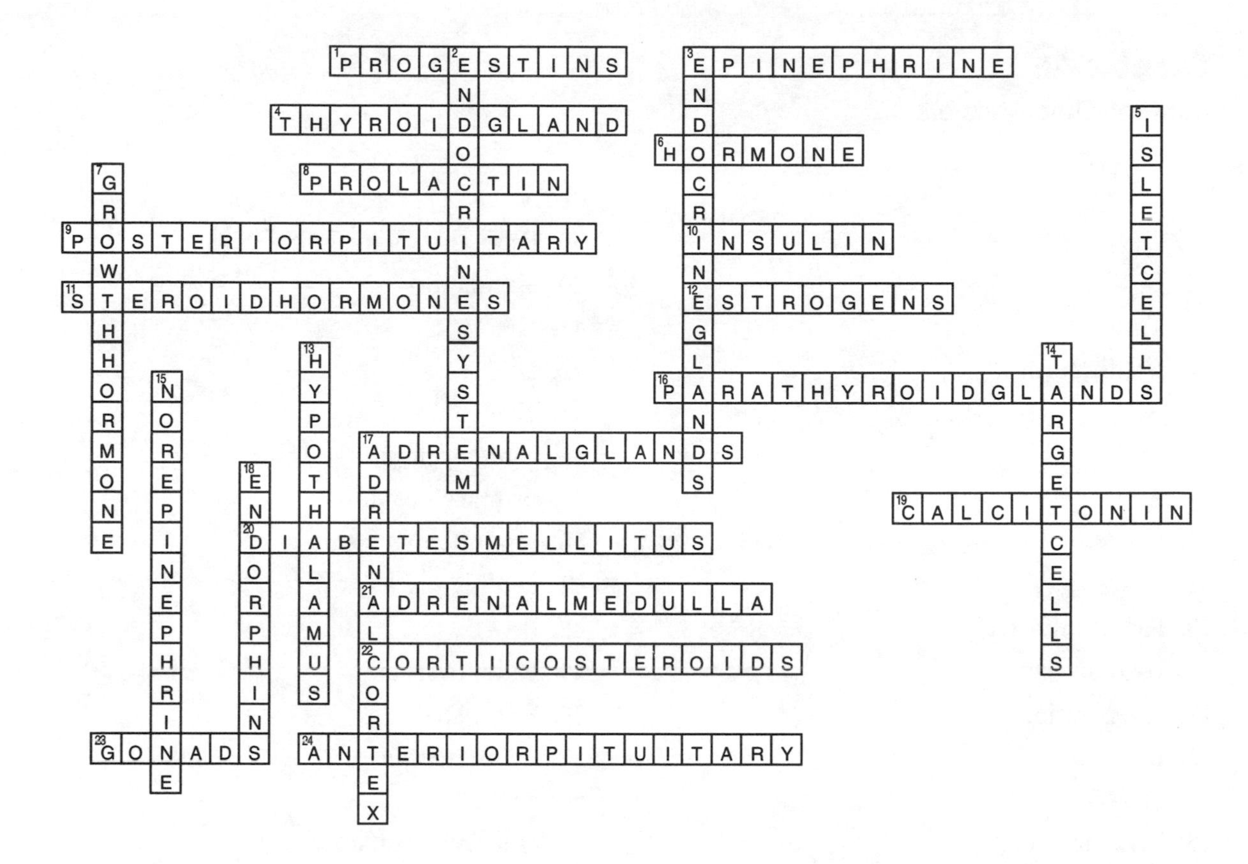

PROGESTINS
ENDOCRINESYSTEM
EPINEPHRINE
ENDOCRINEGLANDS
THYROIDGLAND
ISLETCELLS
HORMONE
GROWTHHORMONE
PROLACTIN
POSTERIORPITUITARY
INSULIN
STEROIDHORMONES
ESTROGENS
HYPOTHALAMUS
TARGETCELLS
NOREPINEPHRINE
PARATHYROIDGLANDS
ADRENALGLANDS
ADRENALCORTEX
ENDORPHINS
CALCITONIN
DIABETESMELLITUS
ADRENALMEDULLA
CORTICOSTEROIDS
GONADS
ANTERIORPITUITARY

Chapter 26

Content Quiz Answers

1. C
2. B
3. True
4. fertility drugs
5. B
6. D
7. A
8. C
9. C
10. False, uniform
11. asexual
12. regeneration
13. D
14. True
15. False, internal
16. sperm; ovum
17. zygote
18. hermaphrodites
19. A
20. D
21. F
22. B
23. H
24. G
25. E
26. C
27. E
28. True
29. gonads
30. ovulation
31. E
32. C
33. F
34. A
35. D
36. B
37. G
38. A
39. False, do not suggest
40. ejaculation
41. C
42. D
43. True
44. gametogenesis
45. primary oocytes
46. seminiferous tubules
47. epididymis
48. B
49. False, ovarian
50. False, FSH
51. menstruation
52. 14
53. anterior pituitary gland
54. HCG, human chorionic gonadotropin
55. D
56. B
57. F
58. A
59. G
60. C
61. H
62. E
63. D
64. True
65. progestin
66. RU-486 (mifepristone)
67. morning after pills
68. B
69. True
70. True
71. bacteria
72. fungi
73. viruses
74. E
75. B
76. False, man

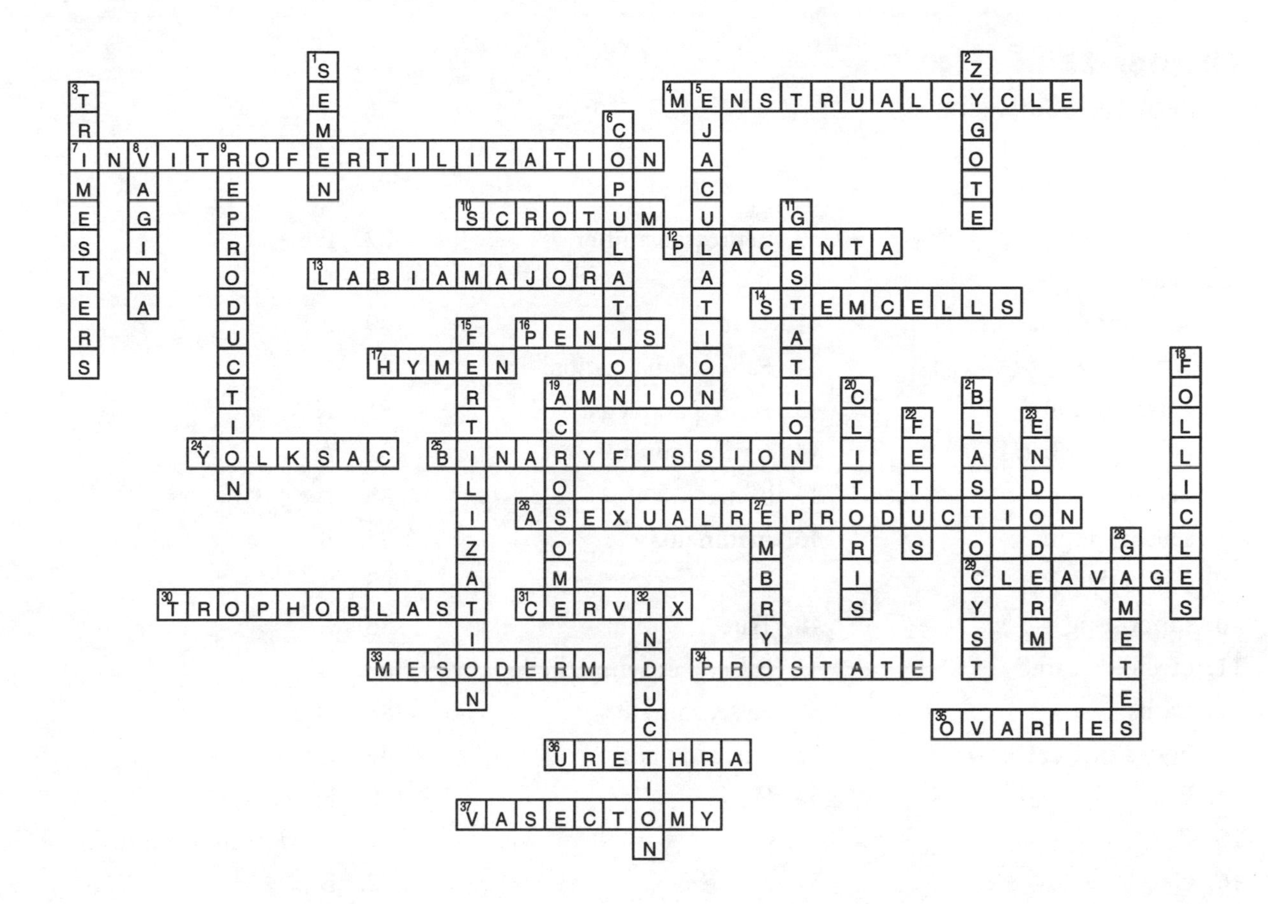

77. impotence
78. impotence
79. C
80. True
81. “in glass”
82. E
83. False, haploid
84. acrosome
85. mitochondria
86. C
87. A
88. F
89. G
90. H
91. D
92. B
93. E
94. C
95. True
96. A
97. C
98. B
99. D
100. A
101. False, can
102. trophoblast
103. placenta
104. stem
105. D
106. second
107. third
108. first
109. first
110. E
111. False, prolactin
112. expulsion
113. labor
114. placenta
115. dilation
116. A
117. False, retain
118. menopause

Chapter 27

Content Quiz Answers

1. E
2. True
3. serotonin
4. A
5. False, motor
6. False, peripheral
7. nerve
8. sensory input
9. cell body
10. supporting
11. myelin sheath
12. axons
13. nodes of Ranvier
14. B
15. C
16. C
17. D
18. True
19. False, negatively
20. resting potential
21. stimulus
22. threshold potential
23. F
24. E
25. C
26. H
27. I
28. B
29. A
30. D
31. G
32. B
33. D
34. True
35. False, electrical
36. synaptic cleft
37. synapse
38. chemical
39. neurotransmitter
40. B
41. True
42. False, cephalization
43. False, paraplegia
44. centralization
45. brain
46. meningitis
47. E
48. True
49. False, parasympathetic
50. False, somatic
51. referred pain
52. G
53. C
54. B
55. E
56. F
57. A
58. D
59. B
60. False, cerebellum
61. lateralization
62. biological
63. E
64. D
65. A
66. B
67. C
68. B
69. False, transduction
70. True
71. receptor potential
72. prostaglandins
73. F
74. M
75. J
76. O
77. P
78. L
79. E
80. C
81. N
82. A
83. I
84. B
85. K
86. G
87. D
88. H
89. C
90. False, aqueous
91. rhodopsin; photopsin
92. B
93. A
94. E
95. C
96. D
97. F
98. 8
99. 5
100. 7
101. 1
102. 3
103. 9
104. 2
105. 6
106. 4
107. C
108. E
109. True
110. tinnitus
111. C
112. A
113. False, appendicular
114. True

115. False, yellow
116. pivot
117. ball-and-socket
118. hinge
119. A
120. B
121. D
122. C
123. D
124. True
125. False, Thick
126. True
127. tendons
128. sarcomeres
129. myofibrils
130. motor unit
131. neuromuscular
132. A
133. False, away from
134. eyespots

Chapter 28

Content Quiz Answers

1. E
2. F
3. True
4. fungus
5. C
6. D
7. False, two leaves
8. cotyledons
9. A
10. B
11. False, stems
12. True
13. False, leaves
14. taproots
15. shoot
16. nodes; internodes
17. apical dominance
18. axillary
19. runner
20. root; stem (called a rhizome)
21. roots
22. root hairs
23. leaves
24. tendril
25. F
26. D
27. B
28. H
29. A
30. E
31. C
32. G
33. False, cell wall
34. cellulose
35. tracheids; vessel elements
36. I
37. E
38. A
39. B
40. J
41. F
42. D
43. H
44. C
45. G
46. True
47. True
48. vascular
49. ground
50. complex
51. E
52. C
53. F
54. H
55. A
56. I
57. D
58. G
59. B
60. False, leaves
61. True
62. False, asexual
63. flower
64. ovule
65. D
66. A
67. C
68. B
69. E
70. F
71. False, stigma
72. double fertilization
73. D
74. seed coat

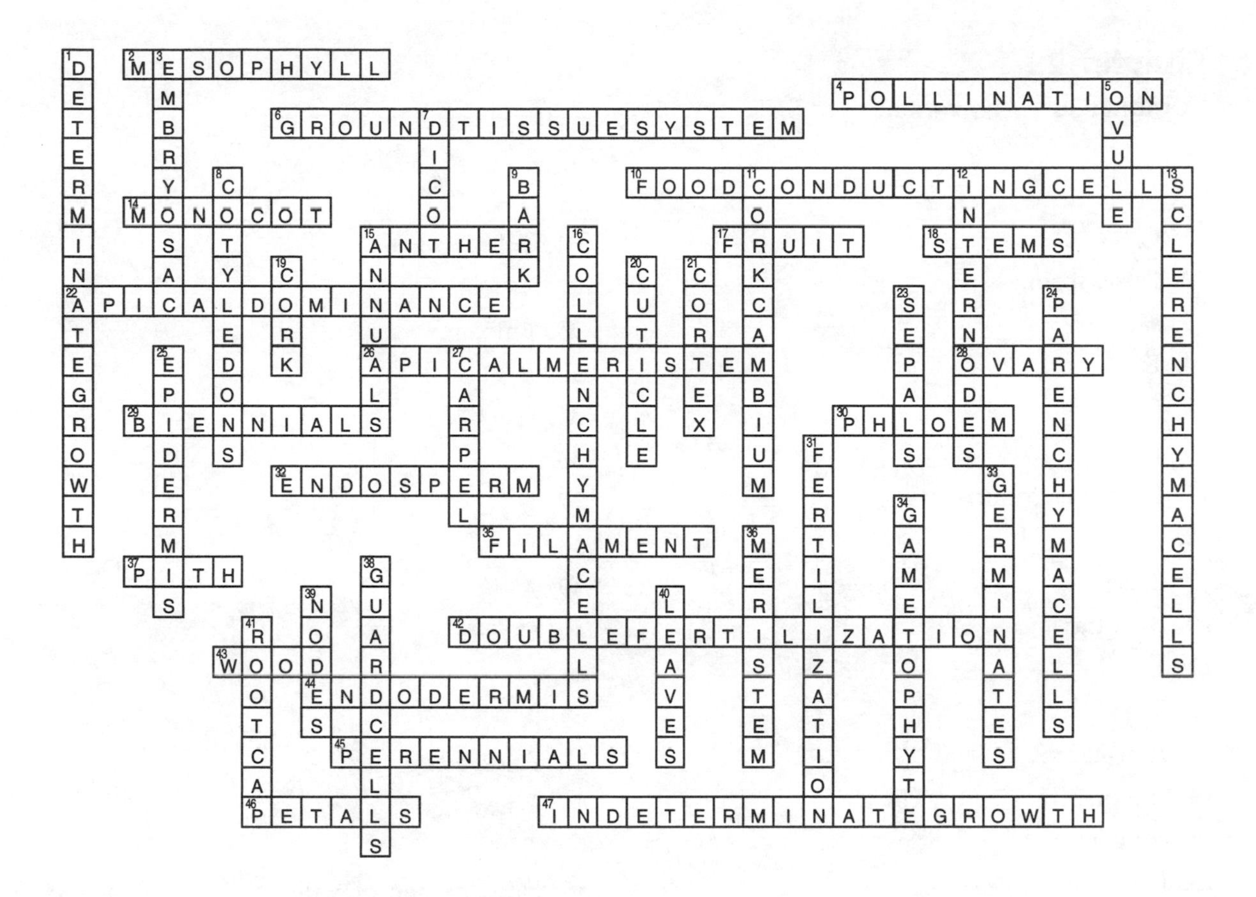

75. endosperm
76. D
77. E
78. False, root
79. True
80. seeds
81. B
82. False, indeterminate
83. perennials
84. annuals
85. B
86. False, primary
87. root cap
88. apical meristems
89. primary
90. A
91. B
92. D
93. False, secondary
94. True
95. vascular cambium
96. xylem
97. lignin
98. cork cambium
99. bark
100. B
101. False, bees
102. nectar

Chapter 29
Content Quiz Answers

1. C
2. False, spring
3. sap
4. E
5. C
6. B
7. False, macronutrients
8. True
9. True
10. essential
11. E
12. True
13. root hairs
14. mycorrhiza
15. root
16. B
17. A
18. E
19. C
20. F
21. D
22. False, cannot
23. nitrogen
24. A
25. C
26. B
27. False, cohesion
28. True
29. xylem
30. C
31. False, source
32. True
33. phloem
34. B
35. C
36. E
37. A
38. D
39. A
40. C
41. D
42. A
43. False, faster
44. True
45. True
46. False, carbon dioxide
47. hormones
48. auxin
49. abscisic acid
50. auxins
51. phototropism
52. C
53. False, spring or early summer
54. thigmotropism
55. gravitropism
56. E
57. True
58. aphids
59. aphids

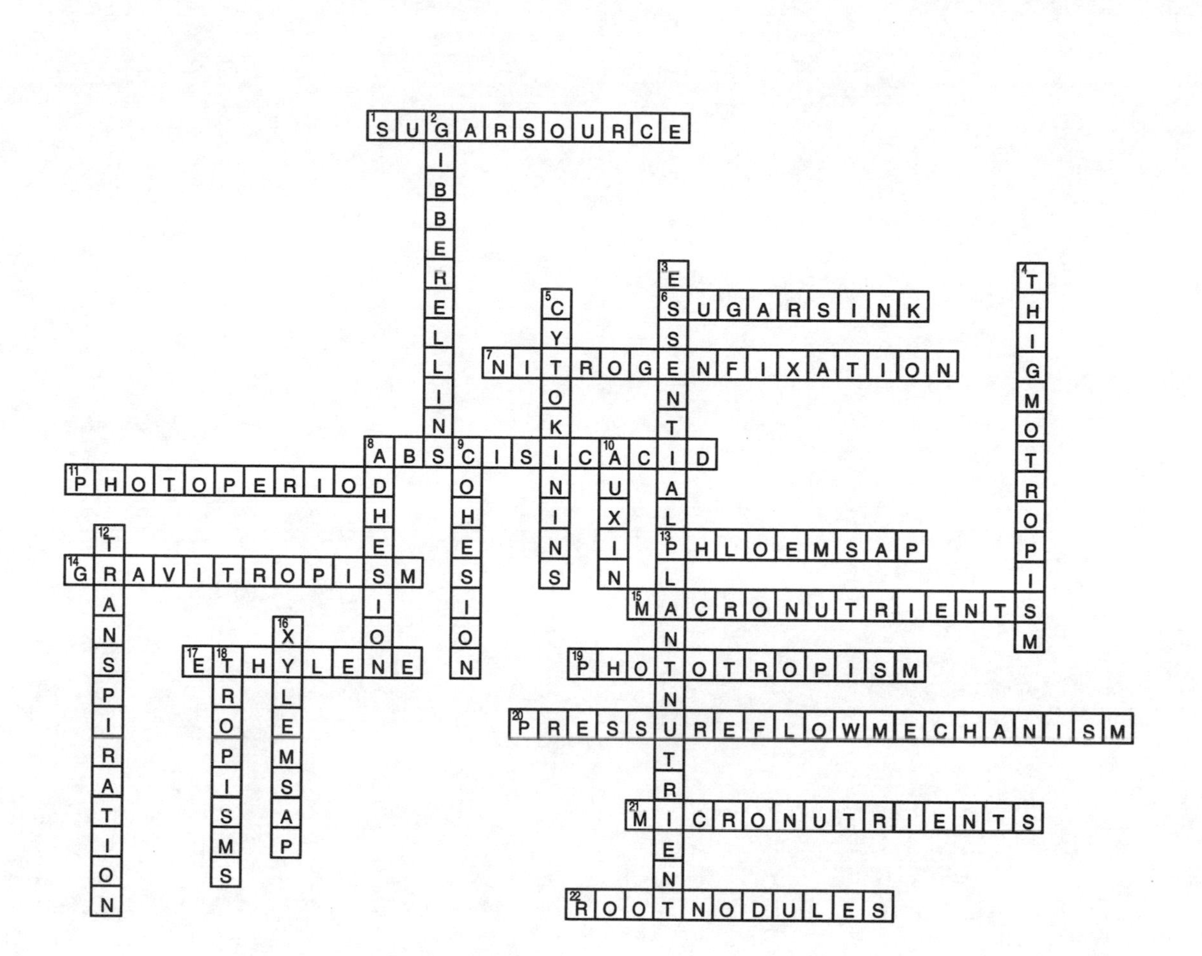

1 SUGARSOURCE
2 GIBBERELLIN
3 ESSENTIALELEMENT
4 THIGMOTROPISM
5 CYTOKININS
6 SUGARSINK
7 NITROGENFIXATION
8 ABSCISICACID
8 ADHESION
9 COHESION
10 AUXIN
11 PHOTOPERIOD
12 TRANSPIRATION
13 PHLOEMSAP
14 GRAVITROPISM
15 MACRONUTRIENTS
16 XYLEMSAP
17 ETHYLENE
18 TROPISMS
19 PHOTOTROPISM
20 PRESSUREFLOWMECHANISM
21 MICRONUTRIENTS
22 ROOTNODULES